AF358823

Sustainable Mining as the Key for the Ecological Transition: Current Trends and Future Perspectives

Sustainable Mining as the Key for the Ecological Transition: Current Trends and Future Perspectives

Editors

Pierfranco Lattanzi
Elisabetta Dore
Fabio Perlatti
Hendrik Gideon Brink

 Basel • Beijing • Wuhan • Barcelona • Belgrade • Novi Sad • Cluj • Manchester

Editors

Pierfranco Lattanzi
The Institute of Geosciences
and Earth Resources
CNR
Firenze
Italy

Elisabetta Dore
Department of Chemical and
Geological Sciences
University of Cagliari
Cagliari
Italy

Fabio Perlatti
National Mining Agency
University of São Paulo
Sao Paulo
Brazil

Hendrik Gideon Brink
Department of Chemical
Engineering
University of Pretoria
Pretoria
South Africa

Editorial Office
MDPI
St. Alban-Anlage 66
4052 Basel, Switzerland

This is a reprint of articles from the Special Issue published online in the open access journal *Minerals* (ISSN 2075-163X) (available at: www.mdpi.com/journal/minerals/special_issues/752XA507Y5).

For citation purposes, cite each article independently as indicated on the article page online and as indicated below:

Lastname, A.A.; Lastname, B.B. Article Title. *Journal Name* **Year**, *Volume Number*, Page Range.

ISBN 978-3-7258-1008-6 (Hbk)
ISBN 978-3-7258-1007-9 (PDF)
doi.org/10.3390/books978-3-7258-1007-9

Contents

About the Editors

Pierfranco Lattanzi

Pierfranco (Piero) Lattanzi was a professor of Mineralogy and Ore deposits at the universities of Firenze and Cagliari, Italy. He is still active as a senior research associate of Consiglio Nazionale delle Ricerche. His research interests include the genesis of mineral deposits and the environmental impact of mining activity.

Elisabetta Dore

Elisabetta Dore was born in 1983 (Cagliari, Italy). She gained a Master's Degree in Geological Sciences and Technologies in 2011 and a PhD in Earth Science in 2015. She was a Research Fellow from 2015 to 2021. Her current position is Junior Professor in Geochemistry at the Department of Chemical and Geological Sciences (DSCG), University of Cagliari, Italy.

She specializes in environmental geochemistry and mineralogy. Her main research topics are focused on the interaction between the hydrosphere and the other geospheres, especially the processes ruling the mobilization and dispersion of inorganic contaminants (mainly metals and metalloids in (abandoned) mining areas), bio-mineralization and bio-geochemical barriers, and hydrologic tracers. Her other topics of interest include the determination of geochemical background, sorption processes (involving both natural and synthetic minerals), water stable isotopes, and biofertilization. Her research activities are developed using a multiscale and multidisciplinary approach through the study of natural processes and via laboratory tests, involving both field activities and analytical studies (XRD, SEM, IC, ICP-OES, ICP-MS, etc.).

Fabio Perlatti

Fabio Perlatti is a distinguished Agronomic Engineer and Ecologist with an extensive academic and professional trajectory in the field of environmental science and mineral resource management. He earned both his Master's and PhD in Ecology from the Federal University of Ceará, Brazil. During his PhD, he also spent a year as a research fellow at the University of Santiago de Compostela, Spain, enriching his research perspective and international experience. Dr. Perlatti further advanced his expertise by undertaking a four-year Postdoctoral Fellowship at the University of São Paulo's Soil Science Department. There, he specialized in the biogeochemistry of mine tailings, soils, and plants, contributing to several esteemed international scientific journals. His research has provided significant insights into soil remediation, technosols, phyto/agromining, and sustainable mining practices.

Professionally, Fabio has been pivotal in his role at the National Mining Agency of Brazil - ANM, where he has served for over 18 years as an Environmental and Mineral Resources Specialist. His work primarily revolves around the analysis and approval of mining permits, focusing on sustainable mining practices. He is particularly noted for his expertise in the rehabilitation of degraded and abandoned mine sites and in navigating the complex issues associated with mine closure and post-mine closure transitions. Dr. Perlatti's contributions to the field have positioned him as a key figure in shaping policies and practices that aim to balance resource extraction with environmental sustainability. His dedication to the field is evident in his ongoing efforts to ensure that mining activities are conducted responsibly, promoting ecological recovery and community benefits.

Hendrik Gideon Brink

Since obtaining his PhD in Chemical Engineering in 2015, with a specialization in Bioreaction Engineering, Professor Hendrik Brink (PhD, PrEng) has made significant strides in academia. After commencing his tenure at the University of Pretoria's Department of Chemical Engineering as a Senior Lecturer in 2015, and later advancing to the position of Associate Professor in January 2022, he has firmly solidified his reputation as an eminent presence within the field. Throughout his tenure, Prof. Brink has demonstrated a prolific scholarly output, with 85 publications garnering 659 citations and culminating in an H-index of 13. His academic contributions extend beyond research, as he has effectively supervised or co-supervised 22 master's degree candidates and 6 PhD candidates, with the ongoing supervision of 8 Master's degree and four PhD candidates. Active in the academic community, Prof. Brink has participated in 26 national and international conferences, furthering knowledge exchange and collaboration. Notably, his research endeavors are centered around waste valorization, aiming to contribute to the establishment of a circular economy—a vital pursuit in sustainable development efforts. Prof. Hendrik Brink stands as a dedicated scholar committed to advancing chemical engineering towards a more sustainable future.

Editorial

Editorial for the Special Issue "Sustainable Mining as the Key for the Ecological Transition: Current Trends and Future Perspectives"

Hendrik Gideon Brink [1], **Elisabetta Dore** [2], **Fabio Perlatti** [3] and **Pierfranco Lattanzi** [4,*]

1 Department of Chemical Engineering, University of Pretoria, Pretoria 0002, South Africa; deon.brink@up.ac.za
2 Department of Chemical and Geological Sciences, Università degli Studi di Cagliari, 09124 Cagliari, Italy; elisabetta.dore@unica.it
3 National Mining Agency of Brazil (ANM), São Paulo 04040-033, Brazil; fperlatti@gmail.com
4 Istituto di Geoscienze e Georisorse (UOS Firenze), Consiglio Nazionale delle Ricerche, 50121 Firenze, Italy
* Correspondence: pierfrancolattanzi@gmail.com

Citation: Brink, H.G.; Dore, E.; Perlatti, F.; Lattanzi, P. Editorial for the Special Issue "Sustainable Mining as the Key for the Ecological Transition: Current Trends and Future Perspectives". *Minerals* **2024**, 14, 389.
https://doi.org/10.3390/min14040389

Received: 1 April 2024
Accepted: 2 April 2024
Published: 8 April 2024

A crucial aspect in the pursuit of sustainable development is the necessary shift toward an "ecological transition", a transformation in societal paradigms to align human activities with the global ecosystem. This change stems from the urgent need to tackle challenges such as climate change and growing social disparities.

A key component of the ecological transition involves moving away from fossil-fuel-dominated energy production to renewable and less environmentally harmful sources like solar, wind, and geothermal energy. However, this shift to alternative energy sources increases the demand for minerals, surpassing historical needs associated with fossil fuel energy generation. This includes essential materials such as iron, aluminum, and copper, which are necessary for any kind of engine (from wind turbines to electric motors), as well as scarce metals like lithium, cobalt, and rare earth elements (REE) used in technological devices, batteries, solar panels, magnets, etc.

Paradoxically, the ecological transition intensifies our reliance on mining, which is historically linked to environmental and societal issues. Addressing these challenges falls under the concept of "Sustainable Mining," requiring diverse expertise in geology and eco-friendly extraction, and an understanding of the long-term environmental impacts. It also involves waste recycling, improved reclamation planning, and strategies to minimize permanent effects on the mining areas and their surroundings.

In this Special Issue, we present a collection of papers addressing various facets of sustainable mining, spanning from the characterization of new resources to the assessment of the environmental impact of dismissed mines, as well as the development of new solutions aimed at the reuse of wastes, to offer an updated snapshot of the current state of the art and upcoming challenges.

The introductory paper by Richard Herrington provides a comprehensive overview of the impending needs and challenges for a low-carbon future. While recycling remains a priority, new mined resources will be indispensable until at least 2050 due to the escalating demand for metals in emerging technologies. This may involve re-evaluating old deposits, repurposing abandoned mining wastes, and exploring unconventional new deposits, all while adhering to environmental and social governance standards.

The work by Sedda et al. epitomizes the theme of this Special Issue by investigating a segment of the Montevecchio vein system located in Sardinia, Italy. This former mine district is recognized for its substantial production, exceeding 2 million metric tons of lead (Pb) and zinc (Zn) metals. The research integrates conventional resource characterization with an assessment of the environmental consequences and the prospective retrieval of crucial materials, including REE, from tailings, wastes, and biomineralization.

Grieco et al. describe the formation of ophiolite-hosted magnesite deposits in Evia Island, Greece, where the magnesite ores are in close spatial association with chromite and olivine. As both magnesium (Mg) and chromium (Cr) are critical metals, the authors suggest their potential for an efficient combined exploitation.

Fornasaro et al. and Onnis et al. investigate the long-term environmental impacts of past mining activities. Fornasaro et al. show how tree rings in chestnut grown in the former mercury (Hg) district of Monte Amiata, Italy, clearly document the evolution of atmospheric Hg concentrations during Hg production, and after the cessation thereof in the district. A notable aspect behind this study is that Hg was, in the past century, a true "critical metal" due to its many applications. Onnis et al. investigate how fluvial geomorphology and soil geochemistry drive zinc and lead dispersion along a mine-impacted stream in Wales, UK. This study pinpoints the fluvial geomorphological zones (erosional, transport, and depositional areas) where metals were released and trapped across different streamflow conditions and identifies the phases with which the metals are associated.

Further papers explore the efficient handling of waste produced by mining and related industrial activities. Muedi et al. and Mogashane et al. present innovative methods for utilizing acid mine drainage (AMD) as a resource, while Fugazzotto et al. illustrate the transformation of waste material from the ceramic industry into precursors for building blocks/binders through an alkaline activation process.

In the context of ecological transition, coal and fossil fuels will gradually be phased out. However, they will remain in use for some time. The paper by Alexandrova et al. presents an optimized magnetic separation process for iron recovery from coal ash and slag. Lastly, Arunachellan et al. detail how used tires, a bulky waste, can be recycled into carbon-based nanostructures, with various applications including the remediation of copper-contaminated water, a common environmental issue associated with copper mining and processing.

Author Contributions: A preliminary draft was provided by P.L. The current version was reshaped by H.G.B. and revised by E.D. and F.P. All authors have read and agreed to the published version of the manuscript.

Acknowledgments: We extend our sincere thanks to the authors, reviewers, and MDPI staff for their collective efforts in maintaining the high standards expected in *Minerals*.

Conflicts of Interest: The authors declare no conflict of interest.

List of Contributions:

1. Muedi, K.; Masindi, V.; Maree, J.; Brink, H. Rapid Removal of Cr(VI) from Aqueous Solution Using Polycationic/Di-Metallic Adsorbent Synthesized Using Fe3+/Al3+ Recovered from Real Acid Mine Drainage. *Minerals* **2022**, *12*, 1318. https://doi.org/10.3390/min12101318.
2. Grieco, G.; Cavallo, A.; Marescotti, P.; Crispini, L.; Tzamos, E.; Bussolesi, M. The Formation of Magnesite Ores by Reactivation of Dunite Channels as a Key to Their Spatial Association to Chromite Ores in Ophiolites: An Example from Northern Evia, Greece. *Minerals* **2023**, *13*, 159. https://doi.org/10.3390/min13020159.
3. Mogashane, T.; Maree, J.; Mujuru, M.; Mphahlele-Makgwane, M.; Modibane, K. Ferric Hydroxide Recovery from Iron-Rich Acid Mine Water with Calcium Carbonate and a Gypsum Scale Inhibitor. *Minerals* **2023**, *13*, 167. https://doi.org/10.3390/min13020167.
4. Fornasaro, S.; Ciani, F.; Nannoni, A.; Morelli, G.; Rimondi, V.; Lattanzi, P.; Cocozza, C.; Fioravanti, M.; Costagliola, P. Tree Rings Record of Long-Term Atmospheric Hg Pollution in the Monte Amiata Mining District (Central Italy): Lessons from the Past for a Better Future. *Minerals* **2023**, *13*, 688. https://doi.org/10.3390/min13050688.
5. Onnis, P.; Byrne, P.; Hudson-Edwards, K.; Stott, T.; Hunt, C. Fluvial Morphology as a Driver of Lead and Zinc Geochemical Dispersion at a Catchment Scale. *Minerals* **2023**, *13*, 790. https://doi.org/10.3390/min13060790.
6. Fugazzotto, M.; Mazzoleni, P.; Lancellotti, I.; Camerini, R.; Ferrari, P.; Tiné, M.; Centauro, I.; Salvatici, T.; Barone, G. Industrial Ceramics: From Waste to New Resources for Eco-Sustainable Building Materials. *Minerals* **2023**, *13*, 815. https://doi.org/10.3390/min13060815.

7. Sedda, L.; De Giudici, G.; Fancello, D.; Podda, F.; Naitza, S. Unlocking Strategic and Critical Raw Materials: Assessment of Zinc and REEs Enrichment in Tailings and Zn-Carbonate in a Historical Mining Area (Montevecchio, SW Sardinia). *Minerals* **2024**, *14*, 3. https://doi.org/10.3390/min14010003.
8. Aleksandrova, T.; Nikolaeva, N.; Afanasova, A.; Chenlong, D.; Romashev, A.; Aburova, V.; Prokhorova, E. Increase in Recovery Efficiency of Iron-Containing Components from Ash and Slag Material (Coal Combustion Waste) by Magnetic Separation. *Minerals* **2024**, *14*, 136. https://doi.org/10.3390/min14020136.
9. Herrington, R. The Raw Material Challenge of Creating a Green Economy. *Minerals* **2024**, *14*, 204. https://doi.org/10.3390/min14020204.
10. Arunachellan, I.; Bhaumik, M.; Brink, H.; Pillay, K.; Maity, A. Efficient Aqueous Copper Removal by Burnt Tire-Derived Carbon-Based Nanostructures and Their Utilization as Catalysts. *Minerals* **2024**, *14*, 302. https://doi.org/10.3390/min14030302.

Article

The Raw Material Challenge of Creating a Green Economy

Richard Jeremy Herrington 

Centre for Resourcing the Green Economy, The Natural History Museum, London SW75BD, UK;
r.herrington@nhm.ac.uk

Abstract: Clean technologies and infrastructure for our low-carbon, green future carry intense mineral demands. The ambition remains to recycle and reuse as much as we can; however, newly mined resources will be required in the near term despite the massive improvements in the reuse and recycling of existing end-of-use products and wastes. Growth trends suggest that mining will still play a role after 2050 since the demand for metals will increase as the developing world moves toward a per capita usage of materials comparable to that of the developed world. There are sufficient geological resources to deliver the required mineral commodities, but the need to mine must be balanced with the requirement to tackle environmental and social governance issues and to deliver sustainable development goals, ensuring that outcomes are beneficial for both the people and planet. Currently, the lead time to develop new mines following discovery is around 16 years, and this needs to be reduced. New approaches to designing and evaluating mining projects embracing social, biodiversity, and life cycle analysis aspects are pivotal. New frontiers for supply should include neglected mined wastes with recoverable components and unconventional new deposits. New processing technologies that involve less invasive, lower energy and cleaner methodologies need to be explored, and developing such methodologies will benefit from using nature-based solutions like bioprocessing for both mineral recovery and for developing sustainable landscapes post mining. Part of the new ambition would be to seek opportunities for more regulated mining areas in our own backyard, thinking particularly of old mineral districts of Europe, rather than relying on sources with potentially and less controllable, fragile, and problematic supply chains. The current debate about the potential of mining our deep ocean, as an alternative to terrestrial sources needs to be resolved and based on a broader analysis; we can then make balanced societal choices about the metal and mineral supply from the different sources that will be able to deliver the green economy while providing a net-positive deal for the planet and its people.

Keywords: green economy; mining; new frontiers; social governance; biodiversity; nature-based solutions; net positive; circular economy

Citation: Herrington, R.J. The Raw Material Challenge of Creating a Green Economy. *Minerals* **2024**, 14, 204.
https://doi.org/10.3390/min14020204

Academic Editors: Pierfranco Lattanzi, Elisabetta Dore, Fabio Perlatti and Hendrik Gideon Brink

Received: 19 November 2023
Revised: 26 January 2024
Accepted: 6 February 2024
Published: 17 February 2024

1. Introduction

There are three key factors driving the current need for the growth in mineral use and resultant renewed mining activity [1]. The first and biggest factor is the use of so-called 'critical minerals' to decarbonize our energy generation, transportation, and industry; these goals are to be achieved largely by employing mineral-hungry technologies [2]. Secondly, the general world economic growth and consumerism, particularly in the economies of the BRIC nations and other less developed nations, stimulate demand for materials as the growing and developing world population rises to per capita material consumption to match that of the more developed world [3]. Thirdly, mining can be shown to be directly implicated in achieving several of the UN's 17 sustainable development goals [4]. Dealing with the last two factors, the World Economic Forum [4] recognized that mining has very direct relevance to a number of the UN's Sustainable Development goals (SDGs), both in a positive and a negative sense. On the positive side, mining is important to providing materials for technologies delivering renewable energy (Goal 7), economic growth (8), and innovation and infrastructure (9). However, negative impacts are noted with respect to goals focused on clean water and sanitation (6) as well as life on land (15).

The metal copper, as an example, is a key enabler of sustainable development since it is an essential metal in housing, transportation, electrification, and appliances, and it underpins both global economic growth and human development [5]. The International Copper Association of India recently announced an 11% increase in the amount of copper used per m^2 in new housing, simply reflecting development and the increased use of domestic consumer electrical appliances in that country. This figure matches the general annual worldwide growth in copper per capita [6], even before the recent rapid increase in demand due to the green energy transition.

2. Mineral Usage Growth Due to the Green Energy, Transport, and Industrial Transition

The key factor driving the growth in the use of minerals is due to the decarbonizing of energy, transport, and industry. Climate change is recognized to be largely driven by the increase in greenhouse gases associated with humankind's use of carbon bound in fossil fuels. Although CO_2 from the burning of carbon is not the only culprit, volumetrically, it is the most important driver of global warming with 2022 reaching a record level of 417.06 parts per million [7], so the reduction of this figure is an essential aspect of tackling global warming. The largest producers of CO_2 are power generation and heating, followed by transportation and then industry, which are collectively responsible currently for around 85% of the global annual production of CO_2 from fossil fuels, as shown in Figure 1.

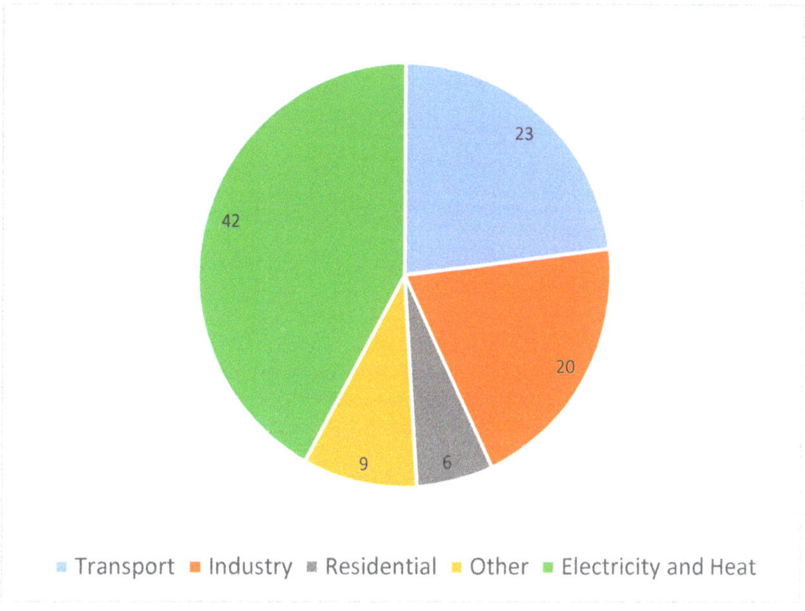

Figure 1. Relative contributions (in percentage terms) of different sectors to the measure of global CO_2 emissions in 2021 [8].

The 2015 Paris Agreement set goals to keep the rise in the mean global temperature below 2 °C (preferably to below 1.5 °C), recognizing that CO_2 emissions need to be cut by roughly 50% by 2030. Many governments have therefore made pledges to drastically reduce emissions, and 44 countries plus the European Union have made 'net-zero' pledges: these countries are collectively responsible for around 70% of current CO_2 emissions [2]. Despite the pledges, global CO_2 emissions grew by 0.9% in 2022 to their highest levels of 36.8 Gt collectively [8]. This increase largely reflected a growth in emissions from energy generation, resulting from a switch to the burning of coal due to the squeeze on gas supplies because of the Ukraine war; the switch added 423 Mt of CO_2 to the emissions budget. Nevertheless, solar and wind energy generation did grow to around 550 TWh of installed capacity, and further bright news is that emissions from industrial processes dropped by 103 Mt, with the deployment of renewable technologies (renewables, EVs, and heat pumps) avoiding the addition of a further 550 Mt of CO_2 release. As a result, the EU saw a significant (70 Mt or 2.5%) reduction in collective CO_2 emissions.

3. How Will We Decarbonize the Economy?

The bulk of industry CO_2 emissions is clearly linked to the generation of heat, and thus, decarbonizing heating processes are identified as key targets [9]. The total world energy consumption in 2022 was around 165,000 TWh with only around 11,000 of this from renewables [10], amounting to 7% of global usage. For electrical production, renewables (including nuclear) accounted for around 38.5%, which was very minimally changed from 2021, partly due to a decline in nuclear power but also a growth in electricity needs due to economic recovery and growth. The renewable sector needs to grow significantly. However, this growth in renewables will put further pressure on the mineral supply. Table 1 shows some of the metal demands for diverse energy sources and metal demands for wind energy that are an order of magnitude or more increased for a range of metals.

Table 1. Calculated metal needs for contrasting energy technologies per megawatt of installed capacity.

Kg/MW	Copper	Nickel	Manganese	Cobalt	Chromium	Molybdenum	Zinc	REE
Offshore Wind	8000	240	790	0	525	109	5500	239
Onshore Wind	2900	404	780	0	470	99	5500	14
Solar PV	2822	1.3	0	0	0	0	30	0
Nuclear	1473	1297	148	0	2190	70	0	0.5
Coal	1150	721	4.63	201	308	66	0	0
Natural Gas	1100	16	0	1.8	48.34	0	0	0

Source of data: [2]; PV = photovoltaics; REE = rare earth elements.

However, the increases in demands for metals and minerals is not limited to so-called 'critical' elements. It is also recognized that the energy revolution will demand increases in other major commodities like steel, glass, and concrete [11]. Steel demands are also significantly increased with two times as much metal used in a wind power array and up to three times as much for photovoltaic systems per MW. To put the estimates for increased copper use into a volumetric perspective, a recent study estimated that humankind will need to mine as much copper between now and 2050 as has been mined throughout history [12].

Decarbonizing transportation will also demand increased usage of materials as shown in Table 2. Using those figures, just to replace the entire 31.5 million UK-based private internal combustion engine vehicles today with electric vehicles, assuming they use the most resource-frugal next-generation NMC 811 batteries, would take 207,900 tonnes of cobalt, 264,600 tonnes of lithium carbonate (LCE), and at least 7200 tonnes of neodymium and dysprosium, in addition to 2,362,500 tonnes of copper.

Table 2. Calculated 'critical mineral' demands for an average internal combustion engine car versus a battery electric alternative (with industry standard battery chemistry).

Kg/Vehicle	Cu	Li	Ni	Mn	Co	Graphite	Zn	REE	Others
BE car	53.2	8.9	39.9	24.5	13.3	66.3	0.1	0.5	0.31
ICE car	22.3	0	0	11.2	0	0	0.1	0	0.3

Source of data: [2]; REE = rare earth elements; BE = battery electric; ICE = internal combustion engine.

Another important note is that creating a battery electric car fleet will have serious implications for the electrical power generation needed to recharge these vehicles. Based on figures published for the current generation of battery electric vehicles [13] and the average of 328.2 billion miles driven by car owners [14], there will be a demand for an additional 80 TWh or 25% increase in the UK-generated electrical capacity (hopefully to be powered from renewable resources like wind and PV that will further demand metals).

4. Can Recycling or Waste Recovery Deliver Everything We Need?

The long-term ambition would be to bring the 'cradle-to-cradle' philosophy of a truly circular economy into the human use of natural resources. However, the following analysis will show that this is not currently possible and unlikely to happen in the foreseeable future [15]. Firstly, in the case of many metals and minerals we use, the end-of-life recycling exceeds 50%, but for some important commodities, it is currently less than 1%, as shown in Figure 2.

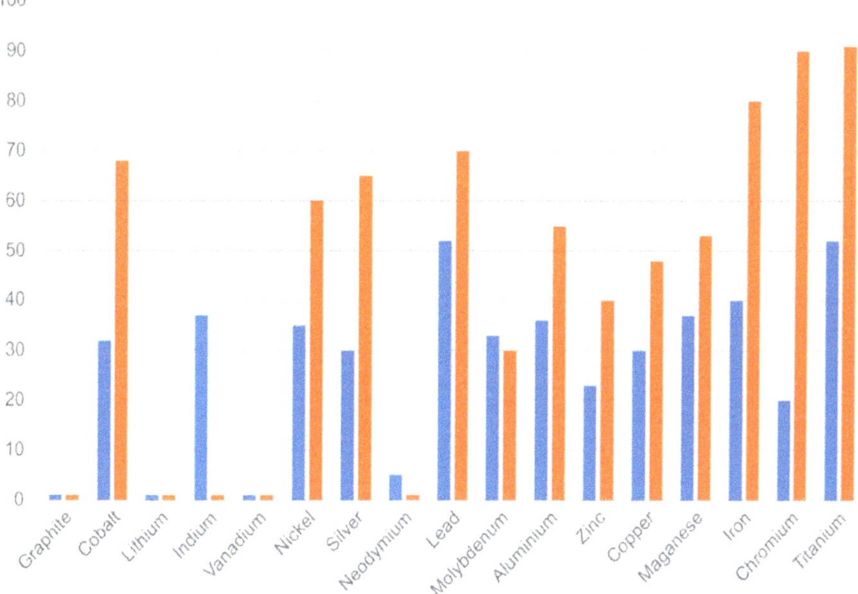

Figure 2. The graph shows (in orange) the rates of the end-of-life recycling for a range of metals and (in blue) the percentages of current material demands that can be met from recycled stock. Author's own compiled figures—various sources incl. [1].

It is acknowledged that it will not be until at least 2035 that stocks of the newly demanded metals like lithium and cobalt will be available as significant stocks to fuel the secondary recycling market, and even then, these will account for less than 50% of the demand for those two metals [16]. Recycled nickel, graphite, and manganese will similarly fall well short of supplying 50% of needs.

Mining annually produces 72 billion tonnes of rock waste and more than 8.8 billion tonnes of processing tailings, and 46% of the volume of tailings comes from the mining of copper where 99% of the mined material goes to waste [17]. The total volume of accumulated tailings worldwide is now more than 282 billion tonnes, and given the inefficiency of mineral processing history, many tailings facilities could have recoverable metals locked within [18,19]. For example, tailings stored at the Bor copper mine in Serbia alone contain more than 200,000 t of Cu, 55,000 t of Mo, and 390,000 t of Zn, sitting, poorly processed, in a tailing storage facility [20]. That material is running at 0.4% Cu, 1100 ppm Mo, and 0.79%

Zn and has a current in situ value of more than $8 billion dollars and compares favourably to the grades of some active mines [21]. It is estimated that tailings from copper mines worldwide could contain up to 43 million tonnes of copper [22]. Nevertheless, even if all the contained copper was all recovered, it would provide only two years of the current world copper production, not enough for what the energy transition demands. However, there are attractions to the processing of tailings since the process of mining and grinding ore itself currently accounts for a significant 2–3% of the current world CO_2 emissions [23], yet mine tailings are already finely grounded, which would significantly reduce the energy consumption of onward processing.

Many of the new metals we need for the green economy are by-products of the mining of another major metal and are therefore what are termed 'companion' metals [24]. Companion metals are often recovered at the processing stage, although in some cases the companion metals may be only partially recovered or not at all and therefore can be found in the waste (see Table 3). The economics of mining operations are largely determined by the primary or host commodity, so the supplies of some critical metals, cobalt being a good example, are at the mercy of the economics of another metal (for cobalt, this could be copper or nickel). More than a third of the EU's 2020 'critical metals and minerals' are recovered as companion metals or mineral by-products. Increasingly, the recognition that some of these companion metals have found their way onto the waste dumps has turned companies toward reprocessing former waste for their contained metals [25].

Table 3. Table showing the geological relationships between the main recovered metals and their metal 'companions'.

Main Host Metal	Companion Elements
Ni	Sc, Co, Ru, Rh, Pd, Os, Ir
Cu	Co, As, Se, Mo, Ag, Te, Re, Au
Fe	V, Sc, La, Ce, Pr, Nd
Zn	Ge, Ag, Cd, In, Tl
Pb	Ag, Sb, Tl, Bi
Al	V, Ga
Ti	Zr, Hf
REE	Y, Th
Mo	Re
Au	Ag, Te

Modified from [25].

While these sources can mitigate the need for newly mined resources, even if combined with optimized recycling and the complete remining of existing mine wastes, new mining will be necessary for many of the larger volume commodities on the critical minerals list [1].

5. The Future for Mining

5.1. Mineral Supply

A number of pessimistic views concerning long-term global mineral supplies have appeared in the literature in the modern era, beginning with the Club of Rome treatise on 'The Limits to Growth' [26] and periodically by other authors since then, e.g., [27–29]. However, it is economics, not geology, that define what companies report as 'reserves' (these are legally defined as 'economically extractable' bodies of mineral resources), and it is these figures upon which pessimistic perspectives declaring that we are 'running out' are erroneously based. Published studies show that geological resources are likely to be much higher than any future demands [30,31] and that the absolute exhaustion of the planet's metals and minerals will not be the major factor limiting the supply of raw materials. Indeed, for copper, the estimates for geologically feasible geological models suggest that currently 'undiscovered' copper deposits are highly likely to constitute more than 40 times the currently identified resources [32,33]. However, the waning success in deposit discovery (Figure 3) suggests that we are not finding these 'undiscovered' resources in a timely fashion.

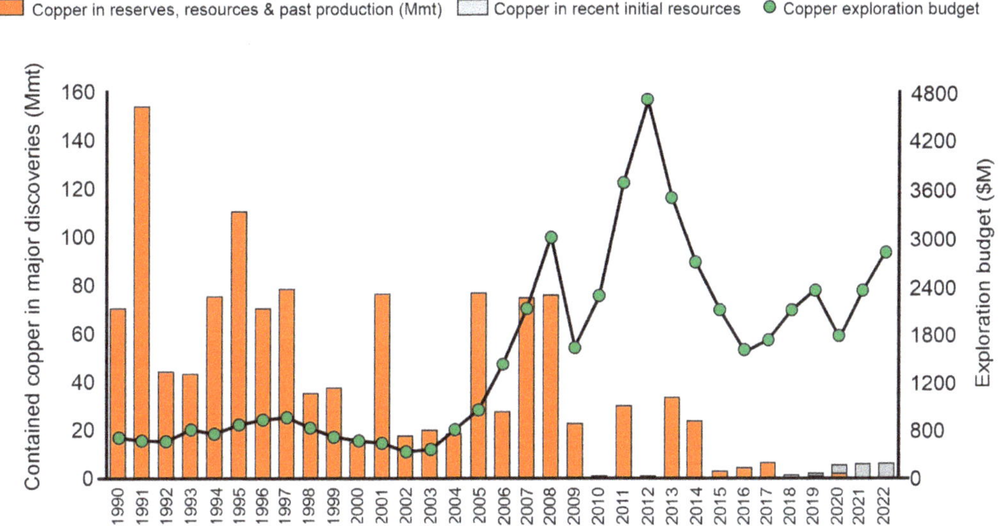

Figure 3. Mineral exploration budgets (green curve) and resultant discovery rates (orange and grey bars) of new copper deposits in the period 1990–2022 [12].

Despite healthy exploration budgets since the early 2000s, discovery rates are disappointingly small, and therefore, discovery rates need to improve rapidly to fill the supply gap as current mines begin to reach their end of life. This probably reflects the need to explore more effectively in 'new frontiers', which is discussed in Section 5.4 below. A further factor that is causing a squeeze in supply is the ever-lengthening timescale for turning discovered resources into producing mines. Using a demand model driven by the modest scenario of restricting the global temperature rise to +2 °C [34], it is apparent that we will need to mine practically all the known copper resources we have at hand. Therefore, there is extreme pressure to replenish those stocks if we are to satisfy ongoing demands. A recently published study by S&P Global [35] reports that for 127 mines opened since 2003, the average lead time from discovery to production is 15.7 years. The average lead time is variable depending on jurisdiction from only 10 years for Cote d'Ivoire to nearly 22 years for Brazil. Taking geological discovery to a bankable feasibility study appears to be the longest step of the process, averaging around 12 years for all mines. Embedded in this process are the acquisition of the necessary permissions and the successful development of a 'social license' to operate.

5.2. Environmental, Social, and Governance Constraints

Sustainable development has three clear dimensions from which John Elkington coined the term 'Triple Bottom Line' [36]; the three stand for the responsibilities that projects have to the economy, society, and biosphere, and are thus a more holistic analysis of the benefits of any project. The language of this analysis focuses on aligning sustainability and the intentions of a business when it comes to the profitability of projects. Given that the current mineral boom has been demanded by the need to arrest climate change and the knock-on effects for human society, holistic valuations need to be applied to any new mining projects if we are to avoid creating new environmental and social issues whilst we are seeking to solve the planetary emergency resulting from climate change [37]. The Triple Bottom Line concept of Elkington was flipped into the concept of a 'Triple Top Line' [38], where it is proposed that the focus should be to align environmental and social sustainability with business profitability from the inception of any product and work to a circular economy for manufacturing. This same philosophy might be translated from the manufacturing industry that Braungart and McDonough [38] studied, toward improving the business of mining, and this will be discussed below.

In building a societal license to operate, the single most important factor is developing trust with the broader society [39–42]. Building trust in the mining industry among the public at large has been most successful where engagement strategies have emphasized dialogue and relationship building [43,44]. Clearly there are lots of cases where mines have failed in either social or environmental aspects, and thus, trust has been lost through the implementation of improper business practices or other environmental or social equity failures [45].

5.3. Minerals versus Other Natural Capital

It is increasingly recognized that mineral resources in a particular locality are only one part of the 'Natural Capital' (NC) of the site. Ekins et al. [46] usefully defined the NC as follows: "Natural capital is a metaphor to indicate the importance of elements of nature (e.g., minerals, ecosystems, and ecosystem processes) to human society. Natural ecosystems are defined by a number of environmental characteristics that in turn determine the ecosystems' capacity to provide goods and services". This includes all the 'ecological capital' [46] that we can currently measure, including stocks like minerals, fossil fuels, forestry and agriculture, fisheries, and water resources. However, NC also includes the ecosystem services that include the site's ability to provide air and water filtration, flood protection, carbon storage, the pollination of crops, and habitats for wildlife. There are four types of NC: (1) the provision of resources (capital stock like minerals or forests) for production; (2) the absorption of wastes through production (either adding to or eroding the ecological capital); (3) basic life support systems (including ecosystem services); and (4) amenity services (e.g., the values of areas as wilderness or for their outstanding beauty). Apart from capital stock, it is difficult to capture the true value of these in a market sense, so we do not really know how much they contribute to the economy. We often take these services for granted and do not know what it would 'cost' if we lost them. The mining industry understands the need to account for this [47], and there are some recent examples that attempt holistic NC accounting in mining projects [48].

5.4. New Exploration Frontiers and Discovery

Encouragingly, geoscience demonstrates that there are still new classes of mineral deposits to be found for the commodities we require. In 2004, during an exploration for borates, Rio Tinto discovered the Jadar deposit in Serbia where more than 100 million tonnes of a Li and B-bearing mineral, jadarite, entirely new to science, was discovered [49]. The Jadar deposit is a member of the emerging new class of ore deposits known as volcano-sedimentary lithium deposits [50]. Explorations for these types of lithium deposits in the last ten years have already successfully discovered more than 60 million tonnes of contained lithium since the discovery of Jadar, and the prognosis for further similar discoveries is good.

Traditionally, exploration budgets have been focused on the Americas and Australia, countries with strong modern mining traditions. Canada is still the number one country for explorers, but new frontiers are opening up too. Recent trends show increased exploration spending in places like Saudia Arabia, where there has been a more than an 155% increase in exploration spending for 2022 [51], and neglected areas like Central Asia have great potential to deliver many of the minerals we need [52]. It is also healthy for the industry to encourage diversity in the geographic sources of critical materials, since diversifying the sources of supply will help to mitigate the security of supply issues linked to the creation of geographic monopolies [53].

The deep ocean floor has enormous untapped potential for metals and other minerals. Massive sulphides formed at seafloor spreading centres hold estimated resources of at least 6×10^8 tonnes of sulphide minerals [54], and the Solwara 1 deposit off Papua New Guinea alone had a signed off resource of more than a million tonnes of sulphide containing 7% Cu and 6 g/t Au. By far, the biggest prize on the deep seafloor is still the enormous potential for polymetallic nodules containing Mn, Ni, Co, Cu, REE, and other minor metals. The Clarion Clipperton Zone alone holds 21 billion tonnes of nodules with an average grade of 27% Mn, 1.3% Ni, 1.05% Cu, and 0.2% Co that amounts to 10 years of the current global Cu production and, respectively, 300 and 450 years of the current Mn and Co productions. In addition to the nodules, the less well-explored polymetallic crusts are also a potential resource of many metals [55].

Asteroid mining has been proposed as an option to avoid terrestrial mining e.g., [56] although the feasibility of returning metals other than very high-value precious metals is questionable [57], particularly in the foreseeable future, and certainly not in time to address the current 2050 net-zero deadline.

5.5. Brownfields and Deep Geological Discovery

Going deeper into existing mining operations to discover further resources offers a clear opportunity too. A review of historical explorations shows that most mineral discoveries have been made within 300 m of the surface [58], and yet mining is possible, even for base metals, below depths of 2 km (Figure 4). Discovery at such depths is possible using new geophysical and other targeting tools like mineral vectoring [59]. A recent example of new geophysics success is the use of seismic methods, a technique normally restricted to hydrocarbon exploration, in the discovery of deep extensions to the Navan Pb–Zn body in Ireland more than 1 km below the previously known mineralization [60]. Good geological reasoning and carefully targeted deep drilling was responsible for the Resolution porphyry discovery in Arizona, discovered around 1 km below an existing mining operation [61].

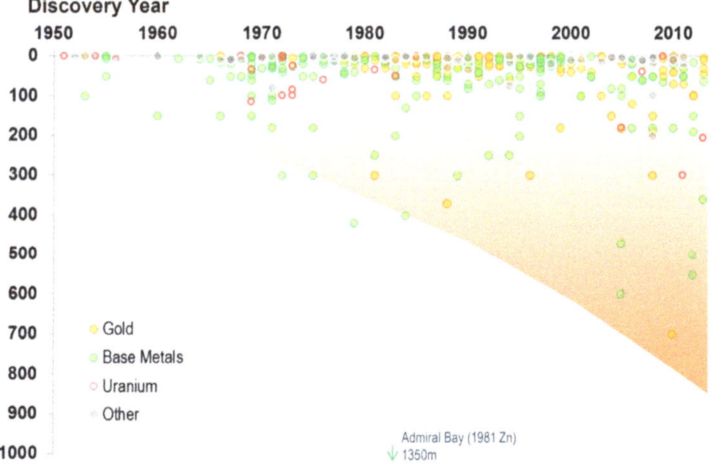

Figure 4. Graphic representation of depths to the tops of new discoveries of significant base and precious metal deposits through time [58].

Former mining camps remain prospective for new deposits. Europe has had a strong history of past mineral production and hosts many significant mineral showings (Figure 5), and there is still much potential with active exploration recorded through the continent [62]. A reassessment of the iron ore deposits of the northern part of Kiruna district resulted in the definition of huge REE resources in apatite-rich iron ores, previously mined only for their iron contents [63]. Likewise, the presence of Li-mica zinnwaldite in Cornish granites was known for many years [64], but the recognition of the distinct Li-rich 'G5' granite phase [65] coupled with a technology breakthrough suggest that it might be possible to process these ores economically. Li has been discovered in significant amounts in other regions of Europe (not shown in Figure 5).

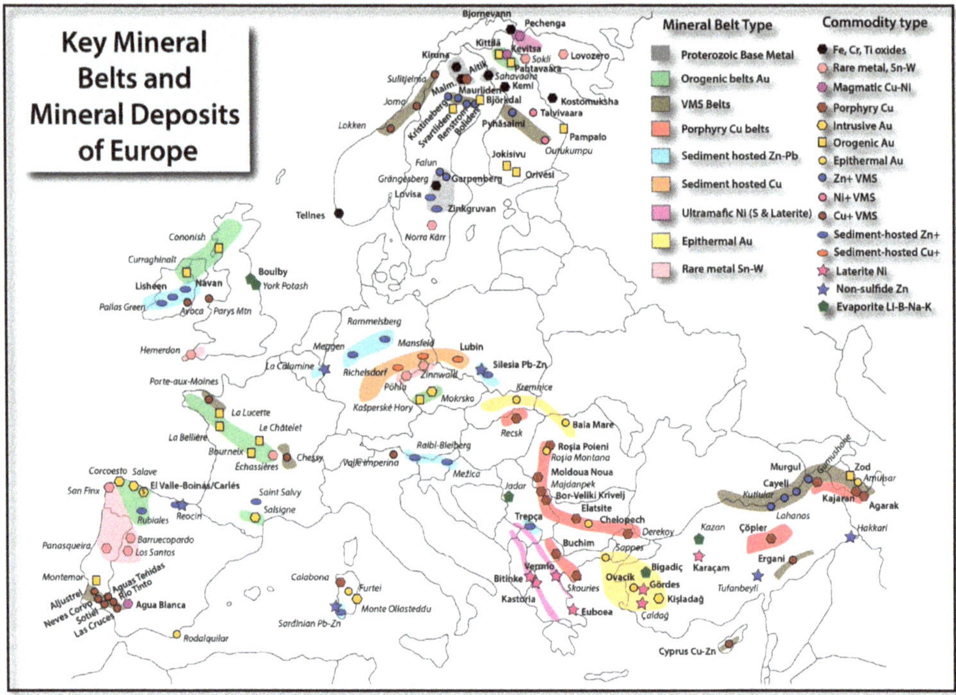

Figure 5. Map of Europe showing the locations of major mineral belts and key deposits (Authors own figure compiled from publically available sources, previously presented in 2013 [66] but unpublished in hard copy).

5.6. Unlocking Potential with Mineral Processing Advances

New processing techniques could be a real game changer for development of new mining projects on existing deposits that remain untapped because of mineral processing issues. It has been suggested there are a number of areas where processing could be improved [67], including using novel solvent extraction/electrowinning, leach processing, bioprocessing, flash smelting, geometallurgy, energy-efficient fine grinding, and introducing underground processing methods.

As an example, ionic liquids and deep eutectic solvents offer an alternative set of lixiviants to those currently used in conventional hydrometallurgy [68,69]. Being both powerful solvents and electrolytes they show great potential to be selective in both dissolution and recovery reactions. Deep eutectic solvents, like choline chloride are environmentally benign compounds, stable, relatively cheap to produce and testwork has shown their direct applicability to extraction of metals from sulfides, tellurides and precious metal minerals [68]. Ionic liquids and deep eutectic solvents have also been proposed to be combined with conventional floatation methods to form new hybrid processing methodologies [69].

Bioleaching is another emerging technology that has the potential to be applied more widely e.g., [70,71]. Overall, around 20% of the world's current copper production is estimated to come from bioleaching [72]. As a technology it is ideally placed for recovery of metals from low grade sulfidic wastes and has successfully been applied to copper, nickel and cobalt bearing sulfide-rich wastes. It also has great potential for the recovery of metals from oxide-rich lateritic ores and valuable oxide wastes [73].

Even conventional mineral processing methods can be improved and there is a strong trend in industry towards decarbonizing existing mining operations with introduction of electrical equipment and renewable energy strategies.

6. Rising to the Challenge—A Way Forward

It must be accepted that the mining industry is not a universally welcomed industry, and in Europe, it probably has the lowest acceptance for any industry sector [62,74]. There are many socially motivated disputes focused on the future development of terrestrial mines [75]. Deep ocean mining, one of the alternatives to terrestrial disturbance, is also a very emotive subject with public opinion largely against allowing operations [76]; although it is currently subject to a de facto moratorium, there is increasing pressure with legal backing to accelerate projects to the mining stage. There are complex and often conflicting arguments that support the choice of mining on land or under the ocean [77]. However, we see the need to mine, and thus, it comes down to a societal choice as to where that mining should be located [1].

What industry clearly needs to establish is an increased level of trust with society, particularly in the assurance that new projects are going to be different from the historical substandard projects and that projects will deliver a positive 'triple top line' as discussed earlier. New mining, therefore, needs to deliver outcomes that are net-positive for people and the planet in addition to being economically viable. A new proposed strategy embraces the concepts of Braungart and McDonough [38] and borrows their terminology in the term 'cradle-to-cradle' mining. This conceptualizes projects that are inherently reconstructive from the start, with project stakeholders embedded from the start, such that impacted communities are part of the decision-making process and thus co-owners of the project, with vested interests in designing something successful for all parties [78]. With such shared equity, the post-mining landscape can be designed as part of the mining process since there will be as many people interested in what happens after mining as there are interested in developing a successful mining venture. Mines are only temporary interventions at a particular site where subsurface minerals of societal need can be recovered at profit; the site itself should have a future that leaves a net-positive nature and people-positive outcome.

There are some positive mine closure examples that have striven to develop sustainable legacies. The Golden Pride operation in Tanzania provides a good case study, where there was strong community and regulator engagement, employee engagement through the closure transition, stakeholder-agreed post-mining land usage, an implemented plan of progressive reclamation, and pit closure that also considered the future needs of small-scale miners [79]. With a new intrinsically regenerative plan for future mining projects, even mines developed closer to home might be more acceptable to European society. Even abandoned and negative legacies have been shown to be able to be repurposed for a net-positive future [80,81].

Even with carefully crafted new project strategies, the 2050 target for net-zero would appear to have material demands that are difficult to deliver, and unless industry becomes more successful at discovering and commissioning new mines in time, it is increasingly recognized that the target of net-zero CO_2 emissions may not be achievable without the use of other interventions such as carbon capture and storage [82,83]. One way to help would be for society to reduce the mineral intensity of its future ambitions by adjusting lifestyles and thereby reducing demands [84]. It seems to be an enormous societal challenge to deliver a net-zero world responsibly, but it is one where geoscientists are clearly front and centre in making sure that its delivery is accomplished in a way that is sustainable, both for the planet and its people.

Funding: This research was funded by NERC research projects CoG[3] NE/M011488/1, LiFT NE/V007068/1, and CuBES NE/T002921/1; EU Framework Project CROCODILE 776473; and GCBC Project Bio+Mine.

Data Availability Statement: Sources for all data used in this paper are included in the references.

Acknowledgments: The author thanks the editors for the invite to write this paper based on a presentation presented at the SIMP congress in 2022 and the patience in waiting for the finished manuscript.

Conflicts of Interest: The author declares no conflicts of interest.

References

1. Herrington, R. Mining our green future. *Nat. Rev. Mater.* **2021**, *6*, 456–458. [CrossRef]
2. IEA. The Role of Critical Minerals in Clean Energy Transitions. 2021. Available online: https://www.iea.org/reports/the-role-of-critical-minerals-in-clean-energy-transitions (accessed on 31 May 2023).
3. Ekins, P.; Hughes, N.; Brigenzu, S.; Clarke, C.A.; Fischer-Kowalski, M.; Graedel, T.; Hajer, M.; Hashimoto, S.; Hatield-Dodds, S.; Havlik, P.; et al. *Resource Efficiency: Potential and Economic Implications*; United Nations Environment Program: Paris, France, 2016.
4. UNDP. Mapping Mining to the SDGs. 2016. Available online: https://www.undp.org/publications/mapping-mining-sdgs-atlas (accessed on 6 November 2023).
5. Klose, S.; Pauliuk, S. Sector-level estimates for global future copper demand and the potential for resource efficiency. *Resour. Conserv. Recycl.* **2023**, *193*, 106941. [CrossRef]
6. Kesler, S.E. Mineral Supply and Demand into the 21st Century, USGS Publication 1294. 2007. Available online: https://pubs.usgs.gov/circ/2007/1294/reports/paper9.pdf (accessed on 6 November 2023).
7. NOOA News April 2023. Available online: https://www.noaa.gov/news-release/greenhouse-gases-continued-to-increase-rapidly-in-2022 (accessed on 24 October 2023).
8. IEA. CO2 Emissions in 2022. 2022. Available online: https://www.iea.org/reports/co2-emissions-in-2022 (accessed on 6 November 2023).
9. Thiel, G.P.; Stark, A.K. To decarbonize industry, we must decarbonize heat. *Joule* **2021**, *5*, 531–550. [CrossRef]
10. BP. Energy Outlook 2022 Edition. 2022. Available online: https://www.bp.com/content/dam/bp/business-sites/en/global/corporate/pdfs/energy-economics/energy-outlook/bp-energy-outlook-2022.pdf (accessed on 6 November 2023).
11. Vidal, O.; Goffé, B.; Arndt, N. Metals for a low-carbon society. *Nat. Geosci.* **2013**, *6*, 894–896. [CrossRef]
12. S&P Capital IQ. Copper Discoveries Still Trending down Amid Increasing Budgets, Higher Prices. S&P Global Market Intelligence, Metals and Mining Research. 2023. Available online: https://www.spglobal.com/marketintelligence/en/news-insights/research/copper-discoveries-declining-trend-continues (accessed on 6 November 2023).
13. Iglesias-Émbil, M.; Valero, A.; Ortego, A.; Villacampa, M.; Vilaró, J.; Villalba, G. Raw material use in a battery electric car—A thermodynamic rarity assessment. *Resour. Conserv. Recycl.* **2020**, *158*, 104820. [CrossRef]
14. GOV.UK. Road Traffic Estimates in Great Britain. Headline Statistics. 2022. Available online: https://www.gov.uk/government/statistics/road-traffic-estimates-in-great-britain-2022/road-traffic-estimates-in-great-britain-2022-headline-statistics (accessed on 13 November 2023).
15. Hagelüken, C.; Goldmann, D. Recycling and circular economy—Towards a closed loop for metals in emerging clean technologies. *Miner. Econ.* **2022**, *35*, 539–562. [CrossRef]
16. Ambrose, H.; O'Dea, J. *Electric Vehicle Batteries: Addressing Questions about Critical Materials and Recycling*; Union of Concerned Scientists: Cambridge, MA, USA, 2021. Available online: https://www.ucsusa.org/resources/ev-battery-recycling (accessed on 6 November 2023).
17. Oberle, B.; Brereton, D.; Mihaylova, A. (Eds.) *Towards Zero Harm: A Compendium of Papers Prepared for the Global Tailings Review*; Global Tailings Review: St Gallen, Switzerland, 2020. Available online: https://globaltailingsreview.org/ (accessed on 6 November 2023).
18. Shaw, R.A.; Petavratzi, E.; Bloodworth, A.J. Resource Recovery from Mine Waste. In *Waste as a Resource*; Hester, R.E., Harrison, R.M., Eds.; The Royal Society of Chemistry: London, UK, 2013. [CrossRef]
19. Araujo, F.S.M.; Taborda-Llano, I.; Nunes, E.B.; Santos, R.M. Recycling and Reuse of Mine Tailings: A Review of Advancements and Their Implications. *Geosciences* **2022**, *12*, 319. [CrossRef]
20. Šajn, R.; Ristović, I.; Čeplak, B. Mining and Metallurgical Waste as Potential Secondary Sources of Metals—A Case Study for the West Balkan Region. *Minerals* **2022**, *12*, 547. [CrossRef]
21. Wanhainen, C.; Broman, C.; Martinsson, O. The Aitik Cu–Au–Ag deposit in northern Sweden: A product of high salinity fluids. *Miner. Depos.* **2003**, *38*, 715–726. [CrossRef]
22. Mining.Com. 2021. Available online: https://www.mining.com/mining-copper-tailings-could-answer-supply-deficits-later-this-decade/0 (accessed on 6 November 2023).

23. Legge, H.; Müller-Falcke, C.; Nauclér, T.; Östgren, E. Creating the Zero-Carbon Mine, McKinsey & Company. 2021. Available online: https://www.mckinsey.com/industries/metals-and-mining/our-insights/creating-the-zero-carbon-mine (accessed on 6 November 2023).
24. Mudd, G.M.; Yellishetty, M.; Reck, B.K.; Graedel, T.E. Quantifying the Recoverable Resources of Companion Metals: A Preliminary Study of Australian Mineral Resources. *Resources* **2014**, *3*, 657–671. [CrossRef]
25. Yilmaz, E.; Koohestani, B.; Cao, S. Recent practices in mine tailings' recycling and reuse. In *Managing Mining and Minerals Processing Wastes, Concepts Design and Applications*; Qi, C., Benson, C.H., Eds.; Elsevier: Amsterdam, The Netherlands, 2023; pp. 271–304. [CrossRef]
26. Meadows, D.H.; Meadows, D.L.; Randers, J.; Behrens, W.W. *The Limits to Growth*; Universe: New York, NY, USA, 1972.
27. Skinner, B.J. Second iron age ahead. *Am. Sci.* **1976**, *64*, 258–269.
28. Laherrere, J. Copper Peak. (Posted by de Sousa, L., 2010). *Oild Rum Eur.* **2010**, *6307*, 1–27. Available online: http://europe.theoildrum.com/node/6307 (accessed on 6 November 2023).
29. Sverdrup, H.; Ragnarsdóttir, K.V. Natural Resources in a Planetary Perspective. *Geochem. Perspect.* **2014**, *3*, 129–341. [CrossRef]
30. Arndt, T.; Fontboté, L.; Hedenquist, J.W.; Kesler, S.E.; Thompson, J.F.H.; Wood, D.G. Future Global Mineral Resources. *Geochem. Perspect.* **2017**, *6*, 1–171. [CrossRef]
31. Jowitt, S.M.; Mudd, G.M.; Thompson, J.F.H. Future availability of non-renewable metal resources and the influence of environmental, social, and governance conflicts on metal production. *Commun. Earth Environ.* **2020**, *1*, 13. [CrossRef]
32. Kesler, S.E.; Wilkinson, B.H. Earth's copper resources estimated from tectonic diffusion of porphyry copper deposits. *Geology* **2008**, *36*, 255–258. [CrossRef]
33. Mudd, G.M.; Weng, Z.; Jowitt, S.M. A Detailed Assessment of Global Cu Resource Trends and Endowments. *Econ. Geol.* **2013**, *108*, 1163–1183. [CrossRef]
34. Seck, G.S.; Hache, E.; Bonnet, C.; Simoën, M.; Carcanague, S. Copper at the crossroads: Assessment of the interactions between low-carbon energy transition and supply limitations. *Resour. Conserv. Recycl.* **2020**, *163*, 105072. [CrossRef]
35. Manalo, P. Discovery to Production Averages 15.7 Years for 127 Mines. 2023. Available online: https://www.spglobal.com/marketintelligence/en/news-insights/research/discovery-to-production-averages-15-7-years-for-127-mines (accessed on 9 November 2023).
36. Adams, C.; Frost, G.; Webber, W. Enter the Triple Bottom Line. In *The Triple Bottom Line*; Routledege: London, UK, 2013; ISBN 9781849773348. [CrossRef]
37. Pell, R.; Tijsseling, L.; Goodenough, K.; Wall, F.; Dehaine, Q.; Grant, A.; Deak, D.; Yan, X.; Whattoff, P. Towards sustainable extraction of technology materials through integrated approaches. *Nat. Rev. Earth Environ.* **2021**, *2*, 665–679. [CrossRef]
38. Braungart, M.; McDonough, W. *Cradle to Cradle: Remaking the Way We Make Things*; North Point Press: New York, NY, USA, 2002.
39. Thomson, I.; Boutilier, R.G. Social license to operate. In *SME Mining Engineering Handbook*; Elsevier: Amsterdam, The Netherlands, 2011; Volume 1, pp. 1779–1796.
40. Prno, J.; Slocombe, D.S. Exploring the origins of 'social license to operate' in the mining sector: Perspectives from governance and sustainability theories. *Resour. Policy* **2012**, *37*, 346–357. [CrossRef]
41. Moffat, K.; Zhang, A. The paths to social licence to operate: An integrative model explaining community acceptance of mining. *Resour. Policy* **2014**, *39*, 61–70. [CrossRef]
42. Suopajärvi, L.; Umander, K.; Jungsberg, L. Social license to operate in the frame of social capital: Exploring local ac-ceptance of mining in two rural municipalities in the European North. *Resour. Policy* **2019**, *64*, 101498.
43. Prno, J. An analysis of factors leading to the establishment of a social licence to operate in the mining industry. *Resour. Policy* **2013**, *38*, 577–590. [CrossRef]
44. Mercer-Mapstone, L.; Rifkin, W.; Louis, W.R.; Moffat, K. Company-community dialogue builds relationships, fairness, and trust leading to social acceptance of Australian mining developments. *J. Clean. Prod.* **2018**, *184*, 671–677. [CrossRef]
45. Herrington, R.; Gordon, S. Delivering Critical Raw Materials: Ecological, Ethical and Societal Issues. In *Geoethics for Future: Facing Global Challenges*; Elsevier: Amsterdam, The Netherlands, *under review*.
46. Ekins, P.; Simon, S.; Deutsch, L.; Folke, C.; De Groot, R. A framework for the practical application of the concepts of critical natural capital and strong sustainability. *Ecol. Econ.* **2003**, *44*, 165–185. [CrossRef]
47. Boldy, R.; Santini, T.; Annandale, M.; Erskine, P.D.; Sonter, L.J. Understanding the impacts of mining on ecosystem services through a systematic review. *Extr. Ind. Soc.* **2021**, *8*, 457–466. [CrossRef]
48. Meney, K.; Pantelic, L.; Cooper, T.; Pittard, M. *Natural Capital Accounting for The Mining Sector: Beenup Site Pilot Case Study*; Prepared by Syrinx Environmental PL for BHP; 2023; ISBN 978-0-6456956-0-1. Available online: https://www.bhp.com/-/media/documents/environment/2023/230502_bhpbeenuppilotcasestudynaturalcapitalaccountingreport.pdf (accessed on 6 November 2023).
49. Stanley, C.J.; Jones, G.C.; Rumsey, M.S.; Blake, C.; Roberts, A.C.; Stirling, J.A.; Carpenter, G.J.; Whitfield, P.S.; Grice, J.D.; Lepage, Y. Jadarite, LiNaSiB3O7(OH), a new mineral species from the Jadar Basin, Serbia. *Eur. J. Miner.* **2007**, *19*, 575–580. [CrossRef]
50. Benson, T.R.; Coble, M.A.; Dilles, J.H. Hydrothermal enrichment of lithium in intracaldera illite-bearing clay-stones. *Sci. Adv.* **2023**, *9*, eadh8183. [CrossRef] [PubMed]
51. S&P Global Market Intelligence. World Exploration Trends 2022. 2022. Available online: https://www.spglobal.com/marketintelligence/en/news-insights/blog/world-exploration-trends-2022 (accessed on 17 November 2023).

52. Vakulchuk, R.; Overland, I. Central Asia is a missing link in analyses of critical materials for the global clean energy transition. *One Earth* **2021**, *4*, 1678–1692. [CrossRef]
53. Herrington, R. Road map to mineral supply. *Nat. Geosci.* **2013**, *6*, 892–894. [CrossRef]
54. Hannington, M.D.; Jamieson, J.; Monecke, T.; Petersen, S.; Beaulieu, S. The abundance of seafloor massive sulfide deposits. *Geology* **2011**, *39*, 1155–1158. [CrossRef]
55. Hein, J.R.; Koschinsky, A. Deep-Ocean Ferromanganese Crusts and Nodules. In *Treatise on Geochemistry*, 2nd ed.; Holland, H.D., Turekian, K.K., Eds.; Elsevier: Amsterdam, The Netherlands, 2014; Chapter 11; pp. 273–291.
56. Zacny, K.; Cohen, M.M.; James, W.W.; Hilscher, B. Asteroid mining. In *AIAA Space 2013 Conference and Exposition*; American Institute of Aeronautics and Astronautics: Reston, VA, USA, 2013; p. 5304.
57. Cannon, K.M.; Gialich, M.; Acain, J. Precious and structural metals on asteroids. *Planet. Space Sci.* **2023**, *225*, 105608. [CrossRef]
58. Schodde, R. Uncovering exploration trends and the future: Where's exploration going? In Proceedings of the IMARC Conference, Melbourne, Australia, 22–26 September 2014. Available online: http://minexconsulting.com/wp-content/uploads/2019/04/IMARC-Presentation-by-Richard-Schodde-Sept-2014-FINAL.pdf (accessed on 6 November 2023).
59. Wilkinson, J.J.; Cooke, D.; Baker, M.; Chang, Z.; Wilkinson, C.; Chen, H.; Fox, N.; Hollings, P.; White, N.; Gemmell, J.B.; et al. Porphyry indicator minerals and their mineral chemistry as vectoring and fertility tools. In *Application of Indicator Mineral Methods to Bedrock and Sediments*; McClenaghan, M.B., Layton-Matthews, D., Eds.; Geological Survey of Canada Open File No. 8345; Natural Resources Canada, Geological Survey of Canada: Ottawa, ON, Canada, 2017; pp. 67–77.
60. Ashton, J.H.; Beach, A.; Blakeman, R.J.; David Coller, D.; Henry, P.; Lee, R.; Hitzman, M.; Hope, C.; Huleatt-James, S.; O'Donovan, B.; et al. Discovery of the Tara Deep Zn-Pb Mineralization at the Boliden Tara Mine, Navan, Ireland: Success with Modern Seismic Surveys. In *Metals, Minerals, and Society*; Arribas, A.M., Mauk, J.L., Eds.; Society of Economic Geologists (SEG): Littleton, CO, USA, 2018; Volume 21. [CrossRef]
61. Manske, S.L.; Paul, A.H. Geology of a Major New Porphyry Copper Center in the Superior (Pioneer) District, Arizona. *Econ. Geol.* **2002**, *97*, 197–220. [CrossRef]
62. Pellegrini, M. Fostering the mining potential of the European Union. *Eur. Geol.* **2016**, *42*, 10–14.
63. Martinsson, O. Genesis of the Per Geijer apatite iron ores, Kiruna area, northern Sweden. In Proceedings of the 13th Biennial Meeting of The SGA, Nancy, France, 23–27 August 2015.
64. Siame, E.; Pascoe, R. Extraction of lithium from micaceous waste from china clay production. *Miner. Eng.* **2011**, *24*, 1595–1602. [CrossRef]
65. Simons, B.; Andersen, J.C.; Shail, R.K.; Jenner, F.E. Fractionation of Li, Be, Ga, Nb, Ta, In, Sn, Sb, W and Bi in the peraluminous Early Permian Variscan granites of the Cornubian Batholith: Precursor processes to magmatic-hydrothermal mineralisation. *Lithos* **2017**, *278–281*, 491–512. [CrossRef]
66. Herrington, R.J. European Mineral Belts: A review of past and current production—Highlighting the future potential. *PDAC Shortcourse New Mines in the Old World*, 1 March 2013.
67. Hoal, K.O.; McNulty, T.P.; Schmidt, R. Metallurgical Advances and Their Impact on Mineral Exploration and Mining. In *Wealth Creation in the Minerals Industry: Integrating Science, Business, and Education*; Doggett, M.D., Parry, J.R., Eds.; Special Publications of The Society of Economic Geologists; Society of Economic Geologists: Littleton, CO, USA, 2005; Volume 12. [CrossRef]
68. Jenkin, G.R.T.; Al-Bassam, A.Z.M.; Harris RCAbbott, A.P.; Smith, D.J.; Holwell, D.A.; Chapman, R.J.; Stanley, C.J. The application of deep eutectic solvent ionic liquids for environmentally-friendly dissolution and recovery of precious metals. *Miner. Eng.* **2016**, *87*, 18–24. [CrossRef]
69. Tian, G.; Liu, H. Review on the mineral processing in ionic liquids and deep eutectic solvents. *Miner. Process. Extr. Met. Rev.* **2022**, *45*, 130–153. [CrossRef]
70. Johnson, D.B. Biomining—Biotechnologies for extracting and recovering metals from ores and waste materials. *Curr. Opin. Biotechnol.* **2014**, *30*, 24–31. [CrossRef]
71. Johnson, D.B.; Roberto, F.F. Evolution and Current Status of Mineral Bioprocessing Technologies. In *Biomining Technologies*; Johnson, D.B., Bryan, C.G., Schlömann, M., Roberto, F.F., Eds.; Springer: Cham, Switzerlands, 2022; 314p. [CrossRef]
72. Roberto, F.F.; Schippers, A. Progress in bioleaching: Part B, applications of microbial processes by the minerals industries. *Appl. Microbiol. Biotechnol.* **2022**, *106*, 5913–5928. [CrossRef]
73. Santos, A.L.; Schippers, A. Reductive Mineral Bioprocessing. In *Biomining Technologies*; Johnson, D.B., Bryan, C.G., Schlömann, M., Roberto, F.F., Eds.; Springer: Cham, Switzerlands, 2022; pp. 261–274. [CrossRef]
74. Lesser, P.; Poelzer, G.; Tost, M. Perceptions of Mining in Europe, Summary Report, MIREU Survey Results. 2020. Available online: https://mireu.eu/documents/mireu-survey-results-perceptions-mining-europe (accessed on 31 May 2023).
75. Kivinen, S.; Kotilainen, J.; Kumpula, T. Mining conflicts in the European Union: Environmental and political perspectives. *Fenn.—Int. J. Geogr.* **2020**, *198*, 163–179. [CrossRef]
76. Kim, R.E. Should deep seabed mining be allowed? *Mar. Policy* **2017**, *82*, 134–137. [CrossRef]
77. Lèbre, É.; Kung, A.; Savinova, E.; Valenta, R.K. Mining on land or in the deep sea? Overlooked considerations of a reshuffling in the supply source mix. *Resour. Conserv. Recycl.* **2023**, *191*, 106898. [CrossRef]
78. Herrington, R.; Tibbett, M. Cradle-to-cradle mining: A future concept for inherently reconstructive mine systems? In Proceedings of the Mine Closure 2022: 15th Conference on Mine Closure, Perth, Australia, 4–6 October 2022; pp. 19–28. [CrossRef]

79. Stevens, R.; Hartnett (Sinclair), J.; Kasege, G. Inclusive Closure and Post-Mining Transition at the Golden Pride Mine, Tanzania. *IGF Case Study*. 2022. Available online: https://www.iisd.org/publications/brief/igf-case-study-golden-pride-mine-tanzania (accessed on 6 November 2023).
80. Finucane, S.; Tarnowy, K. New uses for old infrastructure: 101 things to do with the 'stuff' next to the hole in the ground. In Proceedings of the 13th International Conference on Mine Closure, Perth, Australia, 3–5 September 2019; pp. 479–496. [CrossRef]
81. Smit, T. *Eden*; Corgi Books: London, UK, 2001.
82. Ali, M.; Jha, N.K.; Pal, N.; Keshavarz, A.; Hoteit, H.; Sarmadivaleh, M. Recent advances in carbon dioxide geological storage, experimental procedures, influencing parameters, and future outlook. *Earth-Sci. Rev.* **2021**, *225*, 103895. [CrossRef]
83. Smith, S.M.; Geden, O.; Nemet, G.; Gidden, M.; Lamb, W.F.; Powis, C.; Bellamy, R.; Callaghan, M.; Cowie, A.; Cox, E.; et al. *The State of Carbon Dioxide Removal*, 1st ed.; Mercator Research Institute on Global Commons and Climate Change (MCC): Berlin, Germany, 2023.
84. Marín-Beltrán, I.; Demaria, F.; Ofelio, C.; Serra, L.M.; Turiel, A.; Ripple, W.J.; Mukul, S.A.; Costa, M.C. Scientists' warning against the society of waste. *Sci. Total. Environ.* **2022**, *811*, 151359. [CrossRef]

 *minerals*

Article

Unlocking Strategic and Critical Raw Materials: Assessment of Zinc and REEs Enrichment in Tailings and Zn-Carbonate in a Historical Mining Area (Montevecchio, SW Sardinia)

Lorenzo Sedda, Giovanni De Giudici, Dario Fancello, Francesca Podda and Stefano Naitza *

Dipartimento di Scienze Chimiche e Geologiche, Università degli Studi di Cagliari, 09124 Cagliari, Italy; lorenzo.sedda@unica.it (L.S.); gbgiudic@unica.it (G.D.G.); dario.fancello@unica.it (D.F.); fpodda@unica.it (F.P.)
* Correspondence: snaitza@unica.it

Abstract: Mining wastes are often both a potential source of Strategic and Critical Raw Materials (SRMs and CRMs) and a threat to the environment. This study investigated the potential of mining wastes from the Montevecchio district of Sardinia, Italy, as a source of SRMs and CRMs. The tailings from Sanna mine processing plant were characterized by X-ray diffraction, Scanning Electron Microscopy, and Plasma Mass Spectometry, showing contents of 1.2 wt% of lead, 2.6 wt% of zinc, and about 600 mg/kg of Rare Earth Elements (REEs). White patinas formed in the riverbed, composed by Zn-bearing minerals (hydrozincite and zincite), also contain about 2900 mg/kg of REEs. Characterization of white patinas along the Rio Roia Cani evidenced that their precipitation from water also involves an uptake of Rare Earth Elements, enhancing their contents by an order of magnitude compared with tailings. The process of REEs concentration in Zn-bearing minerals of white patinas is a candidate as a tool for the economic recovery of these elements. These findings suggest that mining wastes from the Montevecchio district could be considered a potential resource for extracting SRMs and CRMs.

Keywords: metal recovery; critical raw materials; green economy

 **check for updates**

Citation: Sedda, L.; De Giudici, G.; Fancello, D.; Podda, F.; Naitza, S. Unlocking Strategic and Critical Raw Materials: Assessment of Zinc and REEs Enrichment in Tailings and Zn-Carbonate in a Historical Mining Area (Montevecchio, SW Sardinia). *Minerals* **2024**, *14*, 3. https://doi.org/10.3390/min14010003

Academic Editor: María Ángeles Martín-Lara

Received: 20 October 2023
Revised: 4 December 2023
Accepted: 14 December 2023
Published: 19 December 2023

1. Introduction

The search for Critical Raw Materials (CRMs) for the energy transition and an ever-greener economy is increasingly important for the European Union in order to secure supplies and reduce dependence on unsafe sources. Since 2011, the EU updates its list of CMRs every three years, now including 35 raw materials of critical and strategic relevance [1]. This prompted the European Commission to support new exploration activities in the old mine districts of the EU members. A specific target of this new research is represented by mining wastes and tailings, whose exploitation may allow the recovery of neglected resources, enhancing at the same time the environmental quality of the involved areas by removing large volumes of contaminated materials [2].

Tailings are different types of waste materials produced by the processing of ore bodies, containing economic metals and minerals, which pose a threat to the environment and various ecosystems. That implies that tailings represent an important source of contamination for the environment for various ecosystems and for the people. A solution could be the re-treatment of these tailings for the extraction of these metals and metalloids if these are contained in economic contents, implicating the removal of a source of contaminations. This solution has several advantages: it allows the recovery of valuable resources that would otherwise have been lost; it reduces the amount of tailings that need to be stored, thus reducing the risk of environmental contamination; and it improves the overall sustainability of the mining sector. Tailings reprocessing must adhere to responsible and sustainable practices, minimizing environmental impact, ensuring worker and community safety, and maintaining cost-effectiveness through market price recovery of metals and metalloids. An essential phase for the definition of these projects consists in the mapping and characterization of the various mining landfills, to define their volumes, surfaces, and concentrations of raw materials [3–5].

Sardinia was a relevant mining region at the European scale, particularly for the extraction of lead and zinc ores. The Montevecchio district, located in SW Sardinia, was one of Italy's most important mining areas for over a century. Mining activities in the district ended in 1991 [6]. Today, it is one of Sardinia's most historically significant mining sites, featuring architecturally valuable abandoned mining buildings and shafts. Beyond this legacy of industrial archaeology, in Montevecchio there are numerous and very large mining dumps that represent a serious environmental problem on a national scale [7]. At the same time, it is increasingly evident from recent studies and geochemical/environmental characterizations [8,9] that these materials may contain significant amounts of base metals and strategic elements, thus becoming a potentially relevant target for CRM research.

The mining area is located between the municipalities of Arbus and Guspini, in Southwestern Sardinia (Figure 1). The intense underground mining activity deeply exploited a huge, NE-SW-directed mineralized vein system, producing about 2 million cubic meters of excavations voids and over 2 million cubic meters of mining dumps. Moreover, the processing plants produced over 6 million cubic meters of tailings [10]. The mining activity was focused on the extraction of minerals containing lead, zinc, and silver, but a large set of associated metals (Cu, Ni, Co, Bi, Sb, In, Ga, Ge, etc.) was recovered in the metallurgical cycles. The mine exploited two types of mineralization, sulfide, and oxidated ores, that required two different treatments to be recovered. Accordingly, two different plants were built in the mine area: the eastern plant (named the Principe Tommaso plant) processed the sulfide minerals, and the western plant (named the Sanna plant) processed the oxidate minerals. Thus, very different tailing materials are currently accumulated in the two old plant areas.

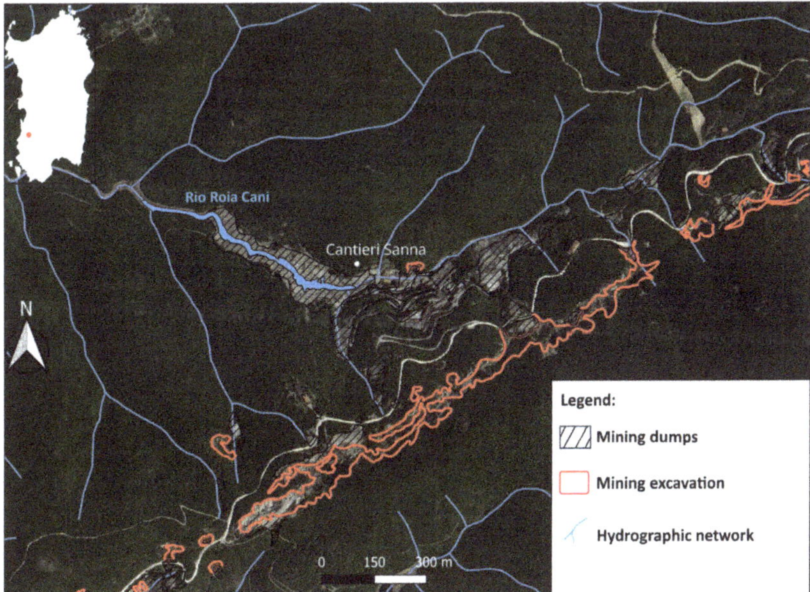

Figure 1. The study area. The main elements related to mining operations (mining dumps and excavation) are evidenced.

This work is focused on the tailings of the Sanna plant, whose composition and current arrangement are a product of the technological evolution of the processing systems for over a century. In detail, oxidate minerals from the whole Montevecchio district were initially processed in the earlier Sanna-Eleonora plant (1869–1936). The process involved a first hammer-sorting, a second hand-sorting, and finally, a separation using jigs and shaking tables. Starting from 1937 and up to 1980, the ores were treated in the new Sanna plant where minerals were first comminuted by jaw and gyratory crushers, then finely grounded by ball mills, and finally enriched by flotation [11]. Both the old and the new plants were set along the valley of the Rio Roia Cani stream, which flows from the mine area towards the west. The resulting tailings, of variable grain size, are now accumulated along the valley and are affected by important erosional phenomena that imply instability of the dumps and produce solid and chemical transport of potentially polluting materials, generating contamination over a large area downstream. Several past studies in the Montevecchio district pointed out that rivers of the area may carry high amounts of metals in solution after having interacted with the mining environments [12–17]. Evidence of the extent of these phenomena are the mineral precipitates (white patinas) that frequently encrust outcropping rocks, pebbles, and sediments at the bottom of the Roia Cani riverbed. They appear analogous to the hydrozincite bio-precipitates induced by bacterial activity, extensively documented in another mining area of the same district (Ingurtosu Mine) along the Rio Naracauli [18,19]. With the dual objective of verifying the possible presence of metals and CRMs in the tailings and the environmental conditions of the area, the study starts from a mineralogical and chemical characterization of the wastes of the Sanna plant but also encompasses some representative minerals of the ore that fed the plant, the Rio Roia Cani waters that directly interact with the wastes, and particularly, the white patinas along the Rio Roia Cani riverbed. Indeed, these latter not only may be considered as markers of the grade of diffusion of pollutants and of their biologically mediated interactions with the natural environment but also may provide interesting information about natural processes of re-concentration of metals and CRMs that may have potential economic implications.

2. Geological Setting

The Montevecchio mine area is located in the external nappe zone (allochthonous Arburese unit) of the Palaeozoic basement of Southwestern Sardinia, close to the contact with the Variscan Foreland (Autochthonous Iglesiente Unit). These low-grade metamorphic units are intruded by the granitoids of the Arbus pluton (304 +/− 1 Ma), which determined only moderate thermometamorphic effects in a relatively thin aureola [20].

The geology of the area is dominated by the Upper Cambrian-Lower Ordovician rocks of the Arenarie di San Vito Formation (Arburèse tectonic Unit), consisting of fine metasandstones, metasiltstones, and metargillites [21]. This sequence is unconformably covered by Middle-Ordovician felsic metavolcanic rocks of rhyolitic-rhyodacitic composition [22,23].

The Palaeozoic rocks are in turn unconformably covered by Tertiary clastic sediments of the Ussana Formation, consisting of conglomerates, breccias, and sandstones with a red-purple clayey-sandstone matrix [21], and by the Oligo-Miocene calc-alkaline sequences of the Monte Arcuentu volcanic complex [23].

3. The Montevecchio Ore Deposits

The large Montevecchio hydrothermal vein system [6] consists of a series of mineralized veins that follow with a "pinch-and-swell" attitude the northern contact of the Arbus pluton for a length of over ten kilometres in a NE-SW direction (Figure 2). The vein field has held great economic importance in the past, having been among the largest in Western Europe; the thickness of the mineralized veins could reach 25–30 m at some points. From east to west, the veins are called: Sant'Antonio, Piccalinna, Sanna, Telle, and Casargiu. Although dominated by lead and zinc sulfides, the composition of the ore may vary significantly from one vein to another, and sometimes even within the same vein, characterized by banded to brecciated textures. In order of abundance or frequency, the economic minerals at Montevecchio are sphalerite [ZnS], galena [PbS], fahlore [($Cu_6[Cu_4(Fe,Zn)_2](As,Sb)_4S_{13}$], chalcopyrite [$CuFeS_2$], bournonite [$PbCuSbS_3$], Ni-Co sulfoarsenides, and pyrite [FeS_2]. The nature of the minerals forming the gangue is equally variable, they are quartz [SiO_2], barite [$BaSO_4$], siderite [$FeCO_3$], Zn carbonate [$ZnCO_3$], calcite [$CaCO_3$], ankerite [$(Ca,Mg,Mn,Fe)(CO_3)_2$], and goethite [$FeO(OH)$] [24–26].

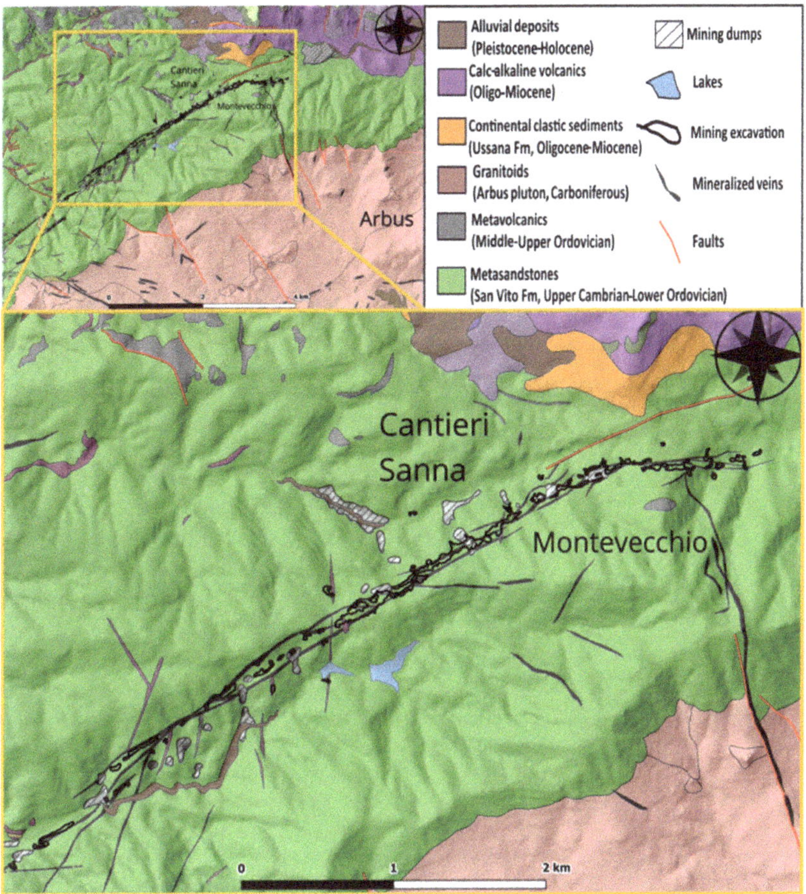

Figure 2. Geological sketch map of the study area.

The Sanna Vein

The Sanna vein was the primary feeder of the Sanna processing plant in Western Montevecchio. The vein displays a smaller longitudinal extension than the other veins of the district (about 300 m), but it has a considerable thickness (15–20 m). A distinctive feature of the Sanna vein is its composition, dominated by supergene minerals. Supergene oxidation processes strongly affected the vein, with an average portion of the oxidated ore reaching about fifty meters in depth and oxidized columnar portions that go down up to 200 m below the topographic surface. The development of supergene processes has deeply transformed the primary sulfide ore (basically: galena, sphalerite, chalcopyrite, and fahlore, with Zn grades of 5–6 wt%, and Pb grades of 1–2 wt%) into secondary phases as anglesite [$PbSO_4$], cerussite [$PbCO_3$], iron hydroxides, zinc, and copper secondary minerals (Figure 3). The gangue is mainly sideritic with subordinate quartz; barite may become relatively abundant in the upper parts of the vein; the lead tends to increase at depth while remaining subordinate to zinc [27].

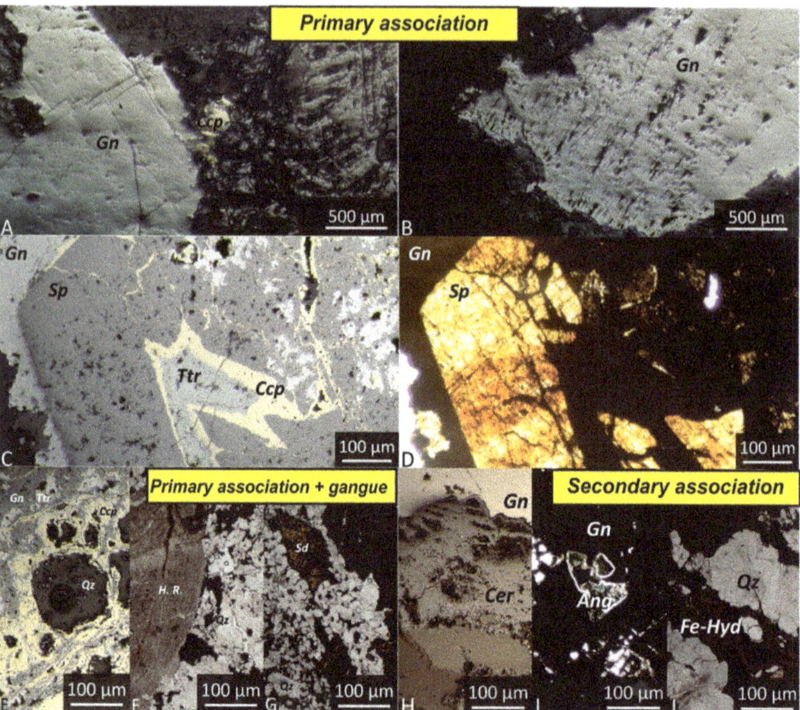

Figure 3. Mineral assemblage of the Sanna vein. (**A**) Yellow chalcopyrite (Ccp) growing at the edge of galena (Gn). (**B**) Galena (Gn) brecciated and deformed (as evidenced by the various orientations assumed by the triangular pits), probably grown during a dynamic phase. (**C**) Tetrahedrite (Ttr) and chalcopyrite (Ccp) included in sphalerite (Sp), in contact with galena (Gn). (**D**) Sphalerite (Sl) in transmitted light, transparency suggests low iron content. (**E**) Quartz crystals (Qz), inside galena (Gn), surrounded by chalcopyrite (Ccp) and tetrahedrite (Ttr). (**F**) Breccia texture with a host rock fragment (H.R.) surrounded by quartz fragments (Qz). (**G**) Siderite (Sd) crystals surrounded by quartz crystals (Qz) forming a mosaic texture. (**H**) Cerussite (Cer) aggregate growing on galena (Gn). (**I**) A crystal of anglesite (Ang) grown within galena (Gn). (J Iron hydroxides (Fe-Hyd) precipitated between quartz crystals (Qz). (**A–C,E,H**): plane-polarized reflected light; (**D,F,G,I,J**): plane-polarized transmitted light. Mineral abbreviations according to Warr (2021) [28].

4. The Sanna Plant Area

The Sanna plant area (about 130,000 m^2) is characterized by the remnants of two past mineral processing plants, consisting of tailings, jigging residues, and waste rocks accumulated along the Rio Roia Cani. During mining operations, tailings were discharged as sludge from the treatment plant to the stream, through a drainage tunnel. Initially, sludge was blocked by a series of dams to form tailing ponds. In past mining operations, dams were opened periodically to make space in the ponds. After the mine closure, the main dam collapsed, and the tailings were partially eroded and washed away. Evidence of a consistent solid transport of these materials may be easily traced several kilometres downstream. Currently, the drainage tunnel is damaged, and tailings are affected by rills and gullies that induce instability and small landslides. Tailings preserve the original flat attitude of the pond surface, whilst the jigging wastes form large heaps along the flanks of the valley (Figure 4), quite unstable and incised by surface waters. In the field, tailings and jigging residues are also easily distinguishable by their granulometry; tailings vary from fine sands to clays, while jigging residues vary from gravel to coarse sand. The tailings cover a surface of about 46.800 m^2, attaining a thickness of more than 10 m. They display a positive gradation that is interpreted as resulting from sedimentation processes into the tailing pond where the waste materials from the plant were discharged. In addition to the solid transport of the waste materials, evidence of solution transport of elements in the water of the Rio Roia Cani is provided by the frequent occurrence of reddish mineral precipitates and white patinas along the riverbed, particularly where the drainage gallery is broken (Figure 4). The area downstream the Rio Roia Cani is low-populated: main potential targets of contamination are surface- and groundwaters and, in general, the flora and fauna present along the river valley, whose habitat and whose development could be conditioned by the presence of pollutants.

Figure 4. The interaction between the flow of the Rio Roia Cani and the mining wastes.

To complete the environmental frame of this old industrial area, also the mine excavations, located upstream close to the plant area must be considered. They constitute an environment of interaction between surface water and mineralized, potentially polluting rocks. Only part of these waters is now conveyed to the Rio Roia Cani, the great part feeds the groundwater system through the underground mining works.

5. Materials and Methods

5.1. Sampling

Eight samples were collected from the tailings of the Sanna processing plant (Figure 5).

Sampling was carried out following the stratigraphic overlap of different materials, distinguished by chromatic differences and different granulometry (Figure 6).

Nine samples of the white patinas present along the bed of the Rio Roia Cani (Figure 7) were also collected. Sampling was realized paying particular attention to avoid the removal of detrital material present under the patinas.

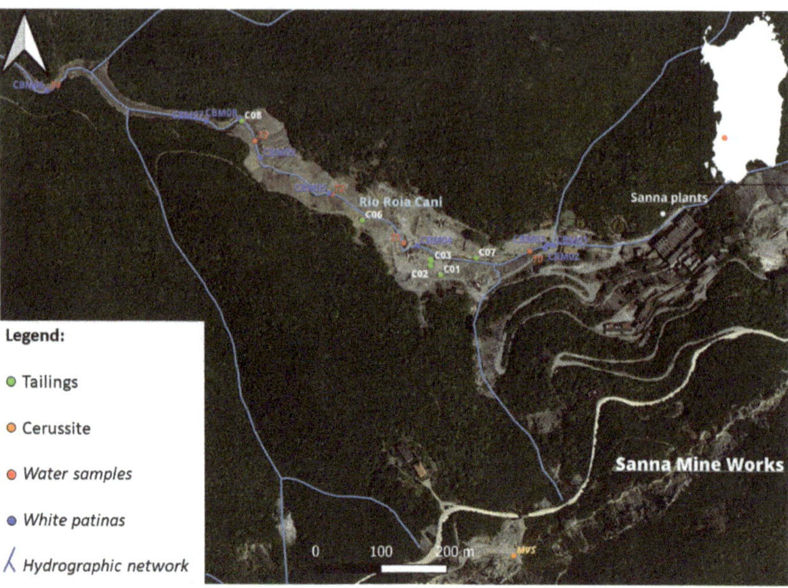

Figure 5. Map with the collected samples. In green, the samples of tailings; in orange, the samples of cerussite; in red, the samples of water; in blue, the samples of white patinas.

Figure 6. The eight sampled layers of tailings along the Rio Roia Cani.

Some cerussite specimens were collected from the oxidate ore in the mine excavations close to the Sanna plant. Moreover, some water samples were taken on the same day of the white patinas sampling, at different points of the Rio Roia Cani, downstream from the processing plant, to have the same environmental conditions for the two sampling sets (Figure 5).

5.2. X-ray Diffraction Analyses

The eight samples of tailings underwent granulometric analyses at first, then they were analysed by an X-Ray Diffractometer (XRD) to detect the main mineralogical phases. An X'Pert Pro PANalytical diffractometer (Malvern Panalytical, Malvern, UK) with an X-ray tube with Cu anticathode (Cu-Kα1 λ = 1.54060 Å), nickel monochromator filter, and X'Celerator detector, was used; data were acquired at 40 kV and 40 mA in an angular range of 5–70 °2θ.

Figure 7. Sampling of the white patinas of the Rio Roia Cani riverbed.

5.3. Scanning Elecron Microscope Analyses

The nine samples of white patinas were analysed by XRD, then by Scanning Electron Microscope (SEM) Quanta Fei 200 unit equipped with a ThermoFischer Ultradry EDS detector (Thermo Fischer Scientific, Waltham, MA, USA) operating under low-vacuum conditions, 25–30 KeV voltage, and variable spot size) to study their structure and assess their composition. Samples were not carbon-coated.

5.4. Inductively Coupled Plasma Spectrometry

Chemical analyses of tailings were performed by Inductively Coupled Plasma Mass Spectrometry (ICP-MS, Perkin-Elmer Elan DRC-e, Shelton, CT, USA), after solubilization through acid digestion by aqua regia and hydrofluoric acid, to detect the possible presence of critical materials, metals, and Rare Earth Elements. Eight tailing samples were analysed, together with a certified reference sample (SRM2711), and a blank. Some samples (C04, C07 and C08) were analysed two times, to verify the correctness of the analytical procedures and analyses performed After acid digestion using aqua regia, cerussite samples were analysed for trace elements by ICP-MS.

White patinas samples were analysed by Inductively Coupled Plasma-Optical Emission Spectroscopy (ICP-OES, Thermo Scientific ARL-3520B Waltham, MA, USA) and ICP-MS, after solubilization by acid digestion by aqua regia and hydrogen peroxide to break down the organic matter component. The water samples were analysed by ICP-MS and by ICP-OES.

6. Analytical Results

Results of the analyses performed on the various types of samples collected on-site are presented below. Samples of tailings were taken in different layers of the old tailing pond, based on chromatic and granulometric variations evidenced in erosional surfaces and gullies along the valley of Rio Roia Cani (Figures 5 and 6). The white patinas were sampled at various points of the riverbed, based on the type of precipitate (Figure 7). The cerussite samples were sampled in the upper part of the Sanna vein, where the primary mineralization was strongly affected by supergene phenomena. The water samples were collected near some inflows and emergencies, trying to maintain a regular interval between the various sampling points.

6.1. Tailings

Mineralogical analyses (X-ray diffraction) on tailing samples (Figure 8 and Table 1), revealed a composition mainly consisting of host rock and gangue minerals, such as quartz and muscovite (present in all the samples), siderite and barite (found in some samples: C04 and C08 siderite, C03, C04 and C05 barite), and montmorillonite. Primary ore phases are represented by galena and sphalerite. Only one sample (C05) shows the presence of zincite, a secondary phase.

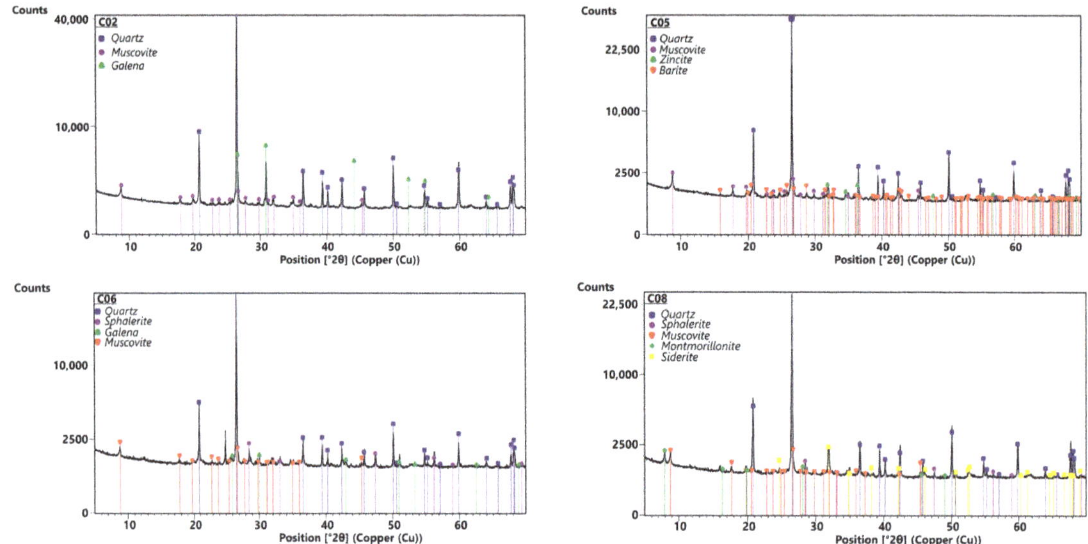

Figure 8. XRD patterns of the mineral phases detected in the different tailing samples. Counts in the *y*-axis are reported in square root to highlight minor peaks.

Table 1. Summary table of the main characters of sampled tailing layers. Qz = quartz, Ms = muscovite, Dol = Dolomite, Gn = galena, Brt = barite, Sd = siderite, Znc = zincite, Sp = sphalerite, Mnt = montmorillonite. Abbreviation as in [28].

Layer	Approximately. Thickness	Sample	Granulometry	Mineralogy
1	50 cm	C01	Moderately graded sand with clays	Qz, Ms, Dol
2	70 cm	C02	Poorly graded sand with clays	Qz, Ms, Gn
3	40 cm	C03	Poorly graded sandy silt with clays	Qz, Ms, Brt
4	50 cm	C04	Poorly graded sand with clays	Qz, Ms, Sd, Brt
5	20 cm	C05	Poorly graded sandy silt with clays	Qz, Ms, Znc, Brt
6	40 cm	C06	Poorly graded gravelly sand with clays	Qz, Ms, Sp, Gn
7	10 cm	C07	Moderately well-graded sand with clays	Qz, Ms
8	50 cm	C08	Poorly graded sandy gravel, with clays	Qz, Ms, Sp, Mnt, Sd

Chemical analyses on the tailing samples show significant amounts of base metals and Rare Earth Elements. A selection of data from the analytical set is reported in Tables 2 and 3. Overall, the dataset displays high Lead and Zinc values, ranging between 760 and 12,300 ppm, and 6600 and 26,500 ppm, respectively. Critical elements once exploited as byproducts in the mine concentrates (Cobalt, Gallium, Indium, Germanium, Antimony, etc.) show low contents (from 15 to 36 ppm of Co, from 8 to 24 ppm of Ga, from 1 to 13 ppm of In, from 4 to 15 ppm of Ge, from 14 to 170 ppm of Sb). On the other hand, the Rare Earth Elements contents show ranges between 69 and 576 ppm. The highest values are shown by Cerium (26–190 ppm), Lanthanum (10–97 ppm), and Neodymium (11–92 ppm).

Table 2. Analytical data for trace elements in tailings. Values are in mg/kg (ppm).

Sample	Al	As	Ba	Bi	Cd	Co	Cu	Fe	Ga	Ge	In	Mn	Ni	Pb	Sb	Sn	V	W	Zn
C01	19,000	78	1400	ND	47	15	140	61,000	ND	ND	ND	3100	25	2400	28	ND	ND	ND	6800
C02	20,100	55	1200	ND	100	15	110	75,400	10	5	2	4300	35	3000	20	2	27	2	7500
C03	19,000	57	1700	ND	110	22	90	80,200	12	4	2	4500	39	2300	19	3	34	3	8700
C04	17,200	58	2000	ND	70	26	65	93,000	8	4	1	5900	34	760	14	2	21	2	6600
C05	22,700	60	3800	ND	140	25	83	88,200	13	4	2	5800	42	1100	19	3	37	3	11,300
C06	18,700	430	270	5	180	24	2070	135,000	24	15	13	6700	23	12,300	170	10	25	2	26,500
C07	16,700	83	5200	ND	150	36	280	154,000	8	5	2	8000	48	5200	40	2	17	2	15,500
C08	17,200	79	2300	ND	57	15	180	64,600	ND	ND	ND	3300	28	3800	34	ND	31	ND	8700

Table 3. Analytical data for Rare Earth Elements in tailings. Values are in mg/kg (ppm).

Sample	Y	La	Ce	Pr	Nd	Sm	Eu	Gd	Tb	Dy	Ho	Er	Tm	Yb	Lu	$\sum$REEs + Y
C01	9	12	30	3	12	3	1	3	ND	2	ND	1	ND	1	ND	77
C02	27	45	81	10	38	9	4	10	1	6	1	2	ND	2	ND	236
C03	65	83	180	20	83	22	10	26	4	17	3	6	1	3	ND	523
C04	64	97	180	22	86	23	10	25	3	15	2	5	1	3	ND	536
C05	76	91	190	23	92	26	11	29	4	19	3	6	1	4	1	576
C06	35	73	130	16	65	16	6	17	2	9	1	3	ND	2	ND	375
C07	48	88	140	19	74	19	8	20	3	11	2	4	ND	3	ND	439
C08	8	10	26	3	11	3	1	3	ND	2	ND	1	ND	1	ND	69

In detail (Tables 2 and 3), some elements such as Cu (65–2070 ppm) and Pb (760–12,300 ppm) vary greatly between the various samples, while others, such as As (55–430 ppm), Ba (270–5200 ppm), Cd (47–180 ppm), Fe (64,600–154,000 ppm), Ga (8–24 ppm), Ge (4–15 ppm), In (1–13 ppm), Sb (14–170 ppm), Sn (2–10 ppm), and Zn (6600–26,500 ppm) vary by only one order of magnitude or zero orders of magnitude, such as Al (16,700–22,700 ppm), Co (15–36 ppm), Mn (3100–8000 ppm), Ni (23–48 ppm), V (17–37 ppm), and W (2–3 ppm). Rare Earths Elements show little variability between samples. The C06 sample is particularly enriched in As, Bi, Cd, Cu, Ga, Ge, In, Pb, Sb, Sn, and Zn, the C02 sample shows the highest contents in Al, the C05 in Ba, W, and REEs, and the C07 in Co, Fe, Mn, and Ni. The samples C01 and C08 display the lowest REEs contents.

6.2. The White Patinas

From XRD mineralogical analyses on white patinas, only slight compositional differences between all the samples are found (Figure 9). They are essentially represented by hydrozincite [$Zn_5(CO_3)_2(OH)_6$], except for the CBM03 and CBM05 samples, consisting of zincite [ZnO].

Based on XRD results, SEM-EDS analyses were performed on the selected CBM01 (hydrozincite) and CBM05 (zincite) representative samples. The morphological study evidenced in the CBM01 sample abundant organic filaments (Figure 10). The EDS spectrum and the EDS compositional maps (Figure 11), show filaments that are made of carbon (C) and oxygen (O). Mineral grains recognizable in the sample are made of zinc (Zn) and oxygen (O), thus confirming the abundance of hydrozincite. Organic filaments are considered to be indicative of biologically mediated formation, as reported for similar occurrences in the same district [18].

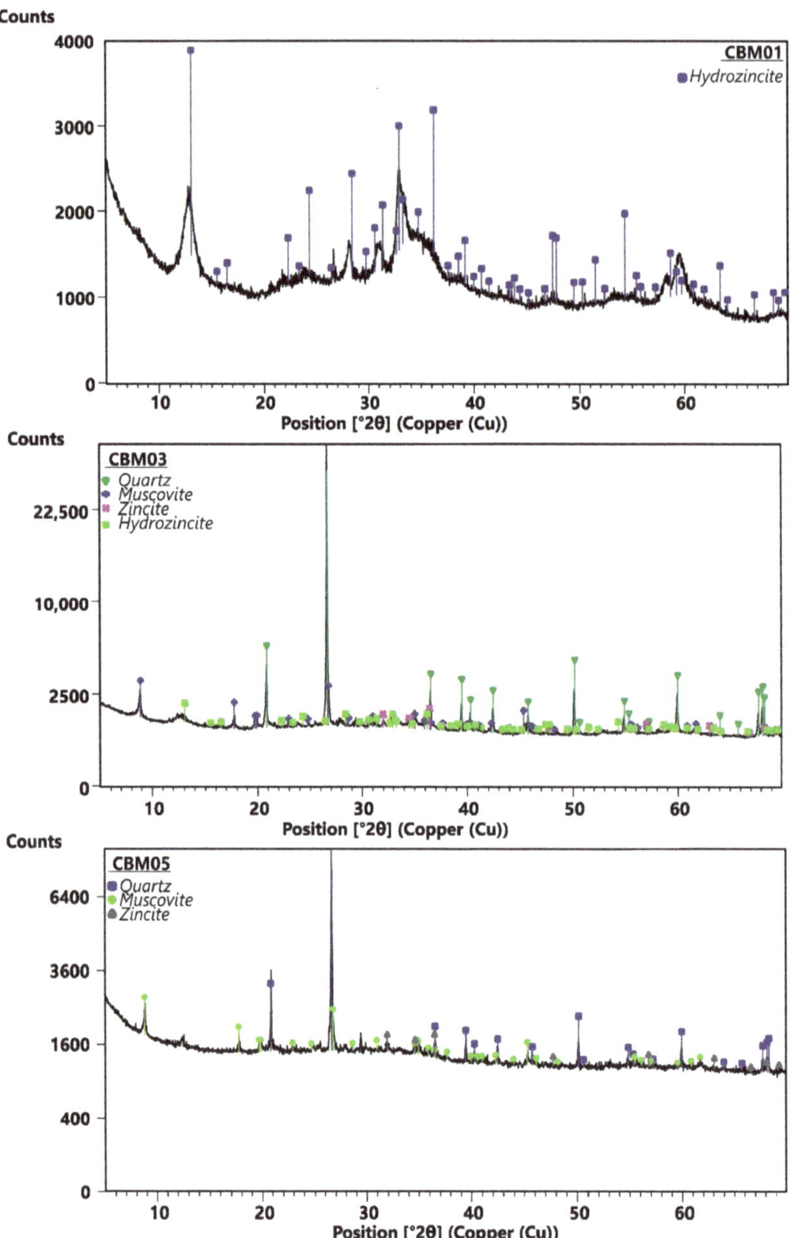

Figure 9. XRD patterns with the mineral phases detected in white patinas. *Y*-axes report counts in square root to highlight the minor peaks, except the first XRD pattern.

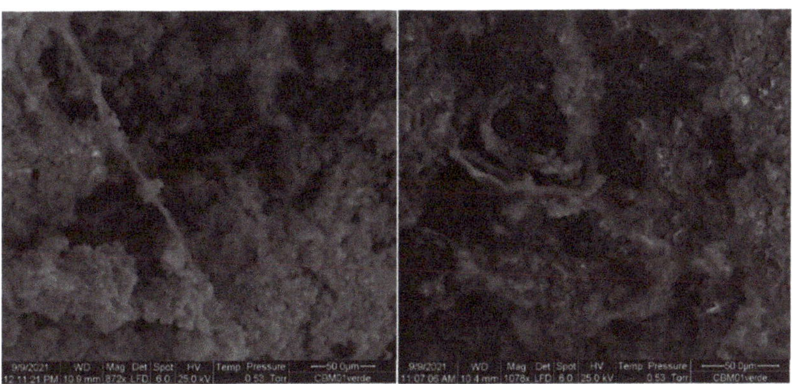

Figure 10. SEM images of CBM01 sample.

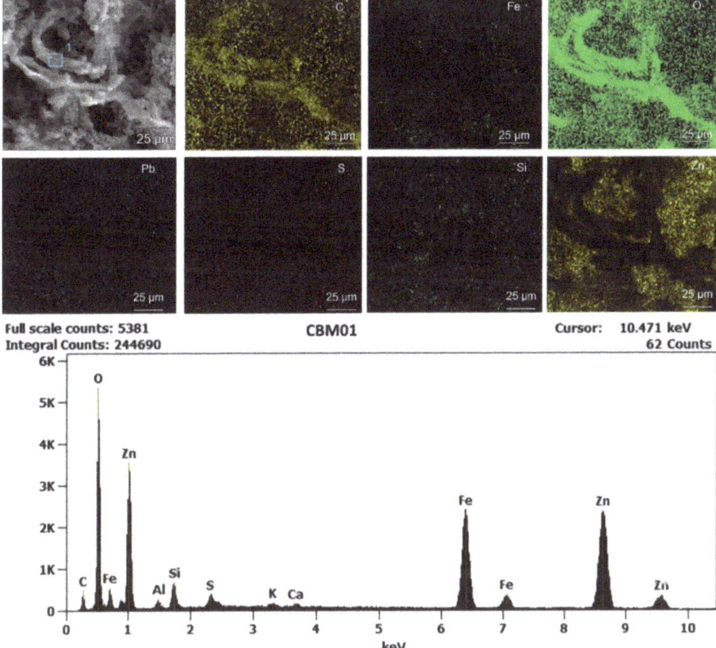

Figure 11. Compositional EDS maps and spectrum for the sample CBM01. The blue square indicates the analysed point.

The CBM05 sample displays a texture dominated by <20 μm elements of globular shape (Figure 12). EDS spectra and compositional maps (Figure 12), confirm the composition reported by the XRD analyses. Moreover, SEM-EDS maps identified grains with traces of arsenic, lead, iron, and oxygen contents, and presence of aluminum-silicates grains.

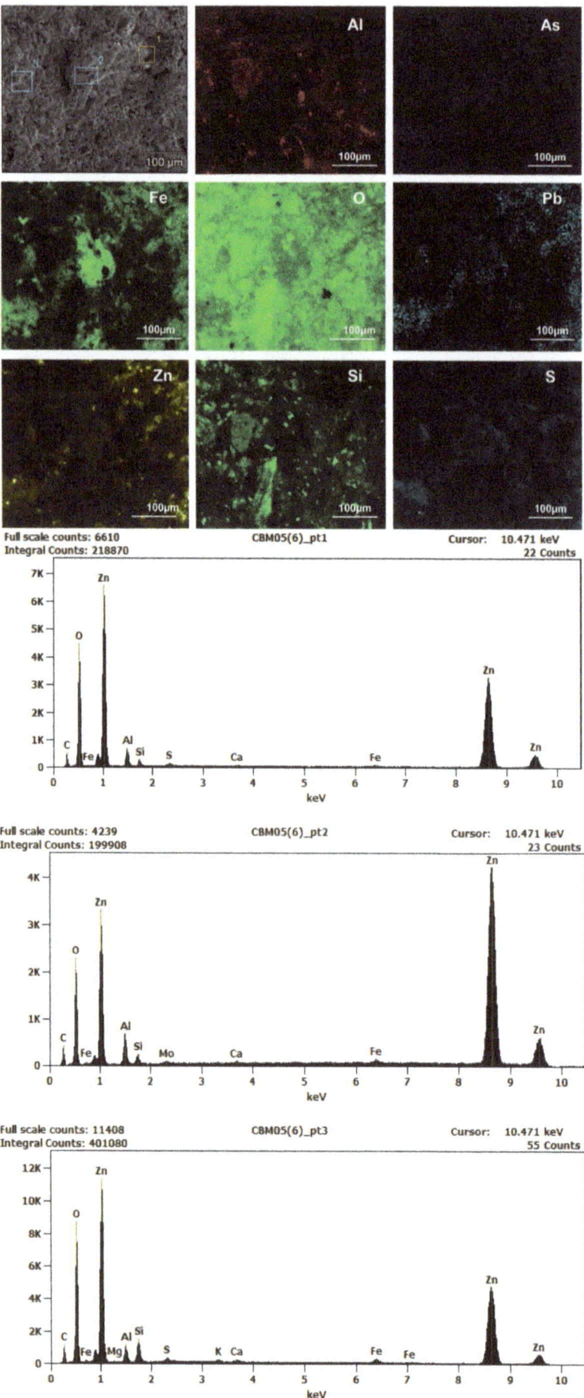

Figure 12. Compositional EDS maps and spectrums for the sample CBM05. The blue and yellow squares indicate the analysed points.

Chemical analyses were performed on each white patinas sample, except for the CBM09, completely utilized for the previous analyses. Trace elements, such as Rare Earth Elements, were analysed by ICP-MS, while major elements were analysed by ICP-OES (Tables 4 and 5).

Table 4. Analytical data for trace elements in white patinas. Values are in mg/kg (ppm).

Sample	Al	As	Ba	Cd	Co	Cu	Fe	Mn	Ni	Pb	Sb	Sr	Zn
CBM01	1770	23	31	610	11	850	27,700	360	150	14,800	8	8	433,200
CBM02	1800	41	310	580	18	730	26,700	650	130	14,600	8	14	415,000
CBM03	10,400	920	16,500	550	84	1430	355,200	14,400	190	33,000	42	200	82,400
CBM04	550	14	120	640	14	300	4870	290	170	6400	5	7	509,000
CBM05	9010	270	1530	400	46	1690	199,800	7450	90	74,500	67	55	90,300
CBM06	10,020	230	11,300	680	29	1490	91,400	4100	92	55,600	110	76	274,300
CBM07	880	23	250	450	10	570	9350	340	130	20,650	13	9	512,200
CBM08	1350	29	860	460	9	660	15,100	400	120	23,370	1	10	489,000

Table 5. Analytical data for Rare Earth Elements in white patinas. Values are in mg/kg (ppm).

Sample	Y	La	Ce	Pr	Nd	Sm	Eu	Gd	Tb	Dy	Ho	Er	Tm	Yb	Lu	$\sum$REEs + Y
CBM01	250	250	140	61	270	63	24	89	11	45	7	15	1	8	1	1235
CBM02	200	200	110	45	200	44	17	65	8	34	6	12	1	6	1	949
CBM03	250	550	970	120	520	130	52	140	18	72	10	20	2	11	2	2867
CBM04	100	86	38	15	61	13	5	22	3	12	2	5	ND	2	ND	364
CBM05	140	180	280	47	220	56	22	68	8	36	5	11	1	6	1	1081
CBM06	91	140	220	33	140	35	14	40	5	2	3	6	1	3	ND	733
CBM07	84	86	41	13	51	11	4	17	2	9	2	4	ND	2	ND	326
CBM08	100	100	54	17	69	15	6	23	3	13	2	5	ND	2	ND	409

Some elements such as Al (550–10,400 ppm), Ba (31–16,500 ppm), Fe (4870–355,200 ppm), Mn (290–14,400 ppm), Sb (1–110 ppm), and Sr (7–200 ppm), vary by two or three orders of magnitude between the various samples, while others, such as As (14–920 ppm), Cd (400–680 ppm), Co (9–84 ppm), Cu (300–1690 ppm), Ni (90–190 ppm), Pb (6400–74,500 ppm), and Zn (82,400–512,200 ppm) vary by only one order of magnitude. The CBM03 sample is particularly enriched in Al, As, Ba, Co, Fe, Mn, Ni, and Sr, the CBM05 sample is the richest in Cu and Pb, the CBM06 in Cd and Sb, and the CBM07 in Zn.

REEs analyses (Table 5) show little variability among the samples, except for the sample CBM03 which appears to be consistently enriched in REEs (2867 ppm).

6.3. Cerussite

To investigate the origin of the REEs contents found in the tailings of the Sanna plant, a study of supergene mineralization of Sanna vein, which was among the main suppliers of the plant, was performed. An interesting aspect of this mineralization is its predominantly carbonate composition, given that from the most recent literature on Sardinian low-temperature hydrothermal ores [29], high concentrations of REEs are found in carbonate phases. Three samples of cerussite ore, prevalently composed of cerussite and goethite, were collected in the upper part of the Sanna vein (Figure 5). Cerussite crystals were therefore selected from the ore by hand picking, and then analysed by ICP-MS, after acid digestion, to investigate their REEs contents (Table 6).

Table 6. Analytical data for Rare Earth Elements in cerussite. Values are in mg/kg (ppm).

Sample	Y	La	Ce	Pr	Nd	Sm	Eu	Gd	Tb	Dy	Ho	Er	Tm	Yb	Lu	$\sum$REEs + Y
MVS_1	ND	110	4	ND	120	20	4	6	1	1	ND	ND	ND	ND	ND	266
MVS_2	ND	96	3	ND	100	17	4	5	ND	1	ND	1	ND	ND	ND	227
MVS_3	ND	61	2	ND	64	10	2	3	ND	ND	ND	ND	ND	ND	ND	142

In Table 6 the results of analyses are reported. Lanthanum and Neodymium show the higher values among the REEs.

6.4. Water Analyses

Water samples collected together with the white patinas were analysed by ICP-MS. Overall, water shows a calcium-magnesian sulphate composition [30], characteristic of sulphide mining areas, and amounts of total REEs plus Y of thousands of ng/L (Table 7), several orders of magnitude below the values reported in the literature for mine drainage waters in Montevecchio area [14].

Table 7. Analytical data for Rare Earth Elements in water. Values are in ng/L.

Sample	T [°C]	pH	Eh [mV]	EC [μs/cm]	TDS [mg/L]	Y	La	Ce	Pr	Nd	Sm	Eu	Gd	Tb	Dy	Ho	Er	Tm	Yb	Lu	∑REEs + Y
T0	16.4	6.3	490	1318	989	1130	730	300	90	340	82	30	130	15	89	17	40	4	19	3	3019
T1	17.5	6.1	491	1301	1071	190	620	750	36	150	34	17	43	4	21	3	5	1	3	ND	1877
T2	16.6	5.7	320	1390	1060	1990	1490	830	240	1040	240	97	400	43	220	35	72	7	45	6	6755
T3	18.6	5.4	471	1365	1030	880	800	290	78	270	64	25	91	10	55	11	22	2	11	1	2610
T4	20.0	4.7	414	1399	1036	450	580	180	57	220	47	25	64	7	34	6	15	1	8	2	1696

From Table 7, the most concentrated REEs in waters are: Y (190–1990 ng/L), La (580–1490 ng/L), Ce (180–830 ng/L), Pr (36–240 ng/L), Nd (150–1040 ng/L), Sm (34–240 ng/L), Gd (43–400 ng/L) and Dy (21–220 ng/L). The sample T0, collected in the stream flow on the tailings immediately downstream to the Sanna processing plant, is among the richest samples in REEs. The sample T1, sampled near a large encrustation of white patina precipitate, is among the poorest samples in REEs. The sample T2 is the richest in REEs, it was sampled where the water leaked off from a collapsed zone of the drainage tunnel of the old tailing pond. The sample T3, among the richest in REEs, was sampled downstream of sample T2. Sample T4, the poorest in REEs, was sampled downstream of other samples and near two tributaries of the Rio Roia Cani (Figure 5).

7. Discussion

Understanding the chemical and mineralogical properties of mining wastes is crucial for the assessment of their economic potential, improving environmental risk assessment, and optimizing the design and execution phases of remediation operations [31]. Raw materials recovery from mining wastes is a specific target for EU policies to improve the supply of CRMs to the Union for the next years [1]. At the same time, these activities provide fundamental contributions to solving environmental issues in abandoned mine areas, as they result in effective remediations of the polluted sites. Effectual recovery of raw materials requires significant technological insight but also requires a strong effort in mineral characterization of waste materials. For the reprocessing of these tailings, it is important to know in detail the initial composition of the ore, the methods used in ore processing, the chemical additives used for mineral beneficiation, how and where tailings were discharged, their current mineralogical composition, and the elements of economic interest contained in them [32]. In environmental studies, tailings characterization generally starts from studies of their granulometry and mineralogy. These characteristics provide essential information on the behaviour of the tailings when they encounter the water [33]. The mineralogy of tailings may change through a single deposit, which may constitute a very heterogeneous system. Minerals in tailings may be very reactive and often soluble in waters under a wide range of environmental conditions; mineralogical characterization of reactive phases (e.g., sulphide and carbonate minerals) allows for predicting the extension and intensity of phenomena such as acid drainage and leaching [31].

7.1. Environmental Issues and Resilience Factors

Several environmental issues emerge from the study in the Sanna plant area. The occurrence in mine wastes, tailings, and secondary precipitates (i.e., the white patinas) of anomalous geochemical contents of metals (i.e., Fe, Al, Zn, Pb, Mn, Cu, Cd, and Hg) and metalloids (As and Sb), represents a supply of several contaminants having various grades of toxicity that imply different threats for the environment. Depending on the physicochemical conditions of the different interactions of these materials with the Rio Roia Cani waters and with atmospheric agents, it may be produced a prevailing dispersion in solution of metalloids (high-pH interaction), or metals and metalloids (low-pH interaction), followed by diffusion in the environment [34–40]. This chemical transport combines with the solid transport of waste constituents due to high erosion rates in poorly consolidated materials.

X-ray diffraction analyses on the tailings of the Sanna plant identified several metallic minerals which must be regarded as fundamental sources of pollution. Indeed, secondary oxidated phases as zincite are more soluble than sulphides, easily releasing zinc in solution, which is confirmed by the frequent occurrence of hydrozincite/zincite mineral precipitates (white patinas) along the riverbed. The equilibrium between the release of contaminants and their precipitation is influenced by the hydrological regime of the Rio Roia Cani, the temperature, the concentration of metals in the solution, and the Eh-pH conditions.

The co-precipitation of Iron and Lead oxy-hydroxides with zincite, suggested by the red colour of the waters and of the riverbed at the sampling point, is confirmed by SEM-EDS in some white patinas samples (i.e., CBM05 sample), which detected Lead, Iron, and Oxygen. Moreover, precipitation of iron-bearing phases is predicted by saturation calculations (Phreeqc Interactive 3.6.2-15100 software) of the water samples collected along the Rio Roia Cani (Figure 5). Where the T2 water sample was collected, the water was oversaturated in ferrite-Zn and goethite, with saturation indexes of 11.54 and 5.94, respectively.

According to Dore et al., 2020 [41], the microscopic processes that control the dissolution and precipitation of minerals in a polluted mining area may act both positively and negatively towards the environment, favouring the natural abatement or, conversely, the dispersion of pollutants. These microscopic processes may be fundamental for the increase or decrease in metal loads in a river; when positive processes dominate, the load decreases or is limited, but, when negative processes dominate, the load increases, and the dispersion of pollutants can become significant. Overall, metal-enriched precipitates such as white patinas may act as resilience factors in a positive process for the natural environment of the area, removing and fixing relevant amounts of pollutants from the Rio Roia Cani waters. This shows the importance of biologically mediated processes as natural remediation agents, as already observed in the waters of Rio Naracauli [42].

7.2. Raw Materials Potential

Chemical analyses of the Sanna plant tailings (Tables 2 and 3) show that they contain high grades of lead and zinc (up to 1.23 wt% and 2.65 wt%, respectively). The highest Pb grades are attained for sample C06, which is also arsenic and antimony-rich (430 and 170 ppm, respectively), representative of a tailing layer which may have been derived from a fahlore-rich primary ore. Although the materials show generally relatively high contents in total REEs, there is no apparent correlation between the metals present in the highest grades (Pb and Zn) and the REEs. On the other hand, the primary source of REEs in the ore has not yet been discovered; this may lead to the conclusion that the REEs are contained in the gangue and/or in the host rock minerals included in the exploited ore, as reported by Moroni et al., 2019 [20] for the five element-type veins of the southern part of the Montevecchio vein system. The analysed cerussite samples (Table 6) show a slight anomaly in lanthanum and neodymium, indicating a possible role of supergene processes in concentrating these elements in secondary phases. Although the role of weathering processes in REEs mobilization and reconcentration is extensively recognized in lateritic and regolith-hosted, ion-adsorbed-type deposits as well as in weathered carbonatite environments [43,44], REEs behaviour in supergene ore zones of hydrothermal vein deposits is less documented [45].

The white patinas in the Rio Roia Cani are made of Zn-carbonate minerals, thus displaying high Zn contents up to 51.22 wt% (Table 4). Remarkably, they show high contents of REEs (up to 2867 ppm $\sum$REEs + Y) (Table 5), which display little correlation with Zn and good correlations with As, Fe, and Mn (Figure 13). The zincite samples (CBM03 and CBM05) show the highest values of REEs. A possible role of biological processes in concentrating REEs from waters in the Rio Roia Cani environment can be inferred. Studies on REEs mineralizing processes in exogenic environments evidenced the role of microorganisms in REEs mobilization from primary minerals (phosphates, carbonates, silicates, etc.) to waters, their capabilities in uptaking REEs from waters and fractionating LREEs from HREEs, and to favour and accelerate REEs concentrations in secondary (bio-) minerals [46,47].

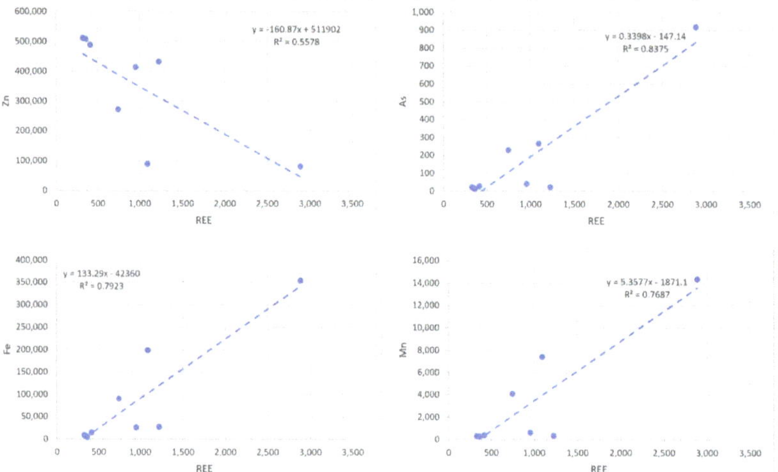

Figure 13. Correlation graphics between some metals and REEs contained in white patinas. Values are in ppm.

The relatively low REEs contents resulting from water analyses of Rio Roia Cani are in good agreement with an effective process of removal of these elements by precipitation of white patinas.

Further insights into the behaviour of REEs in the studied systems are provided by REEs patterns from tailings, white patinas, cerussites and waters, normalized to Post-Archean Australian Shales (PAAS) [48] (Figure 14). Tailings, white patinas, and waters display similar trends: patterns show a flat trend, with a slight enrichment in MREEs and a positive Eu anomaly. They show a good agreement with the patterns reported for the Montevecchio primary ore [6]. Conversely, the cerussite samples show an evident w-type tetrad effect, with strong depletions in Ce, Pr, and HREEs, confirming different paths of REEs fractionation during supergene processes and formation of the secondary ore.

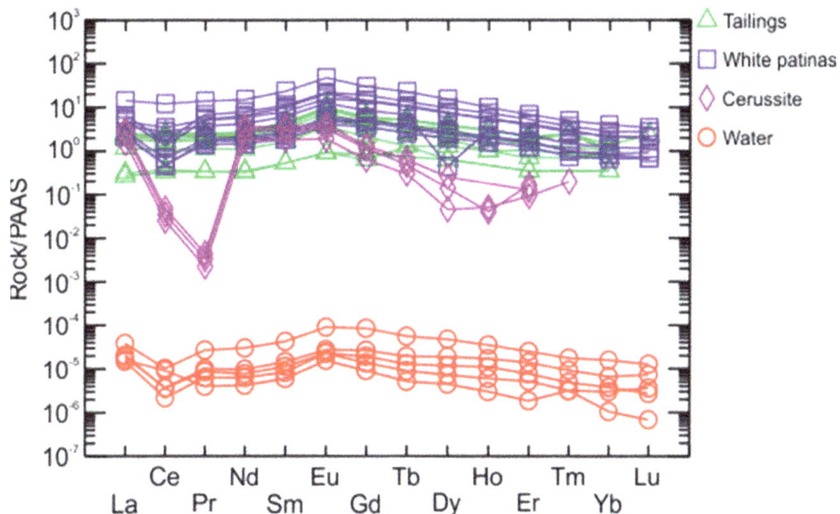

Figure 14. PAAS-normalized REEs patterns for tailings, white patinas, cerussite and waters of the Sanna area.

7.3. Insights for Metals Recovery

Mineral and chemical characterization may also provide some insights into different raw materials recovery from wastes.

The measured contents of Zn and Pb in the tailings may support a project for the recovery of these base metals [49]. In this view, the application of processing techniques essentially depends on different physical (i.e., granulometry) and mineralogical characteristics of waste materials. Due to their fine-grained nature, sulphide-bearing tailings can be effectively treated by flotation, although a large amount of fines increases the consumption of reagents, causing the so-called slime coating phenomenon, hampering the effectiveness of the treatment and, often, making it unfeasible [50]. Successful application of advanced flotation techniques in the processing of the lead and zinc sulphide-bearing tailings of the Principe Tommaso plant in Montevecchio has been recently described in Mercante et al., 2021 [51] and in Sogos et al., 2022 [52]. Different approaches may be required for waste materials in which lead and zinc are contained as oxidate minerals (cerussite, anglesite, smithsonite, hemimorphite, hydrozincite, zincite, etc.), for which metal recovery can be also obtained by leaching techniques (e.g., with sulfuric acid, caustic soda, etc. [53,54]).

The limited presence of economic deposits of REEs and the serious environmental issues related to REE minerals beneficiation [55–59] promoted the studies on alternative and more sustainable treatment techniques, such as biohydrometallurgy, to extract REEs both from primary ores and from secondary resources, as industrial and mining wastes. Recovery of base metals from low-grade ores by bioleaching is a common technique used in the mining industry since the 1970s; it was used with success to leach REEs from mining wastes [60]. As in classical processing techniques, the efficiency of bioleaching is closely linked to the physical and mineralogical composition of the waste materials; actually, the selection of a suitable microorganism depends on the type of REEs minerals (e.g., phosphates, carbonates, etc.) present in the waste, which present different degrees of solubilization in presence of the chemical agents (i.e., organic acids) secreted from microorganisms as chemoheterotrophic bacteria. Biosorption and bioaccumulation processes have been documented for various microorganisms, capable of REEs uptake from leachates [61,62]; conversely, REEs-bearing biomineralization is documented only in a few studies [57,58].

The REEs contents measured in the tailings from the Sanna plant are currently sub-economic, even if with a view to an integral recovery of raw materials from these mine wastes they must certainly be taken into consideration. The current limited knowledge of Rare-Earth-bearing minerals in the Montevecchio system, however, makes treatment hypotheses for the recovery of these metals premature. Nevertheless, the identification of secondary processes of REEs enrichment in the white patinas constitutes a significant incentive to improve biologically mediated techniques for the recovery of REEs elements from suitable mineral phases. To achieve these goals, the characterization of the REEs mineralogical carriers from primary ore to tailings and the identification of the microorganisms associated with hydrozincite forming the white patinas is of primary importance. Until now, a large part of recent investigations on waste materials in Sardinian mine sites was almost exclusively oriented to the environmental aspects [12–19,41,63]; only a few studies faced the wastes as potential sources of raw materials and CRMs [64–66]. The data acquired in this study allow a series of considerations regarding both the raw material potential and the environmental issues of the Sanna-Rio Roia Cani area.

8. Conclusions

The mineralogical and chemical studies on the tailings of the Sanna processing plant area in Montevecchio confirm that these materials constitute sources of environmental risk. At the same time, they must be regarded as potential economic sources of raw materials, primarily Zinc and Lead. The chemical analyses on the various materials present in the site also highlighted specific layers of REEs-enriched fine-grained products along the stratigraphic sequence of the old tailings pond. The mineralogical carrier of REEs is not yet identified, but an origin from gangue and/or host rock minerals may be reasonably inferred. Some indication of supergene REEs-enrichment of the ore mineralization comes from analyses on cerussite from Sanna mine works. Further evidence of remobilisation processes of REEs in the studied area is provided by the analyses on the white patinas precipitated into the riverbed of the Rio Roia Cani, which display 7.45 wt% of Pb, 51.2 wt% of Zn and about 2900 mg/kg of REEs + Y.

Among the various waste materials present on the site, tailings are confirmed as a primary source of contamination, due to their high contents of pollutants (Pb, Zn, Fe, etc.). Contaminants are transported in solution by rainwater and by the flow of the Rio Roia Cani that erodes the mining dumps. Contamination is partially buffered by white patinas that allow the precipitation of Zn in solution as hydrozincite and zincite, a situation similar to the one currently underway in the Rio Naracauli (Ingurtosu mine, west of Montevecchio), where the riverbed is covered by hydrozincite bio-precipitates [18,63]. Furthermore, the analyses indicate that these white patinas retain other metals, preventing them from remaining in solution and from being dispersed into the environment, generating a sort of remediation process.

This study highlights the presence of CRMs in the tailings, suggesting the possibility of considering these materials as sources of secondary raw materials, as also proposed by recent studies in the Eastern Montevecchio area [66]. REEs found in tailings and white patinas are present in sub-economic to economic values, deserving further studies and insights, for defining their origin, and their processes of mobilisation and concentration in the Rio Roia Cani environment.

Overall, the amount of Zinc, Lead, and CRMs found in the solid matrix led to a hypothesis of evaluation of re-treatment of these solids by various techniques of mineralurgical processing. This re-treatment would entail two benefits: one environmental, through the removal of a primary source of contamination, and the other of an economic nature, through the implementation of a circular economy chain linked to the sale and re-use of these raw materials.

Author Contributions: Conceptualization: L.S., S.N. and G.D.G.; Sampling and field activities: L.S., S.N. and G.D.G.; ICP-MS and ICP-OES analyses: L.S. and F.P.; XRD and SEM-EDS analyses: L.S. and D.F.; Writing-original draft preparation: L.S. and S.N.; Writing, review and editing: L.S., S.N., G.D.G., D.F. and F.P.; Funding acquisition: S.N. and G.D.G. All authors have read and agreed to the published version of the manuscript.

Funding: This paper was produced while attending the PhD course in Earth and Environmental Sciences and Technologies at the University of Cagliari, XXXVIII cycle, with the support of a PNRR scholarship, funded by the European Union, NextGenerationEU, Mission 4 "Education and research" Component 2 "From research to business" Investment 3.1 "Fund for the creation of an integrated system of research and innovation infrastructures". Title of the PNRR project: GeoSciences IR. Title of the research project: Update of the Sardinian database of mining and mining-metallurgical dumps and identification of their raw material contents, with priority on critical and strategic raw materials for decarbonization and ecological transition. This work was also supported by Fondazione di Sardegna [grant number CUP F75F21001270007] and by the RETURN Extended Partnership, and received funding from the European Union Next-Generati on EU (National Recovery and Resilience Plan—NRRP, Mission 4, Component 2, Investment 1.3—D.D. 1243 2/8/2022, PE0000005.

Data Availability Statement: The data presented in this study are available on request from the corresponding author. The data are not publicly available as they are part of a larger project currently underway for the mapping of Critical Raw Materials in mining wastes at a regional scale.

Acknowledgments: We thank D. Medas, M.L. Deidda, A. Attardi and E. Musu for their support during the sampling and for the suggestions during the development of the final draft of this paper. We acknowledge the CeSAR (Centro Servizi d'Ateneo per la Ricerca) of the University of Cagliari, Italy, for SEM analysis.

Conflicts of Interest: The authors declare no conflict of interest.

References

1. European Commission. Study on the Critical Raw Materials for the EU. Fifth List. Final Report 2023. Available online: https://op.europa.eu/en/publication-detail/-/publication/57318397-fdd4-11ed-a05c-01aa75ed71a1 (accessed on 20 October 2023).
2. Blengini, G.A.; Mathieux, F.; Mancini, L.; Nyberg, M.; Cavaco Viegas, H.; Salminen, J.; Garbarino, E.; Orveillion, G.; Saveyn, H. *Recovery of Critical and Other Raw Materials from Mining Waste and Landfills*; Publications Office of the European Union: Luxembourg, 2019.
3. Kossoff, D.; Dubbin, W.E.; Alfredsson, M.; Edwards, S.J.; Macklin, M.G.; Hudson-Edwards, K.A. Mine Tailings Dams: Characteristics, Failure, Environmental Impacts, and Remediation. *Appl. Geochem.* **2014**, *51*, 229–245. [CrossRef]
4. Ali, M.A.H.; Mewafy, F.M.; Qian, W.; Alshehri, F.; Almadani, S.; Aldawsri, M.; Aloufi, M.; Saleem, H.A. Mapping Leachate Pathways in Aging Mining Tailings Pond Using Electrical Resistivity Tomography. *Minerals* **2023**, *13*, 1437. [CrossRef]
5. Miler, M.; Bavec, Š.; Gosar, M. The Environmental Impact of Historical Pb-Zn Mining Waste Deposits in Slovenia. *J. Environ. Manag.* **2022**, *308*, 114580. [CrossRef]
6. Moroni, M.; Naitza, S.; Ruggieri, G.; Aquino, A.; Costagliola, P.; De Giudici, G.; Caruso, S.; Ferrari, E.; Fiorentini, M.L.; Lattanzi, P. The Pb-Zn-Ag Vein System at Montevecchio-Ingurtosu, Southwestern Sardinia, Italy: A Summary of Previous Knowledge and New Mineralogical, Fluid Inclusion, and Isotopic Data. *Ore Geol. Rev.* **2019**, *115*, 103194. [CrossRef]
7. Ministero Dell'ambiente e Della Tutela del Territorio. Perimetrazione del Sito di Interesse Nazionale del Sulcis-Iglesiente-Guspinese. GU Serie Generale n. 121 del 27-05-2003. Suppl. Ordinario n. 83. 2003. Available online: https://www.gazzettaufficiale.it/eli/gu/2003/05/27/121/so/83/sg/pdf (accessed on 20 October 2023).
8. Progemisa SpA. *Interventi di Bonifica e Ripristino Ambientale Dell'area Mineraria Dismessa di Montevecchio Ponente. Scheda di Intervento di Emergenza E 3.1—Piano di Caratterizzazione*, Public report; Progemisa SpA: Cagliari, Italy, 2003.
9. Sedda, L. Analisi Ambientale Dell'area di Montevecchio Ponente: Stato Dell'arte E Tecnologie. Master's Thesis, Università degli Studi di Cagliari, Cagliari, Italy, 2021.
10. Piano Regionale di Gestione dei Rifiuti. *Piano di Bonifica Siti Inquinati. Schede dei Siti Minerari. Allegato 5*; Regione Autonoma Sardegna: Cagliari, Italy, 2003.
11. *Studio per il Piano di Recupero dell'area Mineraria Dimessa di Montevecchio—Ingurtosu*; Progemisa SpA: Cagliari, Italy, 2001; pp. 190–203.
12. Cidu, R.; Fanfani, L. Overview of the environmental geochemistry of mining districts in southwestern Sardinia, Italy. *Geochem. Explor. Environ. Anal.* **2002**, *2*, 243–251. [CrossRef]
13. Cidu, R.; Frau, F. Impact of the Casargiu Mine Drainage (SW Sardinia, Italy) on the Mediterranean Sea. In Proceedings of the International Mine Water Conference, Pretoria, South Africa, 19–23 October 2009; pp. 926–931.
14. Cidu, R.; Frau, F.; Da Pelo, S. Drainage at Abandoned Mine Sites: Natural Attenuation of Contaminants in Different Seasons. *Mine Water Environ.* **2011**, *30*, 113–126. [CrossRef]
15. Frau, F.; Medas, D.; Da Pelo, S.; Wanty, R.B.; Cidu, R. Environmental Effects on the Aquatic System and Metal Discharge to the Mediterranean Sea from a Near-Neutral Zinc-Ferrous Sulfate Mine Drainage. *Water Air Soil Pollut.* **2015**, *226*, 55. [CrossRef]
16. De Giudici, G.; Medas, D.; Cidu, R.; Lattanzi, P.; Podda, F.; Frau, F.; Rigonat, N.; Pusceddu, C.; Da Pelo, S.; Onnis, P.; et al. Application of hydrologic-tracer techniques to the Casargiu adit and Rio Irvi (SW-Sardinia, Italy): Using enhanced natural attenuation to reduce extreme metal loads. *Appl. Geochem.* **2018**, *96*, 42–54. [CrossRef]

17. De Giudici, G.; Medas, D.; Cidu, R.; Lattanzi, P.; Rigonat, N.; Frau, I.; Podda, F.; Marras, P.A.; Dore, E.; Frau, F.; et al. Assessment of origin and fate of contaminants along mining-affected Rio Montevecchio (SW Sardinia, Italy): A hydrologic-tracer and environmental mineralogy study. *Appl. Geochem.* **2019**, *109*, 104420. [CrossRef]
18. Medas, D.; Lattanzi, P.; Podda, F.; Meneghini, C.; Trapananti, A.; Sprocati, A.; Casu, M.A.; Musu, E.; De Giudici, G. The amorphous Zn biomineralization at Naracauli stream, Sardinia: Electron microscopy and X-ray absorption spectroscopy. *Environ. Sci. Pollut. Res.* **2014**, *21*, 6775–6782. [CrossRef]
19. Podda, F.; Medas, D.; De Giudici, G.; Ryszka, P.; Wolowski, K.; Turnau, K. Zn biomineralization processes and microbial biofilm in a metal-rich stream (Naracauli, Sardinia). *Environ. Sci. Pollut. Res.* **2014**, *21*, 6793–6808. [CrossRef]
20. Cuccuru, S.; Naitza, S.; Secchi, F.; Puccini, A.; Casini, L.; Pavanetto, P.; Linnemann, U.; Hofmann, M.; Oggiano, G. Structural and metallogenic map of late Variscan Arbus Pluton (SW Sardinia, Italy). *J. Maps* **2016**, *12*, 860–865. [CrossRef]
21. Carmignani, L.; Oggiano, G.; Barca, S.; Conti, P.; Salvadori, I.; Eltrudis, A.; Funedda, A.; Pasci, S. *Geologia Della Sardegna: Note Illustrative Della Carta Geologica Della Sardegna in Scala 1:200.000, Memorie Descrittive Della Carta Geologica d'Italia (Vol. 60)*; Servizio Geologico d'Italia: Rome, Italy, 2001; p. 283.
22. Annino, E.; Barca, S.; Costamagna, L.G.; Annino, E.; Barca, S.; Costamagna, L.G. Lineamenti stratigrafico-strutturali dell'Arburese (Sardegna sud-occidentale). *Rend. Semin. Fac. Sci. Univ. Cagliari.* **2016**, *70*, 403–426.
23. Assorgia, A.; Brotzu, P.; Morbidelli, L.; Nicoletti, M.; Traversa, G. Successione e cronologia (K-Ar) degli eventi vulcanici del complesso calco-alcalino oligo-miocenico dell'Arcuentu (Sardegna centro-occidentale). *Period. di Mineral.* **1984**, *53*, 89–102.
24. Salvadori, I. Studio geo-minerario della zona di Salaponi (Sardegna Sud-occidentale). *Boll. Soc. Geol. Ital.* **1958**, *77*, 91–126.
25. Salvadori, I.; Zuffardi, P. Guida per l'escursione a Montevecchio e all'Arcuentu. Itinerari geologici, mineralogici e giacimentologici in Sardegna. *Ente Min. Sardo* **1973**, *1*, 29–44.
26. Moroni, M.; Rossetti, P.; Naitza, S.; Magnani, L.; Ruggieri, G.; Aquino, A.; Tartarotti, P.; Franklin, A.; Ferrari, E.; Castelli, D.; et al. Factors controlling hydrothermal nickel and cobalt mineralization—Some suggestions from historical ore deposits in Italy. *Minerals* **2015**, *9*, 429. [CrossRef]
27. Zuffardi, P. Fenomeni di ricircolazione nel giacimento di Montevecchio e l'evoluzione in profondità della sua mineralizzazione. *Res. Ass. Min. Sarda* **1962**, *1–2*, 17–73.
28. Warr, L.N. IMA–CNMNC approved mineral symbols. *Miner. Mag.* **2021**, *85*, 291–320. [CrossRef]
29. Mondillo, N.; Boni, M.; Balassone, G.; Spoleto, S.; Stellato, F.; Marino, A.; Santoro, L.; Spratt, J. Rare earth elements (REE)—Minerals in the Silius fluorite vein system (Sardinia, Italy). *Ore Geol. Rev.* **2016**, *74*, 211–224. [CrossRef]
30. Corda, V. Studi Sull'impatto Ambientale Delle Discariche Minerarie a Montevecchio Ponente. Master's Thesis, Università degli Studi di Cagliari, Cagliari, Italy, 2022.
31. Jamieson, H.E.; Walker, S.R.; Parsons, M.B. Mineralogical characterization of mine waste. *Appl. Geochem.* **2015**, *57*, 85–105. [CrossRef]
32. Mulenshi, J.; Gilbricht, S.; Chelgani, S.C.; Rosenkranz, J. Systematic characterization of historical tailings for possible remediation and recovery of critical metals and minerals—The Yxsjöberg case. *J. Geochem. Explor.* **2021**, *226*, 106777. [CrossRef]
33. Wang, C.; Harbottle, D.; Liu, Q.; Xu, Z. Current state of fine mineral tailings treatment: A critical review on theory and practice. *Miner. Eng.* **2014**, *58*, 113–131. [CrossRef]
34. Cidu, R.; Caboi, R.; Fanfani, L.; Frau, F. Acid drainage from sulfides hosting gold mineralization (Furtei, Sardinia). *Environ. Geol.* **1997**, *30*, 231–237. [CrossRef]
35. Jurjovec, J.; Ptacek, C.J.; Blowes, D.W. Acid neutralization mechanisms and metal release in mine tailings: A laboratory column experiment. *Geochim. Cosmochim. Acta* **2002**, *66*, 1511–1523. [CrossRef]
36. Frau, F.; Ardau, C. Geochemical controls on arsenic distribution in the Baccu Locci stream catchment (Sardinia, Italy) affected by past mining. *Appl. Geochem.* **2003**, *18*, 1373–1386. [CrossRef]
37. Moncur, M.C.; Ptacek, C.J.; Blowes, D.W.; Jambor, J.L. Release, transport and attenuation of metals from an old tailings impoundment. *Appl. Geochem.* **2005**, *20*, 639–659. [CrossRef]
38. Concas, A.; Ardau, C.; Cristini, A.; Zuddas, P.; Cao, G. Mobility of heavy metals from tailings to stream waters in a mining activity contaminated site. *Chemosphere* **2006**, *63*, 244–253. [CrossRef]
39. Da Pelo, S.; Musu, E.; Cidu, R.; Frau, F.; Lattanzi, P. Release of toxic elements from rocks and mine wastes at the Furtei gold mine (Sardinia, Italy). *J. Geochem. Explor.* **2009**, *100*, 142–152. [CrossRef]
40. Geng, H.; Wang, F.; Yan, C.; Tian, Z.; Chen, H.; Zhou, B.; Yuan, R.; Yao, J. Leaching behavior of metals from iron tailings under varying pH and low-molecular-weight organic acids. *J. Hazard. Mater.* **2020**, *383*, 121136. [CrossRef]
41. Dore, E.; Fancello, D.; Rigonat, N.; Medas, D.; Cidu, R.; Da Pelo, S.; Frau, F.; Lattanzi, P.; Marras, P.A.; Meneghini, C.; et al. Natural attenuation can lead to environmental resilience in mine environment. *Appl. Geochem.* **2020**, *117*, 104597. [CrossRef]
42. Zuddas, P.; Podda, F. Variations in Physico-Chemical Properties of Water Associated with Bio-Precipitation of Hydrozincite [$Zn_5(CO_3)_2(OH)_6$] in the Waters of Rio Naracauli, Sardinia (Italy). *Appl. Geochem.* **2005**, *20*, 507–517. [CrossRef]
43. Li, M.Y.H.; Zhou, M.F.; Williams-Jones, A.E. The Genesis of Regolith-Hosted Heavy Rare Earth Element Deposits: Insights from the World-Class Zudong Deposit in Jiangxi Province, South China. *Econ. Geol.* **2019**, *114*, 541–568. [CrossRef]
44. Zhukova, I.A.; Stepanov, A.S.; Jiang, S.Y.; Murphy, D.; Mavrogenes, J.; Allen, C.; Chen, W.; Bottrill, R. Complex REE systematics of carbonatites and weathering products from uniquely rich Mount Weld REE deposit, Western Australia. *Ore Geol. Rev.* **2021**, *139 Pt B*, 104539. [CrossRef]

45. Tuduri, J.; Pourret, O.; Chauvet, A.; Gaouzi, A.; Ennaciri, A. Rare earth elements as proxies of supergene alteration processes from the giant Imiter silver deposit. In Proceedings of the 11th Biennal Meeting SGA, Antofagasta, Chile, 2 June 2014.
46. Li, X.; Liang, X.; He, H.; Li, J.; Ma, L.; Tan, W.; Zhong, Y.; Zhu, J.; Zhou, M.F.; Dong, H. Microorganisms Accelerate REE Mineralization in Supergene Environments. *Appl. Environ. Microbiol.* **2022**, *88*, e0063222. [CrossRef]
47. Takahashi, Y.; Hirata, T.; Shimizu, H.; Ozaki, T.; Fortin, D. A rare earth element signature of bacteria in natural waters? *Chem. Geol.* **2007**, *244*, 569–583. [CrossRef]
48. Taylor, S.R.; McLennan, S.M. *The Continental Crust: Its Composition and Evolution*; Blackwell: Oxford, UK, 1985; pp. 1–312.
49. Kaya, M. Assessment of Secondary Zinc and Lead Resources. In *Recycling Technologies for Secondary Zn-Pb Resources*; Kaya, M., Ed.; The Minerals, Metals & Materials Series; Springer: Cham, Switzerland, 2023.
50. Navidi Kashani, A.H.; Rashchi, F. Separation of oxidized zinc minerals from tailings: Influence of flotation reagents. *Miner. Eng.* **2008**, *21*, 967–972. [CrossRef]
51. Mercante, C. Treatment of Mineral-Metallurgical Residues for the Recovery of Useful. Ph.D. Thesis, Università degli Studi di Cagliari, Cagliari, Italy, 2021.
52. Sogos, G. Study of Minerallurgical Processes for the Treatment of Residues from Mining Activities. Ph.D. Thesis, Università degli Studi di Cagliari, Cagliari, Italy, 2022.
53. Espiari, S.; Rashchi, F.; Sadrnezhaad, S.K. Hydrometallurgical treatment of tailings with high zinc content. *Hydrometallurgy* **2006**, *82*, 54–62. [CrossRef]
54. Golik, V.; Komashchenko, V.; Morkun, V. Innovative technologies of metal extraction from the ore processing mill tailings and their integrated use. *Metall. Min. Ind.* **2015**, *7*, 49–52.
55. Massari, S.; Ruberti, M. Rare earth elements as critical raw materials: Focus on international markets and future strategies. *Resour. Policy* **2013**, *38*, 36–43. [CrossRef]
56. Ramos, S.J.; Dinali, G.S.; Oliveira, C.; Martins, G.C.; Moreira, C.G.; Siqueira, J.O.; Guilherme, L.R.G. Rare earth elements in the soil environment. *Curr. Pollut. Rep.* **2016**, *2*, 28–50. [CrossRef]
57. Cotruvo, J.A. The chemistry of lanthanides in biology: Recent discoveries, emerging principles, and technological applications. *ACS Cent. Sci.* **2019**, *5*, 1496–1506. [CrossRef] [PubMed]
58. Castillo, J.; Maleke, M.; Unuofin, J.; Cebekhulu, S. Microbial Recovery of Rare Earth Elements. In *Environmental Technologies to Treat Rare Earth Elements Pollution: Principles and Engineering*; IWA Publishing: London, UK, 2019; Chapter 8; pp. 179–205.
59. Johannesson, K.H.; Zhou, X. Geochemistry of the rare earth elements in natural terrestrial waters: A review of what is currently known. *Chin. J. Geochem.* **1997**, *16*, 20–42. [CrossRef]
60. Fathollahzadeh, H.; Eksteen, J.J.; Kaksonen, A.H.; Watkin, E.L.J. Role of microorganisms in bioleaching of rare earth elements from primary and secondary resources. *Appl. Microbiol. Biotechnol.* **2019**, *103*, 1043–1057. [CrossRef] [PubMed]
61. Park, D.M.; Brewer, A.; Reed, D.W.; Lammers, L.N.; Jiao, Y. Recovery of Rare Earth Elements from Low-Grade Feedstock Leachates Using Engineered Bacteria. *Environ. Sci. Technol.* **2017**, *51*, 13471–13480. [CrossRef] [PubMed]
62. Jin, H.; Park, D.M.; Gupta, M.; Brewer, A.W.; Ho, L.; Singer, S.L.; Bourcier, W.L.; Woods, S.; Reed, D.W.; Lammers, L.N.; et al. Techno-economic Assessment for Integrating Biosorption into Rare Earth Recovery Process. *ACS Sustain. Chem. Eng.* **2017**, *5*, 10148–10155. [CrossRef]
63. De Giudici, G.; Wanty, R.B.; Podda, F.; Kimball, B.A.; Verplanck, P.L.; Lattanzi, P.; Cidu, R.; Medas, D. Quantifying biomineralization of zinc in the rio naracauli (Sardinia, Italy), using a tracer injection and synoptic sampling. *Chem. Geol.* **2014**, *384*, 110–119. [CrossRef]
64. Buosi, M.; Contini, E.; Enne, R.; Farci, A.; Garbarino, C.; Naitza, S.; Tocco, S. Contributo alla conoscenza dei materiali delle discariche della miniera di Monteponi: I "Fanghi Rossi" dell'elettrolisi, caratterizzazione fisico-geotecnica e chimico-mineralogica, definizione del potenziale inquinante e proposte per possibili interventi. *Res. Ass. Min. Sarda* **1999**, *104*, 49–93.
65. Contini, E.; Naitza, S.; Tocco, S.; Garau, A.; Buosi, M.; Sarritzu, R. Fenomeni di contaminazione da discariche minerarie e metallurgiche nel distretto dell'antimonio del Sarrabus-Gerrei (Sardegna Sud-Orientale): L'area di Su Suergiu-Villasalto. *Res. Ass. Min. Sarda* **2008**, *112*, 45–83.
66. Manca, P.P.; Massacci, G.; Mercante, C. Environmental Management and Metal Recovery: Re-processing of Mining Waste at Montevecchio Site (SW Sardinia). In *Proceedings of the 18th Symposium on Environmental Issues and Waste Management in Energy and Mineral Production: SWEMP 2018*; Widzyk-Capehart, E., Hekmat, A., Singhal, R., Eds.; Springer: Cham, Switzerland, 2019.

Article

The Formation of Magnesite Ores by Reactivation of Dunite Channels as a Key to Their Spatial Association to Chromite Ores in Ophiolites: An Example from Northern Evia, Greece

Giovanni Grieco [1], Alessandro Cavallo [2], Pietro Marescotti [3], Laura Crispini [3], Evangelos Tzamos [4] and Micol Bussolesi [2,*]

[1] Department of Earth Sciences, University of Milan, Via Botticelli 23, 20122 Milan, Italy
[2] Department of Earth and Environmental Sciences—DISAT, University of Milan-Bicocca, P.zza della Scienza 1-4, 20126 Milan, Italy
[3] Department of Earth Environment and Life Sciences—DISTAV, University of Genoa, Corso Europa 26, 16126 Genoa, Italy
[4] Ecoresources PC, Giannitson and Santarosa Str. 15-17, 54627 Thessaloniki, Greece
* Correspondence: micol.bussolesi@unimib.it

Abstract: Ophiolite magnesite deposits are among the main sources of magnesite, a raw material critical for the EU. The present work focuses on magnesite occurrences at Kymasi (Evia Island, Greece), in close spatial association with chromitite within the same peridotite massif, and on the relationship between ultramafic rocks and late magnesite veins. Chromitite lenses are hosted within dunite, in contact with a partially serpentinized peridotite cut by magnesite veins. Close to the veins, the peridotite shows evidence of carbonation (forming dolomitized peridotite) and brecciation (forming a serpentinite–magnesite hydraulic breccia, in contact with the magnesite veins). Spinel mineral chemistry proved to be crucial for understanding the relationships between different lithologies. Spinels within partially serpentinized peridotite (Cr# 0.55–0.62) are similar to spinels within dolomitized peridotite (Cr# 0.58–0.66). Spinels within serpentinite–magnesite hydraulic breccia (Cr# 0.83–0.86) are comparable to spinels within dunite and chromitite (Cr# 0.79–0.84). This suggests that older weak zones, such as dunite channels, were reactivated as fluid pathways for the precipitation of magnesite. Magnesite stable isotope composition, moreover, points towards a meteoric origin of the oxygen, and to an organic source of carbon. The acquired data suggest the following evolution of Kymasi ultramafic rocks: (i) percolation of Cr-bearing melts in a supra-subduction mantle wedge within dunite channels; (ii) obduction of the ophiolitic sequence and peridotite serpentinization; (iii) uplift and erosion of mantle rocks to a shallow crustal level; (iv) percolation of carbon-rich meteoric waters rich at shallow depth, reactivating the dunite channels as preferential weak zones; and (v) precipitation of magnesite in veins and partial brecciation and carbonation of the peridotite host rock.

Keywords: magnesite; chromite; critical raw materials; ophiolite; serpentinite

Citation: Grieco, G.; Cavallo, A.; Marescotti, P.; Crispini, L.; Tzamos, E.; Bussolesi, M. The Formation of Magnesite Ores by Reactivation of Dunite Channels as a Key to Their Spatial Association to Chromite Ores in Ophiolites: An Example from Northern Evia, Greece. *Minerals* **2023**, *13*, 159. https://doi.org/10.3390/min13020159

Academic Editors: Pierfranco Lattanzi, Elisabetta Dore, Fabio Perlatti, Hendrik Gideon Brink and Shoji Arai

Received: 18 November 2022
Revised: 17 January 2023
Accepted: 19 January 2023
Published: 21 January 2023

1. Introduction

Magnesium (Mg) is a Critical Raw Material (CRM) for the EU, important for the automotive industry, desulphurization of steel, packaging applications, and construction equipment, as well as some other minor uses [1]. Magnesite ore deposits are the main source of Mg for the industry and are used in the production of various kinds of magnesium compounds. They occur in two main exploitable deposit types: cryptocrystalline magnesite, hosted by ultramafic rocks ("Kraubath type"), and sparry magnesites, hosted by marine platform carbonates ("Veitsch type") [2–4]. While Veitsch-type deposits are more common and have a larger size, Kraubath-type ones usually contain higher-quality magnesite, and are hence preferred for exploitation. Kraubath-type cryptocrystalline magnesite ores usually form irregular bodies frequently associated with the ultramafic rocks of ophiolite suites. The sources of Mg are the Mg-rich minerals of ultramafic rocks, in particular peridotites undergoing serpentinization processes [5]. CO_2, on the other hand, can be ascribed to various sources, including organic ones, carbonate rocks, meteoric waters, or hypogenic hydrothermal fluids [6–8].

Magnesite deposits, moreover, are a natural case study that can provide useful information for the well-known issue of CO_2 sequestration. The study of naturally occurring, ultramafic-hosted magnesites provides a useful playground for understanding the mechanisms of mineral carbonation, both in fully carbonated peridotites and in partially carbonated ones [9].

Greece hosts exploitable ophiolite-related magnesite deposits in two areas: the Evia Island and the Chalkidiki peninsula [10,11]. The ore in Northern Evia consists of a shallow (200 m) stockwork of fracture-filling magnesite veins [12], hosted within peridotites (harzburgites and lherzolites) with various degrees of serpentinization.

Kraubath-type magnesite deposits, due to their geological context, can be spatially associated with other ophiolite-related ore deposits, for example, in the present study, to chrome ores. A similar relation can be found in the Gerakini area (western Chalkidiki), where currently exploited magnesite deposits are found closely associated with chromitites.

Evia Island hosts active magnesite mines operated by TERNA MAG and Grecian Magnesite Mining companies. Several, never-exploited chromitite occurrences have been described in Northern Evia not far from the major magnesite deposits.

This work is focused on minor magnesite occurrences at Kymasi (Evia), that were selected for their very good exposure on an easily accessible horizontal outcrop by the sea, and for their very close association with nearby outcropping chromitite occurrences within the same peridotite massif, hence providing an optimal playground for studying the relationship between two different ores. In this area, magnesite was once extracted in an open pit, and the ore was carried to the nearby port to be transported by sea.

In the present study, we investigate the origin of the spatial relationship between magnesite and chromite deposits. Chromium and magnesite are two georesources considered (chromium) or listed (magnesium) as Critical Raw Materials for the EU. This mineralogical association is hosted within peridotites, the main source of olivine, and can therefore provide an interesting case of sustainable mining. The close spatial relationship between these three commodities could be useful for a reduced-waste, more sustainable exploitation.

2. Geological Setting

Evia is located in central Greece, and it is the second largest island of the country. The island is part of the Hellenides belt, which is divided into two tectonic units: the Internal and External Hellenides. Evia is geologically comprised in the southern part of the Internal Hellenides (Figure 1a), within the Pelagonian Zone (Northern Evia) and the Atticocycladic Zone (Southern Evia). These are two of the many geotectonic domains in which the Internal Hellenides are subdivided. The Atticocycladic Zone exposes a pile of high-pressure tectonic nappes overlain by the Pelagonian Zone [13–15]. The Pelagonian Zone is formed by an assemblage of tectonic units mainly consisting of a Paleozoic metamorphic basement [13,16]. The crystalline basement is overthrust by Late Jurassic ophiolites and Upper Paleozoic and Mesozoic sedimentary sequences [15,17].

The mafic and ultramafic rocks of Evia (Figure 1b) are generally regarded as belonging to the "Western Ophiolite Belt", along with Vourinos, Pindos, and Othris ophiolites, in contrast with the "Eastern Ophiolite Belt", comprising the Vardar Zone and Guevgueli ophiolites, with chromitite occurrences at Vavdos, Gerakini-Ormilya, Vasilika, and Krani [18], and the chromitite occurrences of the Serbo-Macedonian Massif. The ultramafic rocks in the area are known for hosting large magnesite deposits, which are currently exploited [19]. Moreover, chromitite occurrences are reported in the literature in three localities, Papades, Mandoudi, and Limni [20], in small outcrops associated with serpentinites.

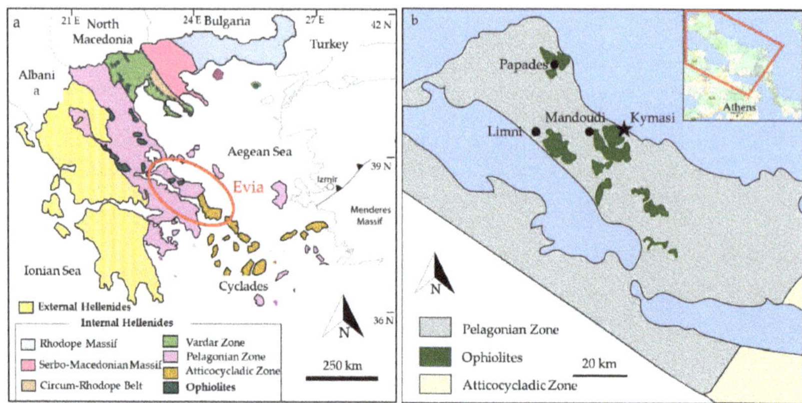

Figure 1. (**a**) Simplified map showing the major structural elements of Greece. Modified after [21,22]; (**b**) Simplified map of Norther Evia showing the major ophiolite bodies of the area. Kymasi mineralization are indicated by the black star. Modified after [23].

3. Materials and Methods

At Kymasi, field work was focused on a well-exposed network of peridotite-hosted magnesite veins (38°48′28.37″ N, 23°31′13.32″ E) and on small chromitite lenses cropping out at a distance of about 120 m (38°48′25.24″ N, 23°31′16.98″ E) (Figure 2).

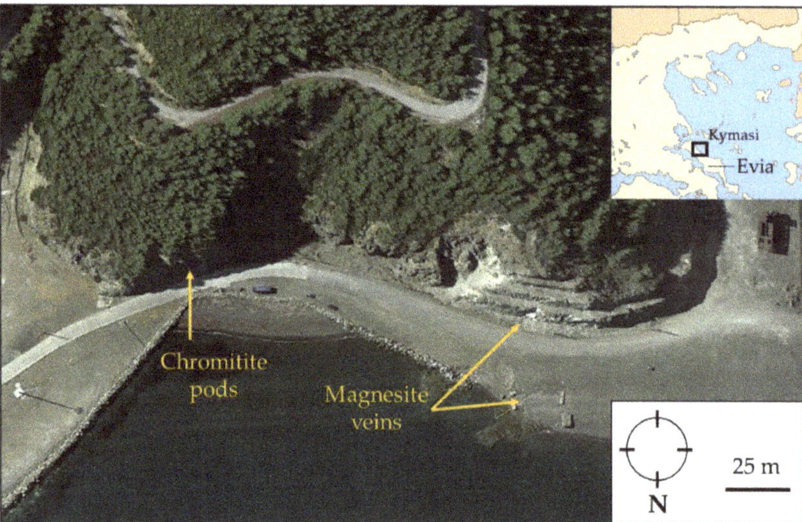

Figure 2. Sampling position (Google Earth®) of chromitite pods, magnesite veins, and their host rock.

A centimetric-to-decimetric chromitite lens at Kymasi is hosted within partially altered dunite, in contact with a harzburgite host rock cut by magnesite veins (Figure 3). Chromitites are densely disseminated, with ~50% chromite modal content. Five chromitite and two dunite specimens were sampled from the outcrop.

Magnesite veins occur as persistent, metric conjugate vein systems (Figure 4a) with orientation NE–SW, NW–SE, and E–W, and with variable thickness, from a few cm up to 50 cm. The center of the veins is composed of white massive magnesite, while at the rim, magnesite shows a reddish secondary alteration (Figure 4b,c).

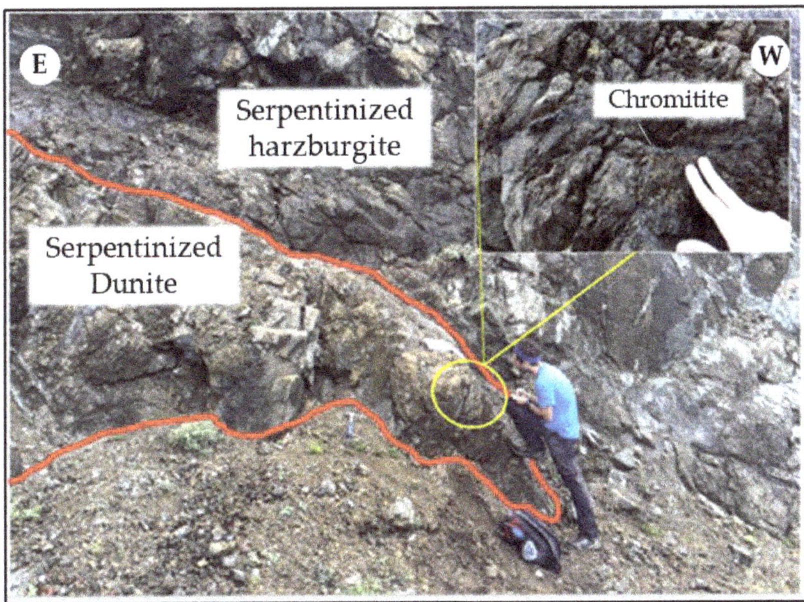

Figure 3. Chromitite pods in serpentinized dunite, hosted by partially serpentinized harzburgite at Kymasi.

Three massive magnesite specimens were sampled from the cores of 3 vein sets (F1, F2, and F3). Vein F2 was then sampled in more detail. From core to rim, the collected samples are: a thick massive magnesite central portion (F4a), a serpentinite breccia with magnesite cement (F4b), and a partially dolomitized peridotite (F4c). The host rock is a partially serpentinized peridotite (F4d). All contacts are irregular and sharp (Figure 4b,c).

The collected samples were studied through optical microscopy in transmitted and reflected light, X-ray powder diffraction (XRPD), scanning electron microscope (SEM-EDS), and electron probe microanalyzer (EMPA). Magnesite samples were also analyzed for their stable carbon and oxygen isotopes.

XRPD analyses were performed at the University of Milan with a PANalytical X'Pert-pro instrument set at the following operating conditions: 40 kV of voltage; 40 mA of current; and Cu anticathode Kα1/Kα2: 1.540510/1.544330 Å. This powder diffractometer is equipped with an incident beam monochromator, which separates the K1 and the K2 and works with the Bragg–Brentano geometry. Analyses were conducted with a step size [°2Th.] of 0.0170 (scan time 30.3607 s). Data were elaborated with software X'Pert Highscore v.2.3.

μ-Raman analyses were performed at the University of Genova with the HORIBA XploRA_Plus Raman Spectrometer (532 nm laser line), equipped with the OLYMPUS BX-41 optical microscope. Raman spectra were acquired using a 2400 grooves mm^{-1}, with an acquisition time of 100 s.

Preliminary SEM analyses were conducted at the University of Milano-Bicocca using a Tescan VEGA TS 5136XM equipped with an EDS analyzer (EDAX Genesis 400), with 200 pA and 20 kV and high-vacuum as standard conditions, on carbon-coated samples (20 nm) using an Edwards 5150B carbon coater.

Mineral chemistry was determined through a JEOL 8200 electron microprobe (JEOL Ltd., Akishima, Japan) equipped with a wavelength dispersive system (WDS) at the Earth Sciences Department of the University of Milan. The microprobe system operated using an accelerating voltage of 15 kV, a sample current on brass of 15 nA, and a counting time of 20 s on the peaks and 10 s on the background. Beam diameter is 1 μm. A series of natural minerals was used as standards: wollastonite for Si, forsterite for Mg, ilmenite for Ti, fayalite for Fe, anorthite for Al and Ca, metallic Cr for Cr, niccolite for Ni, and rhodonite for Mn and Zn. Detection limit is approximately 330 ppm for Ti (6% standard deviation), 460 ppm for Mn (6% standard deviation), 160 ppm for Mg (2% standard deviation), 180 ppm for Si (6% standard deviation), 320 ppm for V (12% standard deviation), 370 ppm for Fe (8% standard deviation), 140 ppm for Ca (15% standard deviation), 135 ppm for Al (4% standard deviation), 370 ppm for Cr (10% standard deviation), 390 ppm for Ni (9%standard deviation), and 800 ppm for Zn (9% standard deviation). Analyses were conducted on carbon-coated samples (20 nm) using an Edwards E306A coating device.

Figure 4. (**a**) Magnesite vein network hosted within variably altered peridotite; (**b**) Sampled magnesite veins (red star), and (**c**) detail of the magnesite vein F2 and sampled host lithologies.

Stable isotope analyses were performed at the University of Milan using a Delta V Advantage mass spectrometer coupled with a GasBench II. Phosphoric acid is added to the solid sample, and the resulting CO_2 is then analyzed.

4. Results

4.1. Mineralogy and Texture

The mineralogy of chromitite and dunite, magnesite veins (sample F1, F2, F3, and F4a), serpentinite–magnesite breccia (sample F4b), dolomitized peridotite (sample F4c), and partially serpentinized peridotite (sample F4d) has been analyzed by means of optical microscopy, XRPD, and μ-Raman.

4.1.1. Chromitite and Dunite

Kymasi densely disseminated chromitites consist of a 40–50 cm-thick schlieren body made up of thin layers of chromitite ~2 cm thick, containing up to 50% chromite, alternated to thin dunite layers up to 10 cm thick. Chromite crystals are submillimetric (50–500 μm), and show rare alteration into ferrian chromite. The interstitial silicate matrix is composed of serpentine, forming a mesh texture, with olivine relicts and rare chlorites associated with ferrian chromite (Figure 5a). Primary phases in dunite are mainly olivine relicts partially replaced by lizardite and chrysotile along the rims (Figure 5b). Cr-spinels are present as accessory phases.

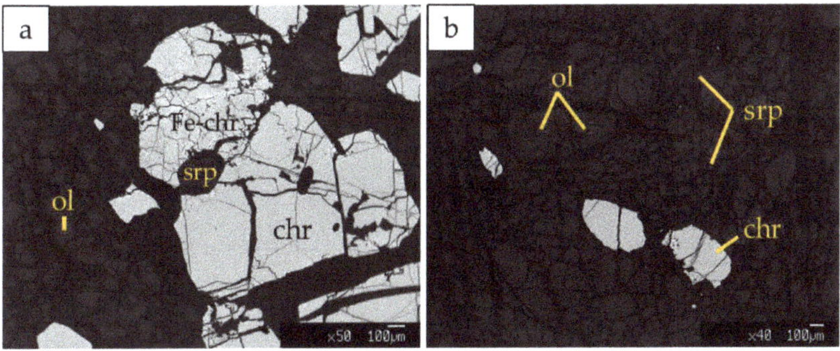

Figure 5. BSE images of (**a**) densely disseminated chromitites at Kymasi; (**b**) Dunite with accessory spinel in contact with chromitites. Fe-chr: ferrian chromite, ol: olivine, chr: chromite, srp: serpentine.

4.1.2. Carbonate-Bearing Peridotites and Magnesite Mineralization

The host rock of magnesite mineralization is a partially serpentinized peridotite (sample F4d), composed of olivine partially replaced by lizardite, minor tremolite, orthopyroxene, clinopyroxene, and accessory spinel. No carbonate minerals were detected (Figure 6a). Chrysotile occurs in very subordinate amounts within sub-millimetric veinlets as well as within rims of mesh texture or cleavage planes of bastites (Figure 7a,b). Chrysotile is often associated with talc and chlorite.

The massive magnesite veins are characterized by secondary halos at the contact with the host peridotite composed of dolomitized peridotite (sample F4c; Figure 6b,c) and a serpentinite–magnesite hydraulic breccia (sample F4b, Figure 6d) moving towards the massive magnesite.

The dolomitized peridotite is mainly composed of dolomite crystals with a euhedral-to-subhedral habit, lizardite, and subordinate clinopyroxene and orthopyroxene relicts (Figures 6b,c and 7a). Spinel is present as an accessory phase.

The serpentinite–magnesite hydraulic breccia consists of poorly sorted angular clasts of chromite-rich serpentinite breccia cemented by magnesite with a cockade texture (Figures 6d and 7a). Lizardite is the main constituent of serpentinite clasts with minor amounts of chrysotile, talc, and chlorite occurring within sub-millimetric veinlets and along rims of mesh texture. The cement of the breccia is composed by cryptocrstalline magnesite with massive texture and, locally, by magnesite and minor dolomite with a microcrystalline polygonal texture (Figure 7b).

Massive magnesite, sampled from the core of the veins, is composed mainly of cryptocrystalline magnesite Figures 6e and 7a), with minor sparitic and spherulitic magnesite. The other mineralogical phases are serpentine relicts and microfracture-filling talc (Figure 6f).

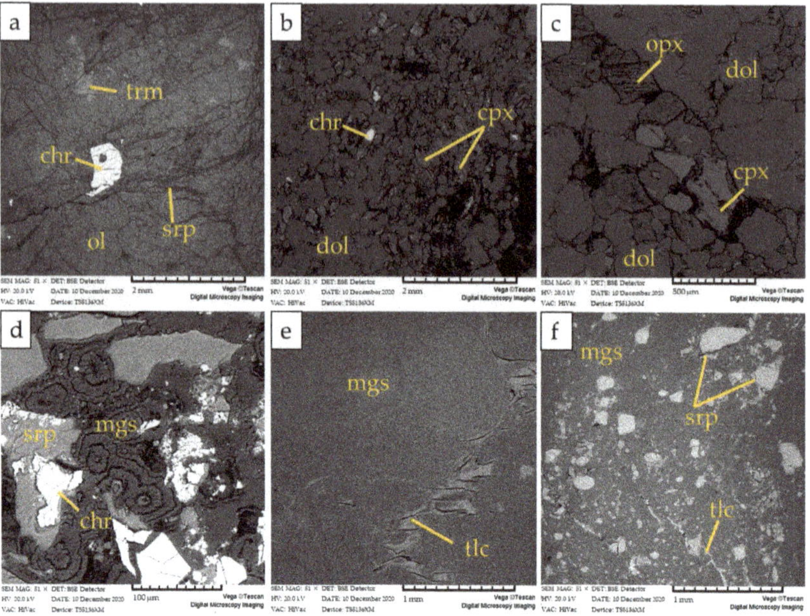

Figure 6. BSE images of (**a**) partially serpentinized peridotite (F4d), showing accessory spinel associated with unaltered olivine, rarely altered into serpentine; (**b**) dolomitized peridotite (F4c) showing euhedral dolomite crystals and rare spinel and clinopyroxene; (**c**) dolomitized peridotite showing euhedral dolomite crystals and subordinate clinopyroxene and orthopyroxene; (**d**) hydraulic breccia (F4b) showing angular serpentine and chromite clasts cemented by magnesite with cockade texture; (**e**,**f**) massive magnesite (F4a) with minor talc and serpentine clasts. Mineral abbreviations are trm: tremolite, chr: chromite; ol: olivine; srp: serpentine; cpx: clinopyroxene; dol: dolomite; opx: orthopyroxene; mgs: magnesite; tlc: talc.

4.2. Mineral Chemistry

The composition of spinels within chromitites, partially serpentinized peridotite, partially dolomitized peridotites, and serpentinite–magnesite hydraulic breccia is reported in Table 1, Figure 8a, and in the Supplementary Materials. The composition of olivine, pyroxene, and carbonates is reported in Figure 8b–d and in the Supplementary Materials. Complete chemical analyses of serpentine, chlorite, and amphibole are reported in the Supplementary Materials.

Table 1. Major and minor elements average composition and standard deviation of chromite cores from Kymasi lithologies; Mg# = [Mg/(Mg + Fe^{2+})]; Cr# = [Cr/(Cr + Al)]. Fe$_2$O$_3$ and FeO are calculated based on stoichiometry.

	Chromitite		Chromitite		Partially Serp. Peridotite (F4d)		Dolomitized Peridotite (F4c)		Serpentinite–Mag. Breccia (F4b)	
	Unaltered Chromite n = 51		Ferrian Chromite n = 41		Chromite n = 15		Chromite n = 5		Chromite n = 5	
Wt%	avg	st.dev.	avg	st.dev.	avg	st.dev.	avg	st.dev.	avg	st.dev.
TiO$_2$	0.10	0.04	0.08	0.04	0.11	0.06	0.06	0.06	0.06	0.04
Al$_2$O$_3$	8.47	0.74	4.30	1.82	24.14	2.67	18.62	1.71	7.96	2.76
Cr$_2$O$_3$	62.84	1.58	64.96	2.12	44.83	4.27	50.17	1.62	61.14	3.84
Fe$_2$O$_3$	2.02	0.75	2.16	1.05	2.13	1.78	1.64	0.98	1.76	1.78

Table 1. *Cont.*

	Chromitite		Chromitite		Partially Serp. Peridotite (F4d)		Dolomitized Peridotite (F4c)		Serpentinite–Mag. Breccia (F4b)	
	Unaltered Chromite n = 51		Ferrian Chromite n = 41		Chromite n = 15		Chromite n = 5		Chromite n = 5	
FeO	15.42	1.44	21.36	1.09	12.86	0.55	17.31	1.07	18.55	3.96
MnO	0.12	0.14	0.31	0.10	0.07	0.15	0.19	0.17	0.27	0.24
MgO	11.92	0.91	7.41	0.92	14.51	0.97	11.30	0.49	9.22	2.73
Total	100.93	1.00	100.62	0.80	98.90	0.56	99.36	0.67	99.24	1.08
Ti	0.00	0.00	0.00	0.00	0.00	0.00	0.00	0.00	0.00	0.00
Al	0.33	0.03	0.17	0.07	0.87	0.08	0.70	0.06	0.31	0.10
Cr	1.62	0.04	1.77	0.09	1.08	0.12	1.26	0.05	1.64	0.14
Fe^{3+}	0.05	0.02	0.06	0.03	0.05	0.04	0.04	0.02	0.04	0.04
Fe^{2+}	0.42	0.04	0.61	0.04	0.33	0.02	0.46	0.03	0.53	0.13
Mn	0.00	0.00	0.01	0.00	0.00	0.00	0.00	0.01	0.01	0.01
Mg	0.58	0.04	0.38	0.04	0.66	0.04	0.54	0.03	0.46	0.12
Cr#	0.812	0.018	0.885	0.042	0.555	0.050	0.631	0.023	0.832	0.049
Mg#	0.579	0.041	0.382	0.041	0.667	0.022	0.538	0.026	0.466	0.125

4.2.1. Chromitite

The Mg#[Mg/(Mg + Fe^{2+})] of unaltered chromitite spinels ranges between 0.51 and 0.64, and the Cr#[Cr/(Cr + Al)] ranges between 0.77 and 0.83. TiO_2 content ranges between 0.02 and 0.23 wt%. They can be classified as chromites (Figure 8a).

Chromite rims altered into ferrian chromite show lower Mg#, comprised between 0.31 and 0.45, and higher Cr#, comprised between 0.80 and 0.96.

Olivine, serpentine, and rare chlorite form the silicate gangue of Kymasi chromitites. The olivine Fo value ranges between 92.28 and 95.36%, and the NiO content ranges between 0.41 and 0.90 wt%. The Mg# of serpentine ranges between 0.93 and 0.96. Rare chlorite, associated with ferrian chromite alteration, shows high Al_2O_3 content (12.32–12.86 wt%) and Cr_2O_3 contents up to 3.45 wt%.

4.2.2. Partially Serpentinized Peridotite (F4d)

Spinels within partially serpentinized peridotite are aluminum chromites characterized by Mg# ranging between 0.63 and 0.66 and lower Cr# (0.55–0.62) with respect to chromitite spinels (Figure 8a). They are classified as Al-chromite.

Olivine is strongly forsteritic (Figure 8b), and shows Fo values ranging between 93.19 and 97.86%, and NiO content up to 0.66 wt%. Pyroxene is augitic, with a composition Wo 34%–42%, En 55%–63%, and Fs 2%–3% (Figure 8c). Serpentine (mainly lizardite) is not widespread, and shows Mg# of 0.97 and 1.00.

4.2.3. Dolomitized Peridotite (F4c)

Spinels within dolomitized peridotite are aluminum chromites, similar to spinels detected within partially serpentinized peridotite (Figure 8a), and show Mg# comprised between 0.51 and 0.54, and Cr# comprised between 0.58 and 0.66.

Primary silicates are enstatite (Figure 8c) with the following ranges: Wo 1.04%–2.23%, En 88.47%–90.85%, and Fs 7.97%–9.53%, and rare diopside relics (Figure 8c) showing the following composition: Wo 46.79–48.34, En 49.30–50.10, and Fs 2.36%–3.28%. Serpentine shows Mg# between 0.85 and 0.94. Dolomite is widespread and shows the following compositional range: MgO (18.38–22.39 wt%) and CaO (28.23–31.19 wt%) (Figure 8d). Tremolite, present as a later alteration phase, has Mg# comprised between 0.97 and 0.99.

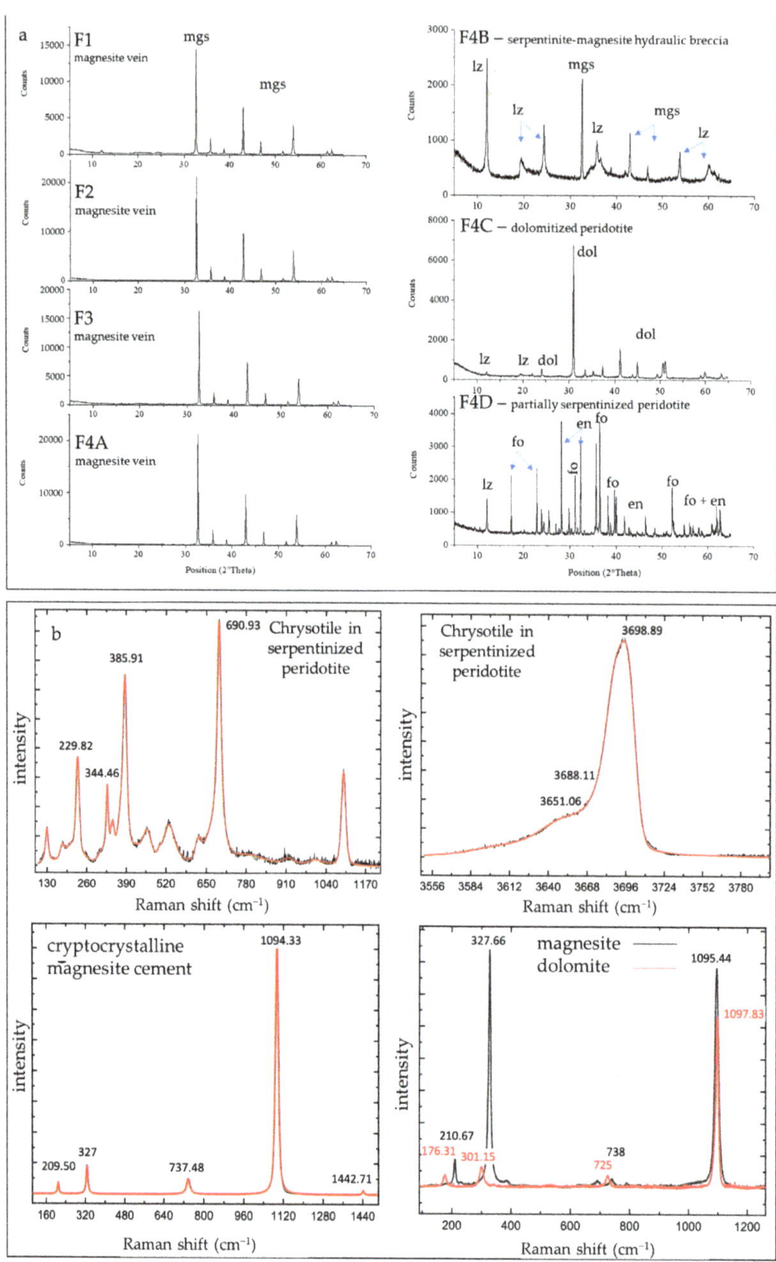

Figure 7. (**a**) XRPD pattern of magnesite samples (F1, F2, F3, F4a), serpentinite–magnesite hydraulic breccia (F4b), dolomitized peridotite (F4c), and partially serpentinized peridotite (F4d). Mgs: magnesite, lz: lizardite, dol: dolomite, fo: forsterite, en: enstatite. (**b**) Selected μ-Raman spectra of: chrysotile within a submillimetric veinlet in a partially serpentinized peridotite for the spectral region 110.86–1097.14 cm^{-1} (**upper left**) and for the spectral region 3599.53–3698.89 cm^{-1} (**upper right**); magnesite with massive cryptocrystalline texture within the cement of serpentinite–magnesite hydraulic breccia (**lower left**); magnesite and dolomite with microcrystalline polygonal texture within the cement of serpentinite–magnesite hydraulic breccia (**lower right**).

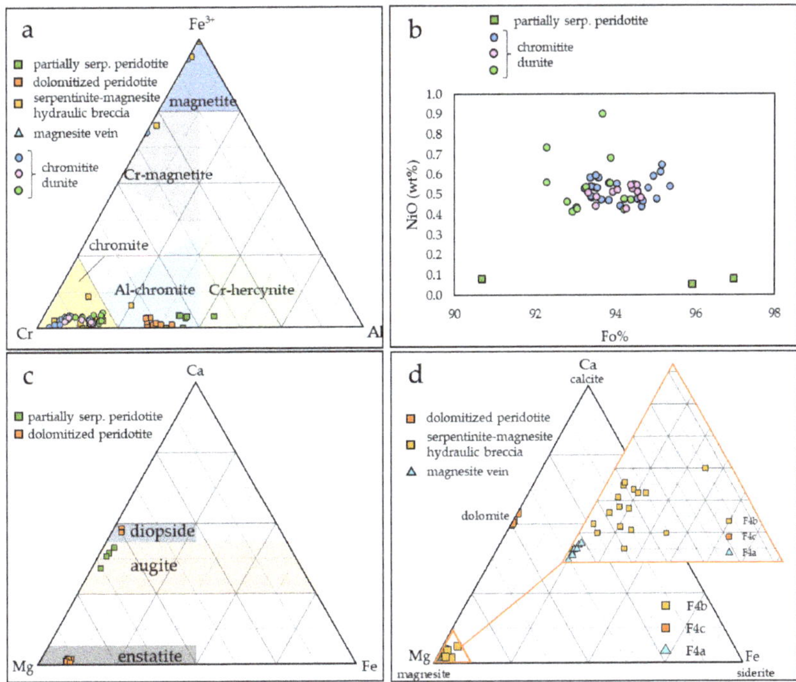

Figure 8. Ternary compositional diagrams of (**a**) spinels from chromitites and peridotites; compositional fields are from [24]; (**b**) olivine from chromitites and peridotites; (**c**) pyroxene from partially serpentinized peridotite and dolomitized peridotite; (**d**) carbonate phases from magnesite vein, serpentinite–magnesite hydraulic breccia and dolomitized peridotite.

4.2.4. Serpentinite–Magnesite Hydraulic Breccia (F4b)

Spinels within the serpentinite–magnesite breccia are chromites, with compositions more similar to those in chromitite than those in partially serpentinized peridotite and dolomitized peridotite (Figure 8a), with high Cr# (0.83–0.93) and Mg# comprised between 0.19 and 0.53.

Lizardite shows Mg# comprised between 0.91 and 0.98. Magnesite filling the fracture network shows MgO varying between 41.35 and 46.57 wt%, highly variable FeO, ranging from 0.12 to 4.07 wt%, and CaO varying between 0.35 and 4.01 wt%.

4.2.5. Magnesite Veins (F4a)

Magnesite filling the veins shows MgO comprised between 33.09 and 46.21 wt%, and low FeO and CaO comprised between 0.16 and 0.33 wt%, and 0.08 and 0.77 wt%, respectively.

4.3. Stable Isotope Geochemistry

Oxygen and carbon isotopic compositions of magnesite in magnesite veins within the serpentinite–magnesite hydraulic breccia, and of dolomite within the dolomitized peridotite are reported in Table 2. The $\delta^{13}C$ values are homogeneous within all magnesite samples, and vary between -10.63 and $-12.68‰$. $\delta^{18}O$ values (VSMOW) are also homogeneous among the different lithologies, varying between 26.33 and 27.72‰. Oxygen and carbon isotopic values of dolomite are comprised in the range of magnesite ones.

Table 2. $\delta^{13}C$ and $\delta^{18}O$ and standard deviation of carbonates from all lithologies.

Sample	$\delta^{13}C$ (VPDB)	$\delta^{18}O$ (VPDB)	$\delta^{18}O$ (VSMOW)	St. Dev. $\delta^{13}C$	St. Dev. $\delta^{18}O$
F1 (mgs)	−12.68	−3.24	27.56	0.03	0.04
F2 (mgs)	−13.11	−3.43	27.37	0.04	0.04
F3 (mgs)	−10.63	−3.09	27.72	0.04	0.03
F4a (mgs)	−12.25	−3.26	27.55	0.04	0.04
F4b (mgs)	−12.53	−4.02	26.76	0.03	0.06
F4c (dol)	−11.74	−3.46	27.33	0.04	0.06

5. Discussion

5.1. Chromitite and Peridotite Genesis

Magnesite veins at Kymasi are hosted within partially altered, chromitite-bearing peridotites. Spinels in the different lithologies show highly variable composition (Figure 9). Chromitites show high Cr# (>0.80) and a wide Mg# range (0.30–0.65). The serpentinite–magnesite hydraulic breccia, in contact with the magnesite vein, hosts spinels with a very similar mineral chemistry. On the contrary, the partially serpentinized peridotite and the dolomitized peridotite host spinels with lower Cr# (0.60–0.70) and narrow Mg# range (0.58–0.66).

Spinels within chromitites are quite homogenous, and are all high-Cr [25,26]. High-Cr chromitites are similar to other ophiolite chromite ores within ultramafic rocks in Greece, such as Vourinos [27–29] and partially Gomati [30,31], while they are quite different from chromitites from the Othris ophiolite [32] (Figure 9a). Disseminated spinels within peridotites at Kymasi are similar to spinels within Vourinos peridotites [27] and Gerakini serpentinites and harzburgites [11].

High-Cr chromitites generally indicate a genesis from high degrees of partial melting (~1200 °C) [33,34], from melts with boninitic affinity (Figure 9b,c). This is due to the fact that boninites commonly contain Cr-rich spinels, with a mineral chemistry similar to ophiolite ones [35], and are hence thought to be the parent melts of high-Cr ophiolite chromitites. Both Kymasi high-Cr chromitites and peridotites were formed in a supra-subduction forearc setting, as indicated by spinel mineral chemistry (Figure 9b). Chromitite ore bodies in supra-subduction settings are thought to be formed within "dunite channels", acting as pathways for the circulation of the parent melts of chromitites [36]. These dunite channels are generated by the dissolution of pyroxenes in supra-subduction zones, leaving a residual, porous dunite. A network of dunite channels provides the ideal conditions for the precipitation of chromitite bodies by the mixing of different melts [36,37].

Disseminated spinels in partially serpentinized peridotite and dolomitized peridotite have lower Cr# than spinels in chromitite, but, in any case, their composition reflects the one of dispersed spinels in restitic harzburgites [38]. On the contrary, disseminated spinels in the serpentinite–magnesite hydraulic breccia are high-Cr, with a mineral chemistry quite similar to the one in chromitites, and to spinels within dunite-hosting chromitites.

Chromites within ophiolites do not always preserve their magmatic imprint, due to sub-solidus re-equilibration, metamorphic, metasomatic, and hydrothermal processes during the obduction of the oceanic lithosphere [39]. Kymasi chromite grains show alteration into ferrian chromite at the rims (Figure 5a), while their cores are unaltered. The transformation of chromite into ferrian chromite involves the reaction of chromite and serpentine to form ferrian chromite and Cr-chlorite [39,40].

Chromite and olivine cores within densely disseminated chromitites are likely to retain their primary composition, and can be used as geothermometers, to estimate a primary temperature [28,41,42]. The major variables of olivine-spinel geothermometers are spinel and olivine Mg# (Mg/(Mg + Fe^{2+}), spinel Cr#*(Cr/Cr + Al + Fe^{3+}), Fe^{3+}# (Fe^{3+}/ΣFe), and Ti moles [43,44] (Table 3). The major uncertainties are Fe^{3+} and Fe^{2+} calculations and the partial resetting of olivine-spinel compositions upon slow cooling at relatively high temperatures [41,45].

Primary temperatures at the core of chromite and olivine grains have been calculated on an average composition for three chromitite samples. They are similar for two samples

(780 and 788 °C) and higher for the third one (900 °C). As primary temperatures represent the temperature below which the compositional homogeneity within chromite and olivine cannot be maintained, it is always lower than the magmatic temperature of formation [42]. Kymasi primary temperatures are consistent with other chromitite temperatures in ophiolite environments, such as Vourinos (Greece) [27,28], Iballe (Albania) [38,42], Kluchevskoy (Russia) [46], and Luobusa (China) [47].

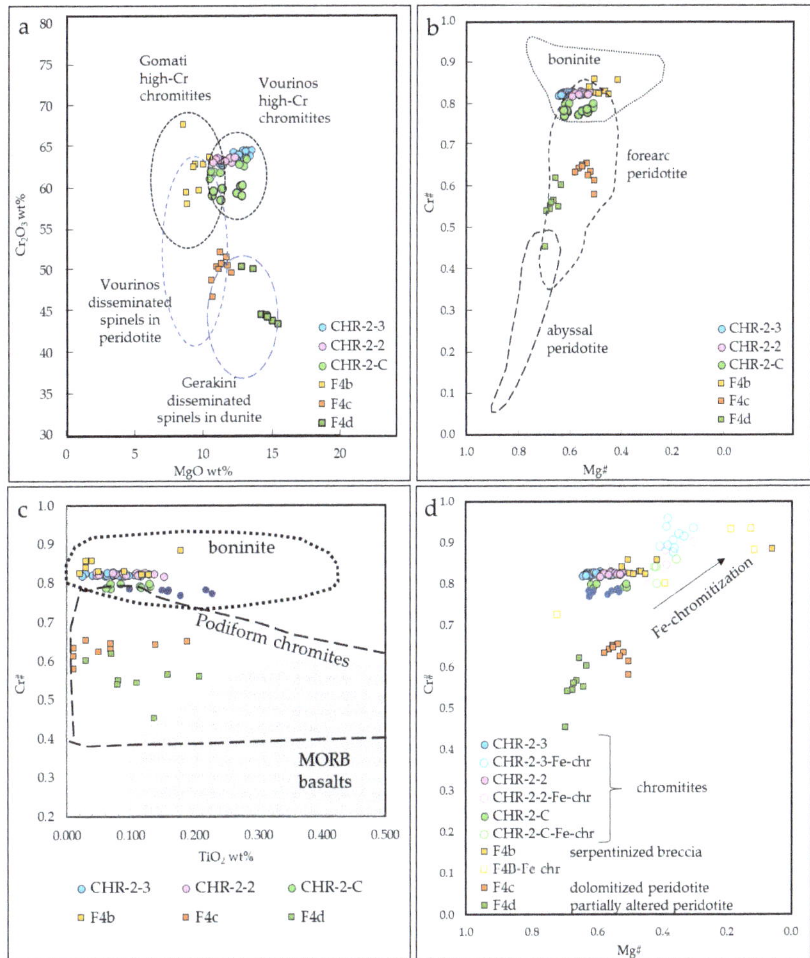

Figure 9. (**a**) Cr_2O_3 vs. MgO (wt%) of chromite cores. Compositional fields: Gomati high-Cr chromitites [30,31], Vourinos high-Cr chromitites [28], Vourinos disseminated spinels in dunite [27] Gerakini disseminated spinels in dunite [11]; (**b**) Cr# vs. Mg# of chromite cores. Compositional fields are from [35,38]; (**c**) Cr# vs. TiO_2 (wt%) of chromite cores. Compositional fields from [48]; (**d**) Cr# vs. Mg# of chromite cores and Fe-chromite rims.

5.2. Peridotite Carbonation and Magnesite Precipitation through Reactivation of Dunite Channels

In general, carbonation of ultramafic rocks is characterized by the replacement of pre-existing minerals by carbonate phases, mainly magnesite, dolomite, and calcite.

Table 3. Estimated primary temperatures and parameters for the olivine-spinel geothermometer according to Ballhaus [43]. The assumed pressure is 0.3 GPa.

Parameters	CHR-2-3 Chromitite	CHR-2C Chromitite	CHR-2-2 Chromitite
Mg# olivine	0.943	0.944	0.942
Fe# olivine	0.057	0.056	0.058
Mg# spinel	0.603	0.554	0.555
Fe# spinel	0.397	0.446	0.445
Cr*# spinel	0.823	0.794	0.824
Fe^{3+}# spinel	0.122	0.096	0.059
Ti moles	0.002	0.002	0.002
KD	10.919	13.481	13.071
T (°C)	900	788	781

At Kymasi, carbonation led to the formation of a thin centimetric layer of dolomitized peridotite close to massive magnesite veins. Magnesite at Kymasi occurs within serpentinite–magnesite hydraulic breccia, as well as within magnesite-filled fractures. Carbonate precipitation is caused by the interaction of serpentinized peridotites with CO_2-rich fluids, usually derived from sea water and the atmosphere [49,50].

Carbonation of ultramafic rocks and circulation of CO_2-rich fluids at Kymasi resulted in the replacement of pyroxene and possibly olivine and serpentine by dolomite within the host peridotite, and in the precipitation of massive magnesite in veins. Progressing from the partially serpentinized peridotite to the magnesite veins, we can find a dolomitized peridotite and a serpentinite–magnesite hydraulic breccia. Dolomite is the main mineralogical phase within the dolomitized peridotite. Its formation can be linked to the alteration of Ca-Mg bearing silicates, such as tremolite or clinopyroxene. Previous studies showed that peridotite serpentinization is characterized by the hydration of olivine and pyroxene, producing serpentinites with mesh textures and bastites, respectively. The excess of Ca^{2+} and Mg^{2+} from the alteration of primary silicates can react with CO_2 to produce calcite or dolomite [51]. Spinel mineral chemistry of dolomitized peridotite is consistent with the nearby partially serpentinized peridotite, with relatively low-Cr spinels.

The serpentinite–magnesite hydraulic breccia, closer to magnesite veins, records a brittle event. Serpentine clasts are angular, and cemented by magnesite with a cockade texture. The low amount of dolomite and the spinel mineral chemistry, more similar to Kymasi chromitites, suggests that carbonation affected a dunite.

The fluids responsible for carbonation and precipitation of magnesites circulated through preferential pathways, possibly reactivating older weakness zones. Chromitite-hosting dunites are formed by percolation of an ascending melt through mantle peridotite in channels where high degree of pyroxene melting weakens the peridotite (Figure 10a), leaving a restitic dunite [36,37,52,53]. Chromitite ores may form in these weakness zones by mixing of two different melts (Figure 10b).

Spinels show compositions similar to those of Kymasi chromitites that cannot be explained by any post-magmatic re-equilibration in spinel composition as post-magmatic processes were limited to minor ferrian-chromitization at rims. The similarity can hence be attributed to the dunitic nature of the carbonated rock.

During obduction, ophiolites undergo strong deformation in a fragile regime, where these weakness zones can be reactivated as faults or shear zones that work as pathways for the circulation of CO_2-rich fluids, precipitating the typical stockwork magnesite veins (Figure 10c). The system of veins sampled at Kymasi exploits an old dunite channel, as suggested by spinel mineral chemistry, and the circulating fluids partially affect also the hosting peridotite, in the specific case inducing dolomitization of the host peridotite (Figure 10d).

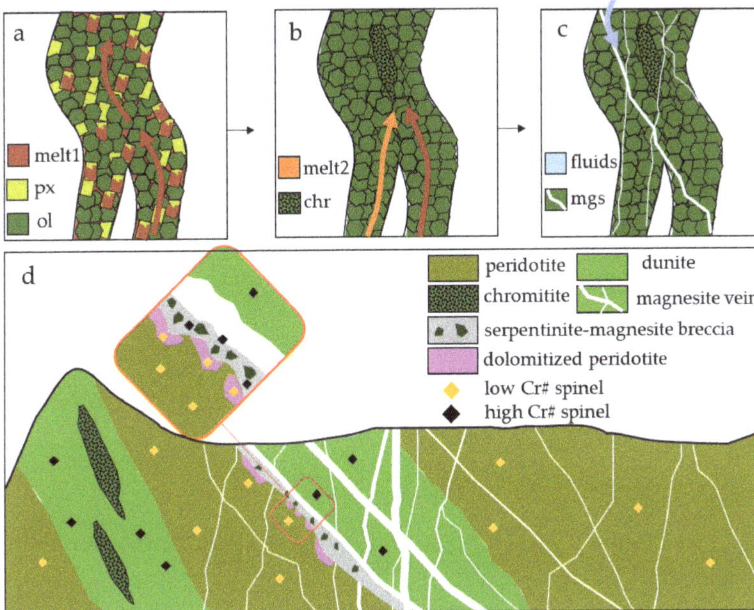

Figure 10. Schematic genetic model showing (**a**) infiltration of melts ascending through peridotite and inducing pyroxene melting, and the formation of a restitic dunite; (**b**) different melts can mix and induce precipitation of chromitite due to chromite oversaturation; (**c**) after the obduction of the oceanic crust, the weaker zones of the dunite channels can be reactivated as CO_2-rich fluid pathways, precipitating magnesite in stockwork veins; (**d**) out-of-scale scheme of Kymasi chromitites and magnesite main veins, showing the difference in spinel mineral chemistry.

5.3. Origin of Magnesite-Forming Fluids

The low-T reaction between surface water and peridotite produces Mg-HCO_3^- waters, also called Type-1 waters [5,6,54–56]. Magnesite, dolomite, serpentine, and clay then precipitate from the reaction between peridotites and Mg-HCO_3^- waters, once they become isolated from the atmosphere [6,55–58]. The resulting waters are enriched in Ca (from the alteration of pyroxene) and OH^-, and depleted in dissolved carbon [5]. These Ca-OH^--rich waters are called Type 2 waters [5,9], and generally form alkaline springs emerging from peridotites. Finally, the alkaline waters can react with the atmosphere to precipitate calcite [9].

Carbon and oxygen stable isotope analyses of magnesite veins are crucial for understanding the origin and source of the mineralizing fluids. The narrow range of $\delta^{18}O$ and $\delta^{13}C$ isotopes, common to other ultramafic-hosted magnesite deposits in the Balkans [59] (Figure 11), suggests that the mineralization was formed through a single process.

$\delta^{18}O$ values of Kymasi magnesite veins, comprised between 26.76 and 27.72 ‰, are comparable to the values of magnesite veins of the California Coast Ranges [60,61] and the Oman magnesite deposits [9], indicating an equilibrium with local waters between 15 and 40 °C. These oxygen isotopic ratios reflect those of meteoric waters, implying a low-T precipitation of the magnesite mineralization very close to the surface.

$\delta^{13}C$ isotopic values can help identify the source of the carbon. Kymasi negative values, comprised between −10 and −15, rule out the possibility of a deep-seated or mantle source of the CO_2 [59,62], which would produce higher isotopic ratios. $\delta^{13}C$ values are more similar to other magnesite ores formed in shallow environments [59,63] and from a biogenic source of the carbon [64]. The organic source of carbon can be provided either through decarboxylation of organic sediments (~−15‰), by thermal contact metamorphism decomposition of limestone [7,59,65], by chemical weathering with involvement of atmospheric CO_2 (~3‰) [4,7,64,66], or by a mixed source. Strongly negative $\delta^{13}C$ PDB values, in any case, point towards an organic carbon source.

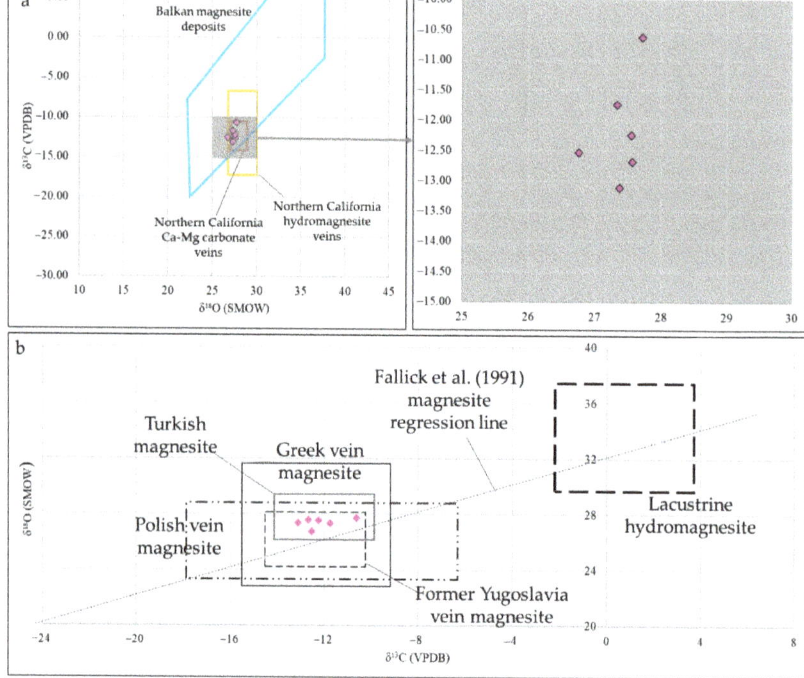

Figure 11. $\delta^{13}C$ vs. $\delta^{18}O$ (**a**) and $\delta^{18}O$ vs. $\delta^{13}C$ (**b**) of Kymasi magnesite. Compositional fields are: Balkan magnesite deposits [59], Northern California hydromagnesite veins [9], Northern California Ca-Mg carbonate veins [9], Turkish magnesite [65], Polish vein magnesite [64], Greek vein magnesite [67], former Yugoslavian vein magnesite, and lacustrine hydromagnesite [59].

6. Conclusions

The coexistence of dunite-hosted chromitite lenses and magnesite vein stockwork within the Kymasi ophiolite mantle offers the opportunity to unravel the sequence of events that affected the rocks during a long time-span, with changes in geotectonic settings: from old deep mantle metasomatism, down to recent carbonation. The following steps can be reconstructed:

Percolation of fluid-rich Cr-bearing melts in a supra-subduction mantle wedge with the formation of dunite channels hosting chromitite lenses;

Obduction of the ophiolite sequence, and partial serpentinization of peridotites;

Uplift and erosion of mantle rocks to a shallow level in the crust;

Percolation at low temperature of meteoric waters rich in biogenic carbon into the mantle rocks at shallow depth, reactivating the dunite channels as preferential weak zones;

Precipitation of magnesite in veins and partial carbonation of peridotite host rock;

Further erosion till exposure at the surface.

The close spatial association of chromitite and magnesite ores in the Gerakini and Vavdos ophiolite mantle rocks of the Halkidiki peninsula envisages the possibility that the genetic relationship between these two kinds of ores, due to their formation within the same weakness zones active in two different dynamic regimes, can be not just a peculiarity of Evia ophiolites, but a more widely occurring process.

Supplementary Materials: The following supporting information can be downloaded at: https://www.mdpi.com/article/10.3390/min13020159/s1, Table S1: sil-chromitite; Table S2: sil-carb-peridotite; Table S3: spi-nel-chromitite; Table S4: spinel-peridotite.

Author Contributions: Conceptualization, G.G. and M.B.; methodology, M.B., P.M., L.C. and E.T.; validation, A.C., P.M. and L.C.; formal analysis, M.B.; investigation, M.B. and G.G.; resources, G.G. and A.C.; data curation, M.B. and G.G.; writing—original draft preparation, M.B.; writing—review and editing, G.G.; visualization, M.B.; supervision, G.G.; project administration, G.G.; funding acquisition, G.G. All authors have read and agreed to the published version of the manuscript.

Funding: This research was funded by the Italian Ministry of Education (MUR) through the project "PRIN2017—Mineral reactivity, a key to understand large-scale processes" and the project "Dipartimenti di Eccellenza 2017".

Data Availability Statement: All data are available and can be found in the article and in the Supplementary Materials.

Acknowledgments: The authors wish to acknowledge Nicola Campomenosi for technical support in the μ-Raman analyses. We also wish to acknowledge MSc students Irene Albieri, Simone Grasso, and Matteo Valentini for their work in the field. We also wish to thank the reviewers for their constructive comments.

Conflicts of Interest: The authors declare no conflict of interest.

References

1. Latunussa, C.E.L.; Georgitzikis, K.; Torres de Matos, C.; Grohol, M.; Eynard, U.; Wittmer, D.; Mancini, L.; Unguru, M.; Pavel, C.; Carrara, S.; et al. *Study on the EU's List of Critical Raw Materials*; Publications Office of the European Union: Luxembourg, 2020.
2. Pohl, W. Genesis of magnesite deposits—Models and trends. *Geol. Rundschau* **1990**, *79*, 291–299. [CrossRef]
3. Pohl, W. Comparative geology of magnesite deposits and occurrences. *Monogr. Ser. Miner. Depos.* **1989**, *28*, 1–13.
4. Schroll, E. Genesis of magnesite deposits in the view of isotope geochemistry. *Bol. Parana. Geociências* **2002**, *50*, 59–68. [CrossRef]
5. Kelemen, P.B.; Matter, J. In situ carbonation of peridotite for CO2 storage. *Proc. Natl. Acad. Sci. USA* **2008**, *105*, 17295–17300. [CrossRef]
6. Barnes, I.; O'Neil, J.R. The relationship between fluids in some fresh alpine-type ultramafks and possible modern serpentinization, Western United States. *Geol. Soc. Am. Bull.* **1969**, *80*, 1947–1960. [CrossRef]
7. Kralik, M.; Aharon, P.; Schroll, E.; Zachmann, D.W. Carbon and oxygen isotope systematics of magnesite: A review. *Monogr. Ser. Miner. Depos.* **1989**, *28*, 197–223.
8. Abu-Jaber, N.S.; Kimberley, M.M. Origin of ultramafic-hosted vein magnesite deposits. *Ore Geol. Rev.* **1992**, *7*, 155–191. [CrossRef]
9. Kelemen, P.B.; Matter, J.; Streit, E.E.; Rudge, J.F.; Curry, W.B.; Blusztajn, J. Rates and mechanisms of mineral carbonation in peridotite: Natural processes and recipes for enhanced, in situ CO_2 capture and storage. *Annu. Rev. Earth Planet. Sci.* **2011**, *39*, 545–576. [CrossRef]
10. Tsirampides, A.; Filippidis, A. Greek mineral resources of Greece: Reserves and value. *Open Geosci.* **2012**, *4*, 641–650. [CrossRef]
11. Tzamos, E.; Bussolesi, M.; Grieco, G.; Marescotti, P.; Crispini, L.; Kasinos, A.; Storni, N.; Simeonidis, K.; Zouboulis, A. Mineralogy and geochemistry of ultramafic rocks from rachoni magnesite mine, Gerakini (Chalkidiki, Northern Greece). *Minerals* **2020**, *10*, 934. [CrossRef]
12. Gartzos, E. On the Genesis of Cryptocrystalline Magnesite Deposits in the Ultramafic Rocks of Northern Evia, Greece. Doctoral Dissertation, ETH Zürich, Zürich, Switerland, 1986.
13. Mountrakis, D. The Pelagonian Zone in Greece: A polyphase-deformed fragment of the Cimmerian continent and its role in the geotectonic evolution of the Eastern Mediterranean. *J. Geol.* **1986**, *94*, 335–347. [CrossRef]
14. Doutsos, T. Kinematics of the Central Hellenides. *Tectonics* **1993**, *12*, 936–953. [CrossRef]
15. Anders, B.; Reischmann, T.; Kostopoulos, D. Zircon geochronology of basement rocks from the Pelagonian Zone, Greece: Constraints on the pre-Alpine evolution of the westernmost Internal Hellenides. *Int. J. Earth Sci.* **2007**, *96*, 639–661. [CrossRef]
16. Papanikolaou, D. The tectonostratigraphic terranes of the Hellenides. *Ann. Geol. Pays Hell.* **1997**, *37*, 495–514.
17. Anders, B.; Reischmann, T.; Kostopoulos, D.; Poller, U. The oldest rocks of Greece: First evidence for a Precambrian terrane within the Pelagonian Zone. *Geol. Mag.* **2006**, *143*, 41–58. [CrossRef]
18. Gartzos, E.; Dietrich, V.J.; Migiros, G.; Serelis, K.; Lymperopoulou, T. The origin of amphibolites from metamorphic soles beneath the ultramafic ophiolites in Evia and Lesvos (Greece) and their geotectonic implication. *Lithos* **2009**, *108*, 224–242. [CrossRef]
19. Kosiari, S. Study of northern Evia's ultramafic rocks as skid-resistant aggregates in pavement construction mountain (central Greece): Correlations and evaluation. *Bull. Geol. Soc. Greece* **2007**, *40*, 1674. [CrossRef]
20. Gartzos, E.; Migiros, G.; Parcharidis, I. Chromites from ultramafic rocks of northern Evia (Greece) and their geotectonic significance. *Schweiz. Mineral. Petrogr. Mitt.* **1990**, *70*, 301–307.

21. Cornelius, N.K. UHP Metamorphic Rocks of the Eastern Rhodope Massif, NE Greece: New Constraints from Petrology, Geochemistry and Zircon Ages. Ph.D. Thesis, University of Mainz, Mainz, Germany, 2008.
22. Zachariadis, P.T. Ophiolites of the Eastern Vardar Zone, N. Greece. Ph.D. Thesis, University of Mainz, Mainz, Germany, 2007.
23. Vavassis, I.; De Bono, A.; Stampli, G.M.; Giorgis, D.; Valloton, A.; Amelin, Y. U-Pb and Ar-Ar geochronological data from the Pelagonian basement in Evia (Greece): Geodynamic implications for the evolution of Paleotethys. *Schweiz. Mineral. Petrogr. Mitt.* **2000**, *80*, 21–43. [CrossRef]
24. Bosi, F.; Biagioni, C.; Pasero, M. Nomenclature and classification of the spinel supergroup. *Eur. J. Mineral.* **2019**, *31*, 183–192. [CrossRef]
25. Uysal, I.; Akmaz, R.M.; Saka, S.; Kapsiotis, A. Coexistence of compositionally heterogeneous chromitites in the Antalya–Isparta ophiolitic suite, SW Turkey: A record of sequential magmatic processes in the sub-arc lithospheric mantle. *Lithos* **2016**, *248–251*, 160–174. [CrossRef]
26. Zhou, M.-F.; Sun, M.; Keays, R.R.; Kerrich, R.W. Controls on Platinum-Group Elemental Distributions of Podiform Chromitites: A Case Study of High-Cr and High-Al Chromitites from Chinese Orogenic Belts. *Geochim. Cosmochim. Acta* **1998**, *62*, 677–688. [CrossRef]
27. Rassios, A.; Tzamos, E.; Dilek, Y.; Bussolesi, M.; Grieco, G.; Batsi, A.; Gamaletsos, P.N. A structural approach to the genesis of chrome ores within the Vourinos ophiolite (Greece): Significance of ductile and brittle deformation processes in the formation of economic ore bodies in oceanic upper mantle peridotites. *Ore Geol. Rev.* **2020**, *125*, 103684. [CrossRef]
28. Grieco, G.; Bussolesi, M.; Tzamos, E.; Rassios, A.E.; Kapsiotis, A. Processes of primary and re-equilibration mineralization affecting chromitite ore geochemistry within the Vourinos ultramafic sequence, Vourinos ophiolite (West Macedonia, Greece). *Ore Geol. Rev.* **2018**, *95*, 537–551. [CrossRef]
29. Tzamos, E.; Filippidis, A.; Michailidis, K.; Koroneos, A.; Rassios, A.; Grieco, G.; Pedrotti, M.; Stamoulis, K. Mineral chemistry and formation of awaruite and heazlewoodite in the Xerolivado chrome mine, Vourinos, Greece. *Bull. Geol. Soc. Greece* **2016**, *50*, 2047–2056. [CrossRef]
30. Bussolesi, M.; Zaccarini, F.; Grieco, G.; Tzamos, E. Rare and new compounds in the Ni-Cu-Sb-As system: First occurrence in the Gomati ophiolite, Greece. *Period. Mineral.* **2020**, *89*, 63–76. [CrossRef]
31. Bussolesi, M.; Grieco, G.; Zaccarini, F.; Cavallo, A.; Tzamos, E.; Storni, N. Chromite compositional variability and associated PGE enrichments in chromitites from the Gomati and Nea Roda ophiolite, Chalkidiki, Northern Greece. *Miner. Depos.* **2022**, *57*, 1323–1342. [CrossRef]
32. Kapsiotis, A.; Rassios, A.E.; Uysal, I.; Grieco, G.; Akmaz, R.M.; Saka, S.; Bussolesi, M. Compositional fingerprints of chromian spinel from the refractory chrome ores of Metalleion, Othris (Greece): Implications for metallogeny and deformation of chromitites within a "hot" oceanic fault zone. *J. Geochem. Explor.* **2018**, *185*, 14–32. [CrossRef]
33. Mysen, B.O.; Kushiro, I. Compositional variations of coexisting phases with degree of melting of peridotite in the upper mantle. *Am. Mineral.* **1977**, *62*, 843–865.
34. Jaques, A.L.; Green, D.H. Anhydrous melting of peridotite at 0–15 Kb pressure and the genesis of tholeiitic basalts. *Contrib. Mineral. Petrol.* **1980**, *73*, 287–310. [CrossRef]
35. Barnes, S.J.; Roeder, P.L. The Range of Spinel Compositions in Terrestrial Mafic and Ultramafic Rocks. *J. Petrol.* **2001**, *42*, 2279–2302. [CrossRef]
36. González-Jiménez, J.M.; Griffin, W.L.; Proenza, J.A.; Gervilla, F.; O'Reilly, S.Y.; Akbulut, M.; Pearson, N.J.; Arai, S. Chromitites in ophiolites: How, where, when, why? Part II. The crystallization of chromitites. *Lithos* **2014**, *189*, 140–158. [CrossRef]
37. Spiegelman, M.; Kelemen, P.B. Extreme chemical variability as a consequence of channelized melt transport. *Geochem. Geophys. Geosyst.* **2003**, *4*, 7. [CrossRef]
38. Saccani, E.; Tassinari, R. The role of morb and SSZ magma-types in the formation of Jurassic ultramafic cumulates in the Mirdita Ophiolites (Albania) as deduced from chromian spinel and olivine chemistry. *Ofioliti* **2015**, *40*, 37–56. [CrossRef]
39. Grieco, G.; Merlini, A. Chromite alteration processes within Vourinos ophiolite. *Int. J. Earth Sci.* **2012**, *101*, 1523–1533. [CrossRef]
40. Merlini, A.; Grieco, G.; Diella, V. Ferritchromite and chromian-chlorite formation in mélange-hosted Kalkan chromitite (Southern Urals, Russia). *Am. Mineral.* **2009**, *94*, 1459–1467. [CrossRef]
41. Bussolesi, M.; Grieco, G.; Tzamos, E. Olivine–Spinel Diffusivity Patterns in Chromitites and Dunites from the Finero Phlogopite-Peridotite (Ivrea-Verbano Zone, Southern Alps): Implications for the Thermal History of the Massif. *Minerals* **2019**, *9*, 75. [CrossRef]
42. Bussolesi, M.; Grieco, G.; Cavallo, A.; Zaccarini, F. Different Tectonic Evolution of Fast Cooling Ophiolite Mantles Recorded by Olivine-Spinel Geothermometry: Case Studies from Iballe (Albania) and Nea Roda (Greece). *Minerals* **2022**, *12*, 64. [CrossRef]
43. Ballhaus, C.; Berry, R.F.; Green, D.H. High pressure experimental calibration of the olivine-orthopyroxene-spinel oxygen geobarometer: Implications for the oxidation state of the upper mantle. *Contrib. Mineral. Petrol.* **1991**, *107*, 27–40. [CrossRef]
44. O'Neill, H.S.C.; Wall, V.J. The Olivine–Orthopyroxene–Spinel Oxygen Geobarometer, the Nickel Precipitation Curve, and the Oxygen Fugacity of the Earth's Upper Mantle. *J. Petrol.* **1987**, *28*, 1169–1191. [CrossRef]
45. Engi, M. Equilibria involving Al-Cr spinel: Mg-Fe exchange with olivine. Experiments, thermodynamic analysis, and consequences for geothermometry. *Am. J. Sci.* **1983**, *283 A*, 29–71. [CrossRef]
46. Zaccarini, F.; Pushkarev, E.; Garuti, G. Platinum-group element mineralogy and geochemistry of chromitite of the Kluchevskoy ophiolite complex, central Urals (Russia). *Ore Geol. Rev.* **2008**, *33*, 20–30. [CrossRef]

47. Yufeng, R.; Fangyuan, C.; Jingsui, Y.; Yuanhong, G. Exsolutions of Diopside and Magnetite in Olivine from Mantle Dunite, Luobusa Ophiolite, Tibet, China. *Acta Geol. Sin. -Engl. Ed.* **2010**, *82*, 377–384. [CrossRef]

48. Pagé, P.; Barnes, S.-J. Using Trace Elements in Chromites to Constrain the Origin of Podiform Chromitites in the Thetford Mines Ophiolite, Québec, Canada. *Econ. Geol.* **2009**, *104*, 997–1018. [CrossRef]

49. Ivan, P.; Jaros, J.; Kratochvil, M.; Reichwalder, P.; Rojkovic, I.; Spisiak, J.; Turanova, L. *Ultramafic Rocks of the Western Carpathians, Czechoslovakia*; Geologicky Ústav Dionyza Stúra: Bratislava, Slovakia, 1985; 258p.

50. Ploshko, V.V. Listvenitization and carbonatization at terminal stages of Urushten igneous complex, North Caucasus. *Int. Geol. Rev.* **1963**, *7*, 446–463. [CrossRef]

51. Boskabadi, A.; Pitcairn, I.K.; Leybourne, M.I.; Teagle, D.A.H.; Cooper, M.J.; Hadizadeh, H.; Nasiri Bezenjani, R.; Monazzami Bagherzadeh, R. Carbonation of ophiolitic ultramafic rocks: Listvenite formation in the Late Cretaceous ophiolites of eastern Iran. *Lithos* **2020**, *352–353*, 105307. [CrossRef]

52. Kelemen, P.B.; Dick, H.J.B. Focused melt flow and localized deformation in the upper mantle: Juxtaposition of replacive dunite and ductile shear zones in the Josephine peridotite, SW Oregon. *J. Geophys. Res. Solid Earth* **1995**, *100*, 423–438. [CrossRef]

53. Cocomazzi, G.; Grieco, G.; Tartarotti, P.; Bussolesi, M.; Zaccarini, F.; Crispini, L. The formation of dunite channels within harzburgite in the wadi tayin massif, oman ophiolite: Insights from compositional variability of cr-spinel and olivine in holes ba1b and ba3a, oman drilling project. *Minerals* **2020**, *10*, 167. [CrossRef]

54. Barnes, I.; LaMarche, V.C.; Himmelberg, G. Geochemical evidence of present-day serpentinization. *Science (80-)* **1967**, *156*, 830–832. [CrossRef] [PubMed]

55. Bruni, J.; Canepa, M.; Chiodini, G.; Cioni, R.; Cipolli, F.; Longinelli, A.; Marini, L.; Ottonello, G.; Vetuschi Zuccolini, M. Irreversible water–rock mass transfer accompanying the generation of the neutral, Mg–HCO3 and high-pH, Ca–OH spring waters of the Genova province, Italy. *Appl. Geochem.* **2002**, *17*, 455–474. [CrossRef]

56. Cipolli, F.; Gambardella, B.; Marini, L.; Ottonello, G.; Vetuschi Zuccolini, M. Geochemistry of high-pH waters from serpentinites of the Gruppo di Voltri (Genova, Italy) and reaction path modeling of CO2 sequestration in serpentinite aquifers. *Appl. Geochem.* **2004**, *19*, 787–802. [CrossRef]

57. Evans, B.W. Control of the products of serpentinization by the Fe2+Mg-1 exchange potential of olivine and orthopyroxene. *J. Petrol.* **2008**, *49*, 1873–1887. [CrossRef]

58. Janecky, D.R.; Seyfried, W.E. Hydrothermal serpentinization of peridotite within the oceanic crust: Experimental investigations of mineralogy and major element chemistry. *Geochim. Cosmochim. Acta* **1986**, *50*, 1357–1378. [CrossRef]

59. Fallick, A.E.; Ilich, M.; Russell, J.K. A Stable isotope study of the magnesite deposits associated with the Alpine-type ultramafic rocks of Yugoslavia. *Econ. Geol.* **1991**, *86*, 847–861. [CrossRef]

60. Barnes, I. *Chemical Composition of Naturally Occurring Fluids in Relation to Mercury Deposits in Part of North-Central California*; US Government Printing Office: Washington, DC, USA, 1973.

61. Blank, J.G.; Green, S.J.; Blake, D.; Valley, J.W.; Kita, N.T.; Treiman, A.; Dobson, P.F. An alkaline spring system within the Del Puerto Ophiolite (California, USA): A Mars analog site. *Planet. Space Sci.* **2009**, *57*, 533–540. [CrossRef]

62. Craig, J.R. Violarite stability relations. *Am. Mineral.* **1971**, *56*, 1303–1311.

63. O'Neil, J.R.; Barnes, I. C13 and O18 compositions in some fresh-water carbonates associated with ultramafic rocks and serpentinites: Western United States. *Geochim. Cosmochim. Acta* **1971**, *35*, 687–697. [CrossRef]

64. Jedrysek, M.O.; Halas, S. The origin of magnesite deposits from the Polish Foresudetic Block ophiolites: Preliminary δ13C and δ18O investigations. *Terra Nov.* **1990**, *2*, 154–159. [CrossRef]

65. Zedef, V.; Russell, M.J.; Fallick, A.E.; Hall, A.J. Genesis of vein stockwork and sedimentary magnesite and hydromagnesite deposits in the ultramafic terranes of Southwestern Turkey: A stable isotope study. *Econ. Geol.* **2000**, *95*, 429–445. [CrossRef]

66. Mirnejad, H.; Aminzadeh, M.; Ebner, F.; Unterweissacher, T. Geochemistry and origin of the ophiolite hosted magnesite deposit at Derakht-Senjed, NE Iran. *Mineral. Petrol.* **2015**, *109*, 693–704. [CrossRef]

67. Gartzos, E. Comparative stable isotope study of the magnesite deposits of Greece. *Bull. Geol. Soc. Greece* **2004**, *36*, 196–203. [CrossRef]

minerals

Article

Tree Rings Record of Long-Term Atmospheric Hg Pollution in the Monte Amiata Mining District (Central Italy): Lessons from the Past for a Better Future

Silvia Fornasaro [1], Francesco Ciani [2], Alessia Nannoni [2], Guia Morelli [3], Valentina Rimondi [2], Pierfranco Lattanzi [3], Claudia Cocozza [4], Marco Fioravanti [4] and Pilario Costagliola [2,*]

1   Department of Earth Sciences, University of Pisa, Via Santa Maria 53, 56126 Pisa, Italy; silvia.fornasaro@unipi.it

2   Department of Earth Sciences, University of Florence, Via G. La Pira 4, 50121 Florence, Italy; francesco.ciani@unifi.it (F.C.); alessia.nannoni@unifi.it (A.N.); valentina.rimondi@unifi.it (V.R.)

3   Consiglio Nazionale delle Ricerche, Istituto di Geoscienze e Georisorse, Via G. La Pira 4, 50121 Florence, Italy; guia.morelli@igg.cnr.it (G.M.); pierfrancolattanzi@gmail.com (P.L.)

4   Department of Agricultural, Food, Environmental and Forestry Sciences and Technologies, University of Florence, Piazzale delle Cascine 18, 50144 Florence, Italy; claudia.cocozza@unifi.it (C.C.); marco.fioravanti@unifi.it (M.F.)

*   Correspondence: pilario.costagliola@unifi.it

Abstract: Trees may represent useful long-term monitors of historical trends of atmospheric pollution due to the trace elements stored along the tree rings caused by modifications in the environment during a tree's life. Chestnut (*Castanea sativa* Mill.) tree trunk sections were used to document the yearly evolution of atmospheric Hg in the world-class mining district of Monte Amiata (MAMD; Central Italy) and were exploited until 1982. An additional source of Hg emissions in the area have been the active geothermal power plants. A marked decrease (from >200 μg/kg to <100 μg/kg) in Hg contents in heartwood tree rings is recorded, likely because of mine closure; the average contents (tens of μg/kg) in recent years remain higher than in a reference area ~150 km away from the district (average 4.6 μg/kg). Chestnut barks, recording present-day Hg pollution, systematically show higher Hg concentrations than sapwood (up to 394 μg/kg in the mining area). This study shows that tree rings may be a good record of the atmospheric Hg changes in areas affected by mining activity and geothermal plants and can be used as a low-cost biomonitoring method for impact minimization and optimal resource and land management.

Keywords: dendrochemistry; tree ring; *Castanea sativa* Mill.; geothermal power plant; mining activity; mercury; Monte Amiata

Citation: Fornasaro, S.; Ciani, F.; Nannoni, A.; Morelli, G.; Rimondi, V.; Lattanzi, P.; Cocozza, C.; Fioravanti, M.; Costagliola, P. Tree Rings Record of Long-Term Atmospheric Hg Pollution in the Monte Amiata Mining District (Central Italy): Lessons from the Past for a Better Future. *Minerals* **2023**, *13*, 688. https://doi.org/10.3390/min13050688

Academic Editor: Saglara S. Mandzhieva

Received: 26 April 2023
Revised: 13 May 2023
Accepted: 17 May 2023
Published: 18 May 2023

1. Introduction

A fundamental requirement of sustainable mining is the ability of predicting potential adverse environmental impacts, even in the long-term. In this study, we present an example of the reconstruction of the long-term record of mercury (Hg) variations in an area affected by Hg mining, metallurgical activities, and the exploitation of geothermal fluids. Mercury is a global and persistent contaminant which generally occurs in the atmosphere in three main species. The dominant species is gaseous elemental mercury (Hg^0, GEM), which comprises >95% of the global atmospheric Hg reservoir. It is a transboundary pollutant due to its low reactivity, long-range transport, and long-term residence time in the atmosphere (0.5–2 years). The other two less abundant Hg species in the atmosphere, reactive gaseous mercury (Hg^{2+}, RGM) and particle-bound mercury (Hgp, PBM), have much shorter residence times (hours to weeks) and are typically deposited near their Hg emission sources [1–8].

Characterizing such anthropogenic sources and quantifying Hg emissions from them are crucial objectives as they may constitute two-thirds of the present global Hg cycle, raising Hg global concentrations four to five times the background levels [9,10]. Such an increase can pose public health concern for workers and local populations through inhalation exposure [11]. Sensitive instrumentation and passive samplers (both abiotic and biotic) have been utilized to quantify Hg air concentrations and observe temporal trends [12,13]. However, these data are not representative of past trends. Data on historical trends of atmospheric Hg deposition, especially on the natural conditions preceding the pollution advent, represent an important gap in the understanding of large-scale Hg pollution and Hg global cycle [14]. At present, the application proxies for investigating Hg contamination over the long-term are lake cores, ice cores, and tree rings [14–17]. Among natural samplers, the dendrochemical analysis of tree rings proved to be an adequate tool to record atmospheric concentrations of Hg^0 ([8] and the references therein, [18–21]). Dendrochronological methods provide a simple, precise, and accurate means of dating with a fine temporal resolution that can be advantageous, despite the short life span of trees relative to longer geochemical archives (i.e., sediment, peat, and ice) [8,22]. On the other hand, dendrochronology is not completely accepted by the scientific community. The debate is still ongoing, because many factors influence the applicability of dendrochemical analyses, including differences in: (i) tree species and physiology (e.g., differences between sapwood and heartwood, tree age, stature, growth rate) [23–25]; (ii) Hg formation in the air; (iii) soil chemistry [26,27]; and (iv) tree location (e.g., proximity to the Hg sources, elevation, proximity to the sea/ocean, wind direction) [18,20,21,23,28,29]. Moreover, attention must be paid to the capability of trees to take up and transport different pollutants into their stem by comparing leaf-to-root and root-to-leaf pathways [30].

The use of dendrochemical techniques has the potential to define the retrospective biomonitoring of pollutants, and, in this context, advances are obtained by study areas well known to support the approach [31]. The Monte Amiata Mining District (MAMD; Central Italy) was chosen for the pollutant history of the study area. It was the third largest mercury (Hg) district worldwide (cumulative production of 102,000 tons; [32]), and currently hosts an important, actively exploited, geothermal field. Mining ceased in the early 1980s but left an impressive Hg contamination legacy in all environmental matrices, i.e., air, soil, sediments, water, and biota, as reported in recent years in several studies ([33] and references therein; [34–39]). Exploitation of geothermal energy in the area started in 1959; currently, there are five active plants equipped with emission control systems (AMIS) to reduce Hg and H_2S discharges from geothermal wells.

This research aims to prove the reliability of dendrochemical analysis to assess past and actual Hg pollution in the MAMD district, reconstructing a long-term record of Hg variations in an area affected by past Hg mining and metallurgical activities and the current exploitation of geothermal fluids. To the best of our knowledge, for the first time in this type of study, chestnut trees (*Castanea sativa* Mill.) were selected to test their ability to store a record of Hg changes in heartwood. Most published tree ring Hg records are developed at relatively coarse resolution (e.g., five-year resolution), and do not take full advantage of the annual resolution of the tree ring record [21]. Conversely, the tree ring analysis in this study provides a high-resolution (i.e., annual) temporal reconstruction of Hg exposure. Although currently Hg extraction, production, and commercialization are banned in most countries [40], it was a true "critical metal" for most of the XX century. The story presented here is, therefore, an instructive example of potential environmental concerns associated with the quest for critical resources.

2. Study Area

2.1. The Monte Amiata Area

The Monte Amiata area (Figure 1) (southern Tuscany, Italy) is characterized by a volcanic activity of dominant trachydacitic composition that occurred between 300 and 190 ka [41,42]. This volcanic activity caused the emplacement of a Pliocene magmatic body at 6–7 km below the sea level. The anomalous heat flux related to this body likely triggered a hydrothermal system that caused the emplacement of Hg ore deposits [32,43,44]. Nowadays, the area hosts two geothermal reservoirs, which feed some CO_2-rich gas manifestations and Ca–SO_4-rich thermomineral waters, mainly located in NE and SW sectors of the volcanic edifice [45–47]. The Monte Amiata district is by extension the second largest geothermal field in Italy after Larderello.

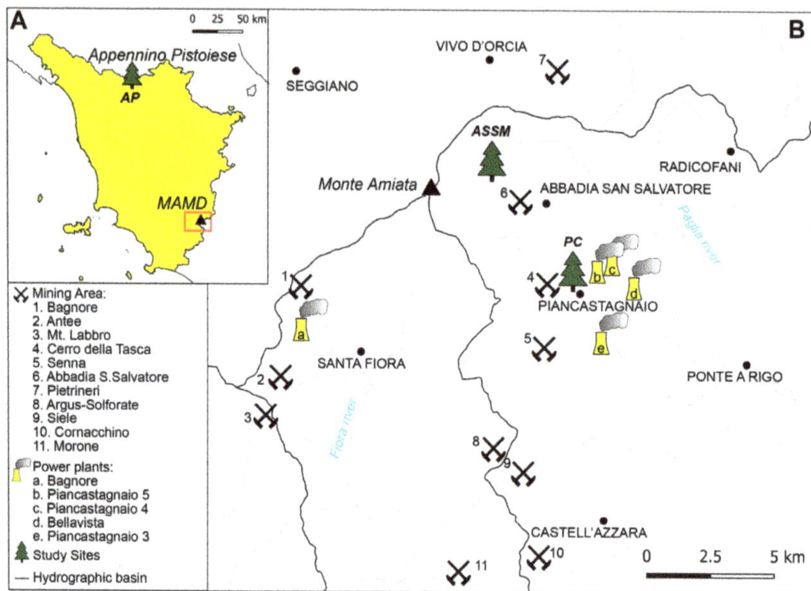

Figure 1. Study area and location of sampling sites. (**A**) Location of MAMD and AP sites, the red box indicates the position of Figure 2B. (**B**) The sampling location in the MAMD (ASSM and PC sites).

The climate of the area is temperate, with hot and dry summer, and cold rainy winter. In the last decade (2010–2020), the weather station of Abbadia San Salvatore-Laghetto Verde (TOS 11000114) reports an average annual temperature of 10.9 °C, and an average annual precipitation of 1543 mm (http://www.sir.toscana.it/consistenza-rete (accessed on 2 June 2022)). About two-thirds of the total annual precipitation are concentrated in the autumn–winter season [48]. Prevailing winds come from the W direction (280°), while subordinate directions are from NE (40°) and SE (140°) [38].

The Monte Amiata area is densely covered by a deciduous forest dominated by chestnut (*C. sativa* Mill.) below 1000 m, and by silver fir (*Abies alba* Mill.), beech (*Fagus sylvatica* L.), and black pine (*Pinus nigra* J.F. Arnold) above 1000 m [49]. Chestnut forests cover 7500 ha; half of the area (3500 ha) is managed for wood production under public ownership. The forests are located mainly in the eastern side of the region at 800 to 1200 m asl. Chestnut forests tend to be monospecific and are of anthropic origin; secondary species are rare or absent.

2.2. Mercury Sources and Mining Activity in the Monte Amiata Area

The main mining and smelting centers of the MAMD are mostly located in the central–eastern and southwestern sectors of the volcanic complex; the principal mining and metallurgical activity was in the municipality of Abbadia San Salvatore (Figure 1). The extraction and metallurgy of cinnabar (HgS) produced metallic Hg from the middle XIX century until the 1970s, when most of the Hg mines in the world were shut down due to environmental concerns.

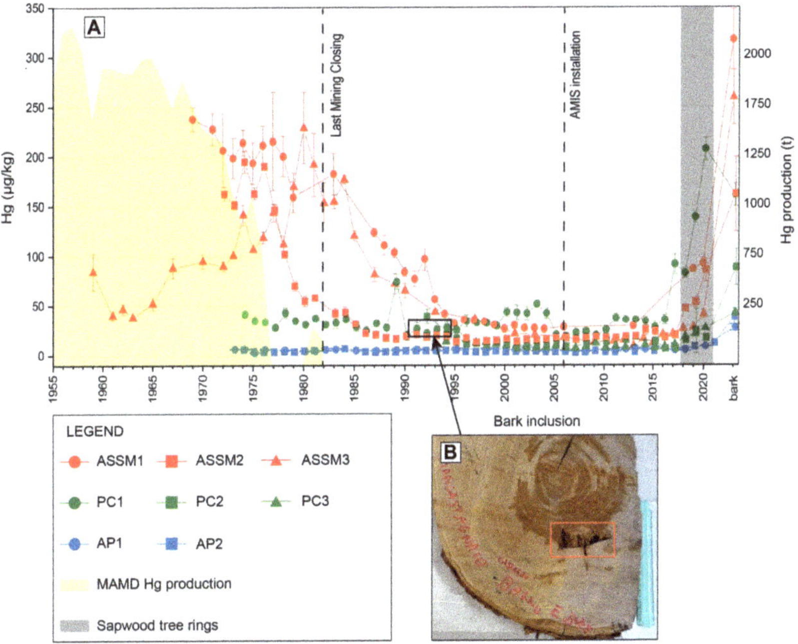

Figure 2. (**A**) Temporal Hg variation in the tree rings of ASSM (red lines), PC (green lines), and background area AP (blue lines). The yellow area represents the Hg production (tons) of the MAMD [50]. The grey area represents the sapwood tree rings. (**B**) Bark inclusion in PC1 sample in PC site (in the red box), dated 1991–1994.

Mining activity began in 1846–1847 with the exploitation of the deposits of Siele and Abbadia S. Salvatore, but, already by 1890, the production of Hg reached the remarkable value of about 450 t, i.e., 10% of the world production of that time [50]. One of the main production peaks occurred during the first decades of the 1900s, driven by the increasing demand of Hg fulminate employed during World War I. In 1917, the production of Hg had exceeded 1000 t, which corresponded to an impressive 26% of world production, overcoming Almadén in Hg flask trading [51]. After the economic crisis in 1930, production decreased and maintained low during World War II, since the district was heavily bombed. After the war, the Hg production had a new pulse during the Korean War up to the mid-1960s [51]; in the 1970s, the Hg demand began a constant decrease, down to a complete halt, with the consequent closure of the mines and plant production sites in 1982 [50]. Among the 42 mining sites and five metallurgical plants, the Abbadia San Salvatore mining area was the last to close its activities in 1981. Remediation of the mining sites occurred at Siele [35] and is ongoing at Abbadia San Salvatore [52].

The exploitation of geothermal energy at Monte Amiata started in 1959. In the study area, the first power plant (Bellavista; Figure 1) was built in 1987; currently, five power plants are active (Figure 1), with an installed capacity of 121 MWe [53]. The main environmental and health concerns of this activity arise from air emissions, represented by non-condensable gases (NCG) vented into the atmosphere, drift emitted from the cooling towers, noise emissions, and visual impact on the landscape. The main pollutants contained in the NCG emitted from geothermal power plants are hydrogen sulfide (H_2S) and Hg [54]. Mercury is present at very low concentrations in geothermal fluids, but its environmental mobility and high toxicity require abating the quantity emitted as much as possible [55]. For this reason, starting from 2002, all the geothermal plants currently operating in the Monte Amiata area are equipped with emission control systems (AMIS) to reduce Hg and H_2S emissions. The installation of AMIS in all the power plants was completed in 2007. However, the efficiency of these systems is not 100% and Hg^0 is still emitted from geothermal plants (from 1 to 4 g/h) [54].

2.3. Sampling Sites

Chestnut trees were sampled at two sites in the MAMD: (i) near the former Abbadia San Salvatore mine (ASSM), the largest mining and smelting center of the district and (ii) near the main geothermal power plant in the Piancastagnaio municipality (PC). At each site, three trees were sampled. Two trees were also sampled in the Appennino Pistoiese (AP; 150 km away from MAMD) that was selected as a reference background area (Figure 1).

3. Materials and Methods
3.1. Sampling and Sample Preparation

Taking profit of routine operations of forest maintenance, three tree disks (cross sections; ~5 cm in thickness) were collected from the stump at a height of 1 m at each sampling site (samples ASSM1, ASSM2, ASSM3 at ASSM; samples PC1, P2, PC3 at PC) in the Monte Amiata area, and two disks at the control site AP (samples AP1 and AP2). Disks were cut within one month from sampling using stainless-steel saws. A portion was polished with progressively finer sandpaper down to 600 grit to expose ring boundaries for identification and remove pollution retained on the rough surfaces. Stem discs allowed us to easily observe the stem profile, enabling easier separation of layers and providing adequate quantities for analysis. In the disk surface, sapwood is clearly recognized from heartwood because of its lighter color; sapwood typically comprises 3–4 growth rings.

After the surfaces of the tree disks were prepared, the tree rings were dated by counting back in time from the outermost ring, which represents the year of sampling (i.e., 2020). All tree rings were measured using a strip of graph paper under the binocular microscope. In addition, any exceptionally large ring or unique ring feature (e.g., impact damage) were noted as marker rings.

Individual disks were cut into one-year segments, starting with the most recent tree ring year, 2020, and moving toward the pith, using a stainless-steel knife; the selected fragments were air-dried. The one-year tree ring splitting was performed in nearly all samples, with the exception where low ring thicknesses made single splitting not possible. In the latter case, several tree rings were grouped and analyzed together: Table S1 in Supplementary Files report the detailed division of the trees. A clean scalpel blade was used to dissect the tree rings into annual growth increments and shave the outer surface to remove any potential superficial contamination that may have accumulated since the time of collection.

Water content was determined in a representative number ($n° = 30$) of dissected tree rings, following the procedure established by [38,56]. A sub-portion of approximately 1 g was dried in the oven at 110 °C for 24 h (or until a constant weight was achieved). The water content was calculated as a fraction of water in the total weight: water content (%) = [(wet weight) − (dry weight)]/(wet weight) × 100. All Hg measurements are reported on a dry weight basis. Water content is about 10% in sapwood, and 7% in heartwood.

The bark was also collected from each tree both at ASSM and PC, and dried. In sample PC1 collected at PC, a bark inclusion in tree stem was sampled and dated approximately between 1991 and 1994 (Figure 2B).

At each site, a representative soil sample was collected at a depth of 1–10 cm, removing the litterfall, i.e., the dead plant material (such as leaves, bark, needles, twigs, and cladodes) above the topsoil. Soils were sampled outside the tree canopy projection, within a distance of a few meters from the trunk. They were subsequently transported into the lab, air dried, sieved to select the fraction <2 mm, and ground.

3.2. Chemical Analysis

Total Hg concentrations in the tree rings sections representing the annual growth increments were quantified with a Milestone tri-cell Direct Mercury Analyzer (DMA-80) by atomic absorption, following thermal decomposition and amalgam preconcentration [57,58]. Tree rings were analyzed using the DMA-80 method recommended by the manufacturer for organic samples: 200 °C maximum starting temperature; 200 °C drying temperature; 50 s drying time; 800 °C decomposition temperature; 90 s decomposition time; and 60 s dwell time.

Soils were analyzed using the DMA-80 method for soil samples: 250 °C maximum starting temperature; 200 °C drying temperature; 40 s drying time; 650 °C decomposition temperature; 60 s decomposition time; and 60 s dwell time.

To estimate the analytical precision, three replicate analyses of the same sample (for both tree samples and soil samples) were routinely performed. Results were generally reproduced within <15% of the average value. Accuracy was evaluated using international standards (NIST 1575a, NIST 1573a, BCR-280R) and was within 10%. Concentrations are reported in µg/kg dry weight for tree rings and soils.

4. Results

The ages of sampled chestnut trees varied by study site (Table 1). At ASSM, trees were mainly 53 years old, ranging between 48 to 62 years, showing 3.5 mm per year as a mean tree ring increment. At PC, trees were younger than ASSM, ranging between 26 to 46 years with an average of 33 years, with a mean incremental growth of 4 mm per year.

Table 1. Characteristics of the study sites. Hg values are reported in µg/kg as average with standard deviation; the min and max values are reported within brackets.

	Abbadia San Salvatore (ASSM)	Piancastagnaio (PC)	Appennino Pistoiese (AP)
Latitude	42.89633	42.85409	44.04356
Longitude	11.64965	11.69055	10.76411
Altitude (m asl)	1070	725	438
Tree species—sample name	*C. sativa* Mill.–ASSM1, ASSM2, ASSM3	*C. sativa* Mill.–PC1, PC2, PC3	*C. sativa* Mill.–AP1, AP2
Tree age average (years)	53 (48–62)	33 (26–46)	46–47
Mean tree ring increments (mm)	3.5	4	-
Exposition	E	E	NW
Forest type	Deciduous forest	Mixed coniferous and deciduous forest	Mixed coniferous and deciduous forest
Hg source	Past mining activity	Geothermal power plant	Global atmospheric pollution
Distance from source (km)	1.5	1	-
Bedrock	Volcanic and volcano-sedimentary successions	Pliocene marine sediments	Flysch
Sample *n°*	3	3	2
Hg in tree rings	71.2 ± 65.8 (9.3–236.9)	26.5 ± 26.8 (6.5–207.2)	4.6 ± 2.1 (1.8–10.9)
Hg in tree rings before 1982	141.2 ± 61.2 (38.2–236.9)	34.6 ± 5 (27.2–41.9)	4.4 ± 1.6 (2.5–7)
Hg in tree rings before AMIS installation (2006)	83.9 ± 69.1 (12.4–236.9)	24.1 ± 13.3 (7.7–73.4)	4.3 ± 1.8 (2.2–9.4)
Hg in bark	245.7 ± 78.7 (161.1–316.9)	97.5 ± 59.3 (43.5–161)	32.8 ± 6.8 (27.9–37.6)
Hg in soil (µg/kg)	3628 ± 2192 (1753–6038)	1730 ± 742 (923–2383)	117 ± 2 (116–119)

In ASSM, Hg in tree rings varied from 9.3 to 236.9 µg/kg. In the mining period (until 1980) Hg concentrations in trees varied between 38.2 and 236.9 µg/kg, whereas in the post-production period after mine closure in the early 1980s, Hg showed a distinct decrease (between 9.3 and 176.7 µg/kg; Figure 2). It is to be noticed that the trend of Hg concentrations in tree rings does not exactly mimic the production trend (yellow area in Figure 2A). In particular, the Hg decreasing trend in trees begins with a slight delay (after 2 years) and extends for some years before reaching an almost constant plateau (post-1994). This fact is not surprising, considering that marked anomalies of atmospheric Hg are to this day recorded in proximity of smelting plants [34,52]. It is noted that samples ASSM1 and ASSM3 recorded a peak in 1980 (181.6 and 228.8 µg/kg, respectively) in correspondence of the last pulse of the Hg production of ASSM.

At PC, tree rings close to the geothermal plants (PC1, PC2, PC3) almost systematically showed a lower Hg content with respect to trees in the mining area (ASSM1, ASSM2, ASSM3), with an average of 26.5 $\pm$ 26.8 (average $\pm$ SD) µg/kg (ranging from 6.5 to 207.2 µg/kg). No significant variations were found in tree rings corresponding to the years after AMIS installation (early 2000s).

In the reference area (AP), the Hg concentrations varied from 1.8 to 10.9 µg/kg, indicating that trees were exposed to a modest local anomaly ([59] report 0.9 µg/kg for a reference site of a rural area).

All samples showed an increase in Hg concentration from heartwood to sapwood (up to an order of magnitude). Chestnut barks systematically showed higher Hg concentrations than sapwood (up to 316.9 µg/kg in ASSM; Figure 2A). The average Hg content in barks of ASSM is 245.7 $\pm$ 78.7 µg/kg, ranging between 161.1 and 316.9 µg/kg. In PC, barks showed an average value of 97.5 $\pm$ 59.3, ranging between 43.5 and 161 µg/kg. The bark inclusion in tree stem in the PC1 sample shows a content of 90 µg/kg. The background value in AP was 32.8 $\pm$ 6.8 µg/kg.

Soil samples collected at ASSM and PC showed concentrations of 3628 $\pm$ 2192 µg/kg and 1730 $\pm$ 74 µg/kg, respectively. These results are consistent with previous work on the presence of geochemical anomalies in the MAMD [32,60]. In the soils of the background area (AP site), Hg varied from 117 $\pm$ 2 µg/kg.

5. Discussion

Studies analyzing past Hg or other potentially toxic elements pollution recorded in tree rings are rare in Italy [31,61–64] and absent in the MAMD area. In this study, long-term reconstruction with an annual resolution of atmospheric Hg^0 in the MAMD district was achieved using chestnut (*C. sativa* Mill.) tree rings from 1958 to 2020.

The primary assumption to perform a dendrochemical analysis is the concept that each annual tree ring should reflect the trace elements available in the environment [65] and be absorbed by the tree through roots, leaves, and barks [30]. However, the pollutants absorption and storage in wood tissues is dependent by species; for instance, some authors ([8] and reference therein) observed in *Pinus* spp. that Hg concentration accumulated in tree rings fails to reconstruct the temporal trend of Hg production, in particular when coniferous species are employed, due to confounded tree physiological and environmental factors (i.e., the radial translocation and tree age effects occurring during the fast growing period).

However, the results presented in this study are broadly consistent with the basic dendrochemical assumption. The changes of Hg concentrations recorded in chestnut tree rings were temporally coherent with the closing of mining activities in the area (i.e., 1982) that would have altered Hg atmospheric concentrations (Figure 2).

On the other hand, although [66] have modeled that the AMIS installation has reduced the emission of Hg^0 by four times in the Piancastagnaio area, this variation has not been registered in our samples.

To the best of our knowledge, this is the first study in relation to tree rings collected in the MAMD which provides a temporal reconstruction of Hg exposure at annual resolutions. Results proved to be a good record of Hg deposition in areas affected by ore mining and smelting [22]. The reliability of the data is probably related to the object of investigation (i.e., Hg) and likely to the atmospheric pollutant that can be best traced by dendrochemical analyses [21,22,67], as well as to the employment of *C. sativa*, which is potentially the ideal arboreal species to track past Hg environmental concentrations.

Mercury concentrations found in the vascular tissues of woody plants ($\cong$90%) are a result of Hg^0 stomatal and cuticular uptake by leaves. After its oxidation to divalent Hg^{2+}, it is incorporated in epidermal and stomatal cell walls [68], while the downward transport into tree rings occurs through the phloem and lateral translocation into the xylem [18,24]. Therefore, Hg concentrations found in tree rings could be conceivably considered as an evidence of past atmospheric Hg pollution. Other possible ways for Hg uptake by tree and translocation into woody tissues are Hg uptake from soil and movements from the outer bark into tree rings. However, both these mechanisms were suggested to be negligible [18,56,58,69,70]. In the present study, the highest Hg concentrations in soil were found at ASSM, where Hg concentrations in wood are the highest. We cannot exclude the fact that root uptake from the soil is a source of Hg to the bole. However, the main species of total Hg in these soils is HgS ([32] and reference therein), a stable and very scarcely soluble Hg phase. Indeed, the translocation factor (Hg_{wood}/Hg_{soil}) is very low (0.019 ASSM; 0.015 PC; 0.039 AP). Thus, a soil Hg uptake seems unlikely, or in any case, minor.

Chestnut tree species are characterized by a relatively low and consistent number of sapwood tree rings able to record changes in atmospheric Hg concentrations more reliably than species with a relatively high and variable number of sapwood tree rings [8]. As previously stated, several authors [59,70,71] highlight that not all of the tree species are suitable for dendrochemical studies, because they clearly do not preserve Hg records. For example, some coniferous trees (e.g., *Pinus*, *Larix*), due to their permeable heartwood and the high number of rings in sapwood, may radially translocate Hg and other contaminants to older tree rings, smearing or otherwise altering Hg time series captured by tree rings [18]. By contrast, the nearly simultaneous response recorded in the collected chestnut (such as the peak observed by ASSM1 e ASSM3 in 1981 connected to the last pulse of Hg production of ASSM) trees indicated little or no translocation of Hg within the tree stem, which is in agreement with other works conducted on other tree species worldwide (e.g., larch in [70]). Among the species of the Fagaceae family and compared to other angiosperms, *C. sativa* wood has the highest concentrations of ellagitannins, a class of hydrolysable tannins, i.e., polyphenol compounds responsible for Hg^{2+} sequestration in barks and involved in several plant reactions, such as redox transformations and cation complexation [72,73]. These compounds generally reach the highest level at the sapwood–heartwood boundary, and after heartwood formation, the no longer active cells become unable to accumulate these compounds [8,74]. In our opinion, the complexation of Hg with these compounds most likely results in the absence of Hg translocation between tree heartwood rings. This evidence is further confirmed by the bark inclusion of the PC1 chestnut (Figure 2B). The concentration of Hg (90 µg/kg) recorded in this bark, dated between 1991 and 1994, exactly followed the Hg positive spike (~73 µg/g) recorded in the previous years (1989) in this tree. No comparable Hg values were recorded in the adjacent rings, underlying the absence of translocation. Moreover, the physiology of heartwood formation could explain the increase in Hg concentration from heartwood to sapwood (up to an order of magnitude), and Hg concentration increase at the sapwood–heartwood boundary (Figure 2A).

Unlike heartwood, in sapwood living cells, in particular in the parenchyma ray cells, tannins are in solution; with the death of the cell, and so at the heartwood formation, tannins solidify and harden in the dead timber [75]. Therefore, radial permeability is enhanced in sapwood and may contribute in reducing temporal precision in the Hg concentrations measured in these rings. Translocation of divalent cations driven by living ray cells has, for example, been documented in oak sapwood [76,77]. Similarly, Hg concentrations found in chestnut trees could be considered not indicative of a single-year exposition, but may be influenced by the radial translocation.

Finally, chestnut barks systematically showed higher Hg concentrations than sapwood (Figure 2A, gray area), as already evidenced in previous studies both for Hg [8,78,79] and other heavy metals [80,81]. Mercury concentrations in barks layers (i.e., inner and outer barks) result from a differential Hg caption and absorption. Arnold et al. [18] and Peckham et al. [82] suggest that Hg translocation from the phloem may be an important source of Hg stored into the inner bark, while the Hg concentration in an outer bark mainly reflects atmospheric Hg depositions, both as gaseous and particulate Hg [56,83]. In this study, no distinction has been made between inner and outer chestnut barks, so Hg concentrations could be assumed as the sum of both bark layers and thus of Hg translocation and deposition.

6. Conclusions

Trunk cross-sections of 26 to 62 year old chestnut trees recorded in their growth rings the yearly evolution of atmospheric Hg in the Monte Amiata mining district and geothermal area (Italy). The oldest trees overlap with the last years of mining activity (1959–1982) and recorded a sharp decrease in Hg content in the rings following mine closure (after 1982). However, Hg contents in tree rings at Monte Amiata remain higher than in the control area 150 km away. This study showed that chestnut tree rings may represent a good record of atmospheric Hg in areas affected by past mining activities and geothermal power plants, and can be used as a low-cost, long-term biomonitoring tool for impact minimization, and the optimal management of environmental and land resources. However, the best results are achieved by selecting tree species with characteristically narrow sapwood and a consistent, low number of sapwood tree rings, such as sweet chestnut.

Moreover, the pathways of Hg uptake and translocation remain uncertain and difficult to study, especially in field settings, as in the MAMD. Further work is needed to assess the radial Hg translocation in more controlled studies with larger sample sizes and thus to satisfactorily interpret this historical Hg record.

Supplementary Materials: The following supporting information can be downloaded at: https://www.mdpi.com/article/10.3390/min13050688/s1, Table S1: Mercury concentration in chestnut tree rings.

Author Contributions: Conceptualization, P.C., S.F., F.C. and V.R.; methodology, S.F., F.C., G.M., C.C. and V.R.; validation, S.F. and F.C.; investigation, P.C., S.F., F.C. and V.R.; data curation, S.F., F.C. and V.R.; writing—original draft preparation, S.F. and F.C.; writing—review and editing, P.C., S.F., F.C., V.R., P.L., G.M., A.N., M.F. and C.C.; visualization, S.F.; supervision, P.C. and P.L. All authors have read and agreed to the published version of the manuscript.

Funding: This research received no external funding.

Data Availability Statement: Not applicable.

Acknowledgments: The authors would like to thank Sauro Visconti and all the staff of the *Unione dei Comuni Amiata Val d'Orcia* for their support during the sampling activity.

Conflicts of Interest: The authors declare no conflict of interest.

References

1. Driscoll, C.T.; Mason, R.P.; Chan, H.M.; Jacob, D.J.; Pirrone, N. Mercury as a global pollutant: Sources, pathways, and effects. *Environ. Sci. Technol.* **2013**, *47*, 4967–4983. [CrossRef] [PubMed]
2. Gustin, M.S.; Lindberg, S.E.; Weisberg, P.J. An update on the natural sources and sinks of atmospheric mercury. *Appl. Geochem.* **2008**, *23*, 482–493. [CrossRef]
3. Lindberg, S.A.; Stratton, W.J. Atmospheric mercury speciation: Concentrations and behavior of reactive gaseous mercury in ambient air. *Environ. Sci. Technol.* **1998**, *32*, 49–57. [CrossRef]
4. Mason, R.P.; Sheu, G.R. Role of the ocean in the global mercury cycle. *Glob. Biogeochem. Cycles* **2002**, *16*, 40–41. [CrossRef]
5. Obrist, D.; Kirk, J.L.; Zhang, L.; Sunderland, E.M.; Jiskra, M.; Selin, N.E. A review of global environmental mercury processes in response to human and natural perturbations: Changes of emissions, climate, and land use. *Ambio* **2018**, *47*, 116–140. [CrossRef]
6. Schroeder, W.H.; Munthe, J. Atmospheric mercury—An overview. *Atmos. Environ.* **1998**, *32*, 809–822. [CrossRef]
7. Selin, N.E. Global biogeochemical cycling of mercury: A review. *Annu. Rev. Environ. Resour.* **2009**, *34*, 43–63. [CrossRef]
8. Novakova, T.; Navrátil, T.; Demers, J.D.; Roll, M.; Rohovec, J. Contrasting tree ring Hg records in two conifer species: Multi-site evidence of species-specific radial translocation effects in Scots pine versus European larch. *Sci. Total Environ.* **2021**, *762*, 144022. [CrossRef]
9. Wihlborg, P.; Danielsson, Å. Half a century of mercury contamination in lake Vänern (Sweden). *Water Air Soil Pollut.* **2006**, *170*, 285–300. [CrossRef]
10. Selin, N.E. Global change and mercury cycling: Challenges for implementing a global mercury treaty. *Environ. Toxicol. Chem.* **2014**, *33*, 1202–1210. [CrossRef]
11. Asano, S.; Eto, K.; Kurisaki, E.; Gunji, H.; Hiraiwa, K.; Sato, M.; Sato, H.; Hasuike, M.; Hagiwara, N.; Wakasa, H. Acute inorganic mercury vapor inhalation poisoning. *Pathol. Int.* **2000**, *50*, 169–174. [CrossRef]
12. Conti, M.E. *Biological Monitoring: Theory and Applications: Biomonitors and Biomarkers for Environmental Quality and Human Exposure Assessment*; WIT Press: Southampton, UK, 2008; p. 512.

13. Lattanzi, P.; Benesperi, R.; Morelli, G.; Rimondi, V.; Ruggieri, G. Biomonitoring Studies in Geothermal Areas: A Review. *Front. Environ. Sci.* **2020**, *8*, 579343. [CrossRef]
14. Bindler, R. Estimating the natural background atmospheric deposition rate of mercury utilizing ombrotrophic bogs in southern Sweden. *Environ. Sci. Technol.* **2003**, *37*, 40–46. [CrossRef]
15. Benoit, J.M.; Gilmour, C.C.; Mason, R.P.; Riedel, G.S.; Riedel, G.F. Behavior of mercury in the Patuxent River estuary. *Biogeochemistry* **1998**, *40*, 249–265. [CrossRef]
16. Schuster, P.F.; Krabbenhoft, D.P.; Naftz, D.L.; Cecil, L.D.; Olson, M.L.; Dewild, J.F.; Abbott, M.L. Atmospheric mercury deposition during the last 270 years: A glacial ice core record of natural and anthropogenic sources. *Environ. Sci. Technol.* **2002**, *36*, 2303–2310. [CrossRef]
17. Kang, S.; Huang, J.; Wang, F.; Zhang, Q.; Zhang, Y.; Li, C.; Guo, J. Atmospheric mercury depositional chronology reconstructed from lake sediments and ice core in the Himalayas and Tibetan Plateau. *Environ. Sci. Technol.* **2016**, *50*, 2859–2869. [CrossRef]
18. Arnold, J.; Gustin, M.S.; Weisberg, P.J. Evidence for nonstomatal uptake of Hg by aspen and translocation of Hg from foliage to tree rings in Austrian pine. *Environ. Sci. Technol.* **2018**, *52*, 1174–1182. [CrossRef]
19. Cooke, C.A.; Martinez-Cortizas, A.; Bindler, R.; Gustin, M.S. Environmental archives of atmospheric Hg deposition—A review. *Sci. Total Environ.* **2020**, *709*, 134800. [CrossRef]
20. Ghotra, A.; Lehnherr, I.; Porter, T.J.; Pisaric, M.F. Tree-ring inferred atmospheric mercury concentrations in the Mackenzie Delta (NWT, Canada) peaked in the 1970s but are increasing once more. *ACS Earth Space Chem.* **2020**, *4*, 457–466. [CrossRef]
21. Clackett, S.P.; Porter, T.J.; Lehnherr, I. The tree-ring mercury record of Klondike gold mining at Bear Creek, central Yukon. *Environ. Pollut.* **2021**, *268*, 115777. [CrossRef]
22. Hojdová, M.; Navrátil, T.; Rohovec, J.; Žák, K.; Vaněk, A.; Chrastný, V.; Svoboda, M. Changes in mercury deposition in a mining and smelting region as recorded in tree rings. *Water Air Soil Pollut.* **2011**, *216*, 73–82. [CrossRef]
23. Wright, G.; Gustin, M.S.; Weiss-Penzias, P.; Miller, M.B. Investigation of mercury deposition and potential sources at six sites from the Pacific Coast to the Great Basin, USA. *Sci. Total Environ.* **2014**, *470*, 1099–1113. [CrossRef]
24. Peckham, M.A.; Gustin, M.S.; Weisberg, P.J. Assessment of the suitability of tree rings as archives of global and regional atmospheric mercury pollution. *Environ. Sci. Technol.* **2019**, *53*, 3663–3671. [CrossRef] [PubMed]
25. Kang, H.; Liu, X.; Guo, J.; Zhang, Q.; Wang, Y.; Huang, J.; Xu, G.; Wu, G.; Ge, W.; Kang, S. Long-term mercury variations in tree rings of the permafrost forest, northeastern China. *Sci. China Earth Sci.* **2022**, *65*, 1328–1338. [CrossRef]
26. Cutter, B.E.; Guyette, R.P. Anatomical, chemical, and ecological factors affecting tree species choice in dendrochemistry studies. *J. Environ. Qual.* **1993**, *22*, 611–619. [CrossRef]
27. Watmough, S.; Brydges, T.; Hutchinson, T. The tree-ring chemistry of declining sugar maple in central Ontario, Canada. *Ambio* **1999**, *28*, 613–618.
28. Stamenkovic, J.; Gustin, M.S. Nonstomatal versus stomatal uptake of atmospheric mercury. *Environ. Sci. Technol.* **2009**, *43*, 1367–1372. [CrossRef]
29. Ahn, Y.S.; Jung, R.; Moon, J.H. Approaches to understand historical changes of mercury in tree rings of Japanese Cypress in industrial areas. *Forests* **2020**, *11*, 800. [CrossRef]
30. Ballikaya, P.; Brunner, I.; Cocozza, C.; Grolimund, D.; Kaegi, R.; Murazzi, M.A.; Schaub, M.; Schönbeck, L.; Sinnet, B.; Cherubini, P. First evidence of nanoparticle uptake through leaves and roots in beech (*Fagus sylvatica* L.) and pine (*Pinus sylvestris* L.). *Tree Physiol.* **2023**, *43*, 262–276. [CrossRef]
31. Perone, A.; Cocozza, C.; Cherubini, P.; Bachmann, O.; Guillong, M.; Lasserre, B.; Tognetti, R. Oak tree-rings record spatial-temporal pollution trends from different sources in Terni (Central Italy). *Environ. Pollut.* **2018**, *233*, 278–289. [CrossRef]
32. Rimondi, V.; Chiarantini, L.; Lattanzi, P.; Benvenuti, M.; Beutel, M.; Colica, A.; Costagliola, P.; Di Benedetto, F.; Gabbani, G.; Gray, J.E.; et al. Metallogeny, exploitation and environmental impact of the Mt. Amiata mercury ore district (Southern Tuscany, Italy). *Ital. J. Geosci.* **2015**, *134*, 323–336. [CrossRef]
33. Nannoni, A.; Meloni, F.; Benvenuti, M.; Cabassi, J.; Ciani, F.; Costagliola, P.; Fornasaro, S.; Lattanzi, P.; Lazzaroni, M.; Nisi, B.; et al. Environmental impact of past Hg mining activities in the Monte Amiata district, Italy: A summary of recent studies. *AIMS Geosci.* **2022**, *8*, 525–551. [CrossRef]
34. Fornasaro, S.; Morelli, G.; Rimondi, V.; Fagotti, C.; Friani, R.; Lattanzi, P.; Costagliola, P. The extensive mercury contamination in soil and legacy sediments of the Paglia River basin (Tuscany, Italy): Interplay between Hg-mining waste discharge along rivers, 1960s economic boom, and ongoing climate change. *J. Soils Sediments* **2022**, *22*, 656–671. [CrossRef]
35. Fornasaro, S.; Morelli, G.; Rimondi, V.; Fagotti, C.; Friani, R.; Lattanzi, P.; Costagliola, P. Mercury distribution around the Siele Hg mine (Mt. Amiata district, Italy) twenty years after reclamation: Spatial and temporal variability in soil, stream sediments, and air. *J. Geochem. Explor.* **2022**, *232*, 106886. [CrossRef]
36. Fornasaro, S.; Morelli, G.; Costagliola, P.; Rimondi, V.; Lattanzi, P.; Fagotti, C. The contribute of the Paglia-Tiber River system (Central Italy) to the total mercury mass load in the Mediterranean Sea. *Toxics* **2022**, *10*, 395. [CrossRef]
37. Rimondi, V.; Bardelli, F.; Benvenuti, M.; Costagliola, P.; Gray, J.E.; Lattanzi, P. Mercury speciation in the Mt. Amiata mining district (Italy): Interplay between urban activities and mercury contamination. *Chem. Geol.* **2014**, *380*, 110–118. [CrossRef]
38. Rimondi, V.; Costagliola, P.; Benesperi, R.; Benvenuti, M.; Beutel, M.W.; Buccianti, A.; Chiarantini, L.; Lattanzi, P.; Medas, D.; Parrini, P. Black pine (*Pinus nigra*) barks: A critical evaluation of some sampling and analysis parameters for mercury biomonitoring purposes. *Ecol. Indic.* **2020**, *112*, 106110. [CrossRef]

39. Chiarantini, L.; Rimondi, V.; Bardelli, F.; Benvenuti, M.; Cosio, C.; Costagliola, P.; Di Benedetto, F.; Lattanzi, P.; Sarret, G. Mercury speciation in Pinus nigra barks from Monte Amiata (Italy): An X-ray absorption spectroscopy study. *Environ. Pollut.* **2017**, *227*, 83–88. [CrossRef]

40. United Nations Environmental Program (UNEP)—Minamata Convention on Mercury. Available online: https://mercuryconvention.org/en (accessed on 29 March 2023).

41. Ferrari, L.; Conticelli, S.; Burlamacchi, L.; Manetti, P. Volcanological evolution of the Monte Amiata, Southern Tuscany: New geological and petrochemical data. *Acta Vulcanol.* **1996**, *8*, 41–56.

42. Cadoux, A.; Pinti, D.L. Hybrid character and pre-eruptive events of Mt Amiata volcano (Italy) inferred from geochronological, petro-geochemical and isotopic data. *J. Volcanol. Geotherm. Res.* **2009**, *179*, 169–190. [CrossRef]

43. Tanelli, G.; Lattanzi, P. Pyritic ores of southern Tuscany. *Italy Geol. Soc. S. Afr. Spec. Publ.* **1983**, *7*, 315–323.

44. Klemm, D.D.; Neumann, N. Ore-controlling factors in the Hg-Sb province of southern Tuscany, Italy. In *Syngenesis and Epigenesis in the Formation of Mineral Deposits: A Volume in Honour of Professor G. Christian Amstutz on the Occasion of His 60th Birthday with Special Reference to One of His Main Scientific Interests*; Springer: Berlin/Heidelberg, Germany, 1984; pp. 482–503.

45. Minissale, A.; Magro, G.; Vaselli, O.; Verrucchi, C.; Perticone, I. Geochemistry of water and gas discharges from the Mt. Amiata silicic complex and surrounding areas (central Italy). *J. Volcanol. Geotherm. Res.* **1997**, *79*, 223–251. [CrossRef]

46. Frondini, F.; Caliro, S.; Cardellini, C.; Chiodini, G.; Morgantini, N. Carbon dioxide degassing and thermal energy release in the Monte Amiata volcanic-geothermal area (Italy). *Appl. Geochem.* **2009**, *24*, 860–875. [CrossRef]

47. Tassi, F.; Vaselli, O.; Cuccoli, F.; Buccianti, A.; Nisi, B.; Lognoli, E.; Montegrossi, G. A geochemical multi-methodological approach in hazard assessment of CO_2-rich gas emissions at Mt. Amiata volcano (Tuscany, central Italy). *Water Air Soil Pollut. Focus* **2009**, *9*, 117–127. [CrossRef]

48. Ciccacci, S.; D'Alessandro, L.; Fredi, P.; Lupia Palmieri, E. Contributo dell'analisi geomorfica quantitativa allo studio dei processi di denudazione nel bacino idrografico del Torrente Paglia (Toscana meridionale-Lazio settentrionale). *Suppl. Geogr. Fis. Dinam. Quat.* **1988**, *1*, 171–188.

49. Selvi, F. Flora and phytogeography of the volcanic dome of Monte Amiata (Central Italy). *Webbia* **1996**, *50*, 265–310. [CrossRef]

50. Segreto, L. *Monte Amiata. Il Mercurio Italiano. Strategie Internazionali e Vincoli Extraeconomici*; Angeli: Milano, Italy, 1991.

51. Caselli, G.; Rosati, C.; Simone, M. La popolazione dei comuni minerari dell'Amiata. *Pop. E Stor.* **2007**, *8*, 63–89. Available online: https://popolazioneestoria.it/article/view/276 (accessed on 30 September 2022). (In Italian).

52. Vaselli, O.; Rappuoli, D.; Bianchi, F.; Nisi, B.; Niccolini, M.; Esposito, A.; Cabassi, J.; Giannini, L.; Tassi, F. One hundred years of mercury exploitation at the mining area of Abbadia San Salvatore (Mt. Amiata, Central Italy): A methodological approach for a complex reclamation activity before the establishment of a new mining park. In *El patrimonio Geológico y Minero: Identidad y Motor de Desarrollo*; Mansilla Plaza, L., Mata Perelló, J.M., Eds.; Instituto Geológico y Minero de España: Madrid, Spain, 2019; pp. 1109–1126.

53. Manzella, A.; Bonciani, R.; Allansdottir, A.; Botteghi, S.; Donato, A.; Giamberini, S.; Lenzi, A.; Paci, M.; Pellizzone, A.; Scrocca, D. Environmental and social aspects of geothermal energy in Italy. *Geothermics* **2018**, *72*, 232–248. [CrossRef]

54. Sabatelli, F.; Mannari, M.; Parri, R. Hydrogen Sulfide and Mercury Abatement: Development and Successful Operation of AMIS Technology. *GRC Trans.* **2009**, *33*, 343–347.

55. Bonciani, R.; Lenzi, A.; Luperini, F.; Sabatelli, F. Geothermal power plants in Italy: Increasing the environmental compliance. In Proceedings of the Conference European Geothermal Congress, Pisa, Italy, 3–7 June 2013.

56. Chiarantini, L.; Rimondi, V.; Benvenuti, M.; Beutel, M.W.; Costagliola, P.; Gonnelli, C.; Lattanzi, P.; Paolieri, M. Black pine (Pinus nigra) barks as biomonitors of airborne mercury pollution. *Sci. Total Environ.* **2016**, *569*, 105–113. [CrossRef]

57. Jeon, B.; Cizdziel, J.V. Determination of metals in tree rings by ICP-MS using ash from a direct mercury analyzer. *Molecules* **2020**, *25*, 2126. [CrossRef]

58. Yanai, R.D.; Yang, Y.; Wild, A.D.; Smith, K.T.; Driscoll, C.T. New approaches to understand mercury in trees: Radial and longitudinal patterns of mercury in tree rings and genetic control of mercury in maple sap. *Water Air Soil Pollut.* **2020**, *231*, 10. [CrossRef]

59. Siwik, E.I.; Campbell, L.M.; Mierle, G. Distribution and trends of mercury in deciduous tree cores. *Environ. Pollut.* **2010**, *158*, 2067–2073. [CrossRef]

60. Protano, G.; Nannoni, F. Influence of ore processing activity on Hg, As and Sb contamination and fractionation in soils in a former mining site of Monte Amiata ore district (Italy). *Chemosphere* **2018**, *199*, 320–330. [CrossRef]

61. Cocozza, C.; Ravera, S.; Cherubini, P.; Lombardi, F.; Marchetti, M.; Tognetti, R. Integrated biomonitoring of airborne pollutants over space and time using tree rings, bark, leaves and epiphytic lichens. *Urban For. Urban Green.* **2016**, *17*, 177–191. [CrossRef]

62. Leonelli, G.; Battipaglia, G.; Cherubini, P.; Morra di Cella, U.; Pelfini, M. Chemical elements and heavy metals in European larch tree rings from remote and polluted sites in the European Alps. *Geogr. Fis. Dinam. Quat.* **2011**, *34*, 195–206.

63. Orlandi, M.; Pelfini, M.; Pavan, M.; Santilli, M.; Colombini, M.P. Heavy metals variations in some conifers in Valle d'Aosta (Western Italian Alps) from 1930 to 2000. *Microchem. J.* **2002**, *73*, 237–244. [CrossRef]

64. Monticelli, D.; Di Iorio, A.; Ciceri, E.; Castelletti, A.; Dossi, C. Tree ring microanalysis by LA–ICP–MS for environmental monitoring: Validation or refutation? Two case histories. *Microchim. Acta* **2009**, *164*, 139–148. [CrossRef]

65. Amato, I. Trapping tree-rings for the environmental tales they tell. *Annal. Chem.* **1988**, *60*, 1103A–1107A.

66. Parisi, M.L.; Ferrara, N.; Torsello, L.; Basosi, R. Life cycle assessment of atmospheric emission profiles of the Italian geothermal power plants. *J. Clean. Prod.* **2019**, *234*, 881–894. [CrossRef]

67. Zhang, L.; Qian, J.L.; Planas, D. Mercury concentration in tree rings of black spruce (*Picea mariana* Mill. BSP) in boreal Quebec, Canada. *Water Air Soil Pollut.* **1995**, *81*, 163–173. [CrossRef]
68. Zhou, J.; Obrist, D.; Dastoor, A.; Jiskra, M.; Ryjkov, A. Vegetation uptake of mercury and impacts on global cycling. *Nat. Rev. Earth Environ.* **2021**, *2*, 269–284. [CrossRef]
69. Bishop, K.H.; Lee, Y.H.; Munthe, J.; Dambrine, E. Xylem sap as a pathway for total mercury and methylmercury transport from soils to tree canopy in the boreal forest. *Biogeochemistry* **1998**, *40*, 101–113. [CrossRef]
70. Navrátil, T.; Nováková, T.; Shanley, J.B.; Rohovec, J.; Matoušková, S.; Vaňková, M.; Norton, S.A. Larch tree rings as a tool for reconstructing 20th century Central European atmospheric mercury trends. *Environ. Sci. Technol.* **2018**, *52*, 11060–11068. [CrossRef]
71. Wang, X.; Yuan, W.; Lin, C.J.; Wu, F.; Feng, X. Stable mercury isotopes stored in Masson Pinus tree rings as atmospheric mercury archives. *J. Hazard. Mater.* **2021**, *415*, 125678. [CrossRef]
72. Hernes, P.J.; Hedges, J.I. Tannin signatures of barks, needles, leaves, cones, and wood at the molecular level. *Geochim. Et Cosmochim. Acta* **2004**, *68*, 1293–1307. [CrossRef]
73. Martínez-Gil, A.; del Alamo-Sanza, M.; Sánchez-Gómez, R.; Nevares, I. Different woods in cooperage for oenology: A review. *Beverages* **2018**, *4*, 94. [CrossRef]
74. Viriot, C.; Scalbert, A.; du Penhoat, C.L.H.; Moutounet, M. Ellagitannins in woods of sessile oak and sweet chestnut dimerization and hydrolysis during wood ageing. *Phytochemistry* **1994**, *36*, 1253–1260. [CrossRef]
75. Chattaway, M.M. The sapwood-heartwood transition. *Aust. For.* **1952**, *16*, 25–34. [CrossRef]
76. Hagemeyer, J.; Breckle, S.W. Cadmium in den Jahrringen von Eichen: Untersuchungen zur Aufstellung einer Chronologie der Immissionen. *Angew. Botanik.* **1986**, *60*, 161–174.
77. Edmands, J.D.; Brabander, D.J.; Coleman, D.S. Uptake and mobility of uranium in black oaks: Implications for biomonitoring depleted uranium-contaminated groundwater. *Chemosphere* **2001**, *44*, 789–795. [CrossRef]
78. Navrátil, T.; Šimeček, M.; Shanley, J.B.; Rohovec, J.; Hojdová, M.; Houška, J. The history of mercury pollution near the Spolana chlor-alkali plant (Neratovice, Czech Republic) as recorded by Scots pine tree rings and other bioindicators. *Sci. Total Environ.* **2017**, *586*, 1182–1192. [CrossRef]
79. McLagan, D.S.; Biester, H.; Navrátil, T.; Kraemer, S.M.; Schwab, L. Internal tree cycling and atmospheric archiving of mercury: Examination with concentration and stable isotope analyses. *Biogeoscience* **2022**, *19*, 4415–4429. [CrossRef]
80. Hampp, R.; Höll, W. Radial and axial gradients of lead concentration in bark and xylem of hardwoods. *Arch. Environ. Contam. Toxicol.* **1974**, *2*, 143–151. [CrossRef]
81. Vaněk, A.; Chrastný, V.; Teper, L.; Cabala, J.; Penížek, V.; Komárek, M. Distribution of thallium and accompanying metals in tree rings of Scots pine (*Pinus sylvestris* L.) from a smelter-affected area. *J. Geochem. Explor.* **2011**, *108*, 73–80. [CrossRef]
82. Peckham, M.A.; Gustin, M.S.; Weisberg, P.J.; Weiss-Penzias, P. Results of a controlled field experiment to assess the use of tree tissue concentrations as bioindicators of air Hg. *Biogeochemistry* **2019**, *142*, 265–279. [CrossRef]
83. Bardelli, F.; Rimondi, V.; Lattanzi, P.; Rovezzi, M.; Isaure, M.P.; Giaccherini, A.; Costagliola, P. Pinus nigra barks from a mercury mining district studied with high resolution XANES spectroscopy. *Environ. Sci. Process. Impacts* **2022**, *24*, 1748–1757. [CrossRef]

minerals

Article

Fluvial Morphology as a Driver of Lead and Zinc Geochemical Dispersion at a Catchment Scale

Patrizia Onnis [1,2,*], Patrick Byrne [1], Karen A. Hudson-Edwards [3], Tim Stott [4] and Chris O. Hunt [1]

1   School of Biological and Environmental Science, Liverpool John Moores University, Liverpool L3 3AF, UK; p.a.byrne@ljmu.ac.uk (P.B.); c.o.hunt@ljmu.ac.uk (C.O.H.)
2   Department of Chemical and Geological Sciences, University of Cagliari, 09042 Monserrato, Italy
3   Environment & Sustainability Institute and Camborne School of Mines, University of Exeter, Penryn TR10 9FE, UK; k.hudson-edwards@exeter.ac.uk
4   Faculty of Education, Health and Community, Liverpool John Moores University, Liverpool L17 6BD, UK; t.a.stott@ljmu.ac.uk
*   Correspondence: patrizia.onnis@unica.it

Abstract: Metal-mining exploitation has caused ecosystem degradation worldwide. Legacy wastes are often concentrated around former mines where monitoring and research works are mostly focused. Geochemical and physical weathering can affect metal-enriched sediment locations and their capacity to release metals at a catchment scale. This study investigated how fluvial geomorphology and soil geochemistry drive zinc and lead dispersion along the Nant Cwmnewyddion (Wales, UK). Sediments from different locations were sampled for geochemical and mineralogical investigations (portable X-ray fluorescence, scanning electron microscope, X-ray diffraction, and electron microprobe analysis). The suspended sediment fluxes in the streamwater were estimated at different streamflows to quantify the metal dispersion. Topographical and slope analysis allowed us to link sediment erosion with the exposure of primary sulphide minerals in the headwater. Zinc and lead entered the streamwater as aqueous phases or as suspended sediments. Secondary sources were localised in depositional stream areas due to topographical obstruction and a decrease in stream gradient. Sediment zinc and lead concentrations were lower in depositional areas and associated with Fe-oxide or phyllosilicates. Streamwater zinc and lead fluxes highlighted their mobility under high-flow conditions. This multi-disciplinary approach stressed the impact of the headwater mining work on the downstream catchment and provided a low-cost strategy to target sediment sampling via geomorphological observations.

Keywords: soil geochemistry; mine waste; metal contamination; geomorphology; monitoring

Citation: Onnis, P.; Byrne, P.; Hudson-Edwards, K.A.; Stott, T.; Hunt, C.O. Fluvial Morphology as a Driver of Lead and Zinc Geochemical Dispersion at a Catchment Scale. *Minerals* **2023**, *13*, 790. https://doi.org/10.3390/min13060790

Academic Editors: Pierfranco Lattanzi, Elisabetta Dore, Fabio Perlatti and Hendrik Gideon Brink

Received: 3 May 2023
Revised: 5 June 2023
Accepted: 6 June 2023
Published: 9 June 2023

1. Introduction

Mining has a long history of exploitation resulting in the accumulation of waste material in river systems [1]. This history provides examples of the soil, water, and biological impacts mining mismanagement and disasters can pose on riverine and coastal systems [2,3]. Over centuries of mining, metal-enriched sediments have been redistributed in catchments via geomorphological processes to achieve morphological equilibrium, eroded and transported along river channels, and deposited in channels or alluvial valleys [4–6]. Metal-bearing minerals dispersed in a river system constitute potential sources of contaminants for the ecosystem via sediments, water, or organisms [7,8]. However, understanding metal dispersion at a catchment scale is difficult due to inherent river geomorphology and sediment geochemistry variabilities.

Previous studies have contributed to understanding the links between metal sediment concentrations, geomorphology, and mineral presence [2,9–11]. In an English mine site, Kincey, et al. [11] quantified 434 t of Zn- and Pb-contaminated mine waste that eroded from mine heaps in the river headwater of Garrigill Burn, a tributary of the upper South

Tyne. Erosional processes such as gullying, bank erosion, and mass movements were observed with laser scanning data acquisition. Large losses of sediments were recorded, particularly during infrequent high-magnitude storms. Foulds, et al. [12] identified the two main processes responsible for mining-affected sediment remobilization and dispersion: (i) the direct erosion of unstable loose sediment mainly characterised by mine waste in headwaters; (ii) bank erosion and the lateral and vertical accretion of floodplain deposits that are common in piedmont areas. Variations in sediment transport velocity, controlled via changes in slope or channel enlargements, have been proposed to create low-energy zones that confine sediments to tributaries, preventing them from reaching main river channels [2]. Therefore, the fluvial geomorphology of mining-impacted catchments enriches the understanding of the erosion, transport, and deposition of metal-enriched sediments.

Mine wastes dispersed into catchments carry ore minerals, such as galena and sphalerite, or secondary metal-bearing minerals, such as sulphates, phosphates, carbonates, and Fe- or Mn-oxyhydroxides [9,10]. These minerals' stability and metal mobility depend on biogeochemical factors including pH, oxygen concentration, metal sediment and water concentrations, water saturation, and bacterial activity [13]. Therefore, geochemical and mineralogical analysis is necessary to identify the presence and stability of metal-bearing phases. Hudson-Edwards, et al. [9] showed that primary minerals, such as lead (Pb) and zinc (Zn) sulphides, were mostly localised at mine tips associated with their weathered products, such as secondary carbonate, silicate, phosphate, and sulphate (CSPS) minerals. Secondary iron- and Mn-oxyhydroxides were abundant downstream, with Mn-oxyhydroxides progressively decreasing in abundance. This study highlights the importance of investigating sediment geochemistry at a catchment scale.

Sediment mineralogical and geochemical analysis must be undertaken to estimate the storage or release capacities of potential metal sources dispersed in a catchment, and geomorphological information can help explain sediment and water geochemistry spatial variations. River headwaters in mining-impacted catchments can host waste tips, in which fluvial processes associated with high slopes continuously erode the wastes, exposing primary minerals to weathering processes. On the other hand, depositional areas provide permeable riverbeds and banks that can potentially promote geochemical processes that are typical of hyporheic zones or clay embankments with low permeability and reductive conditions [14,15]. Statistical approaches have been used to predict downstream metal decay [5] and have shown that downstream concentration decays are metal-specific depending on their biogeochemistry and associated material, while denser primary minerals remain close to mine waste sources [16,17]. Mixing models can account for dilution processes due to clean sediment, multiple source areas, and the attenuation of suspended sediments due to water acidity variations [17,18]. Modelling limitations arise from the variability of geomorphological processes that are not systematic along a channel [2].

Further research is needed to understand the complex geomorphological and sedimentary geochemical processes controlling metal dispersions in mining impacted catchments. This paper contributes to this knowledge gap by combining the study of fluvial morphological and geochemical processes and the categorisation of fluvial zones across a river with defined mineralogical paragenesis and metal release capability. In particular, it will focus on the geochemical behaviour of the mining-related metals Zn and Pb, whose (1) toxicities have been investigated extensively [19–22]; (2) impacts on mining catchments worldwide are well-known [23–27]; and (3) mobility poses management challenges in circum-neutral rivers [14,28–31]. The specific objectives are to (i) map the spatial fluvial morphological variability along the study catchment; (ii) characterise the geochemistry of the metal-enriched sediment; and (iii) demonstrate the fluvial morphological controls on the Zn- and Pb-bearing mineral distribution along the catchment. The findings can link fluvial systems with sustainable land use management for the reclamation and monitoring of ecological status.

2. Materials and Methods

2.1. Study Site

The study catchment is the Nant Cwmnewyddion in central Wales (Figure 1a). This stream drains the Wemyss and Graig Goch Mines, flows west to the Nant Magwar, and then into Afon Ystwyth. A total of 19 km of river length is impacted by Pb and Zn [32]. This area falls into the Ceredigion Uplands, a special landscape area (SLA) of national importance for its outstanding ecological, cultural, and archaeological value [33]. The Graig Goch, Wemyss, and Frongoch Mines exploited the Frongoch mineral load. Wemyss and Frongoch were joined in the 1840s under the same owner, with Wemyss Adit extended to serve the Frongoch Mine, becoming Frongoch Adit. The mining was in operation from the 2nd half of the 19th century until 1904. The geology of the area is characterised by mudstones and sandstones of the Devil's Bridge formation of the Upper Llandovery (Telychian-aged) [34]. More recent units are glacial deposits (diamicton) and alluvium (clay, silt, sand, and gravel). These deposits and the mining-exploited areas provide permeable areas surrounded by the low permeability of the Silurian basement. The mining areas are aligned with the southwest-northeast normal fault containing the Frongoch mineral lode enriched in Cu, Pb, Zn, and Ag. The minerals in the ore veins are sulphides, including galena, sphalerite, occasional pyrite, and gangue minerals comprising mainly quartz, feldspar, illite, and chlorite. The lead-bearing secondary minerals include anglesite and plumbojarosite [35]. Monitoring of the annual hydrographic records at the downstream Ystwyth gauge station 63001 (Ystwyth at Pont Llolwyn) indicated a 'flashy' stream, with high inter- and intra-annual variability with minimum and maximum streamflow values of 0.8 and 74.7 m^3/s in 2016 (NRW station 63001), which are typical of a temperate maritime climate [36].

2.2. Fluvial Geomorphology

The fluvial catchment geomorphology was described using topographical and slope analysis and the grain size distributions of the riverbank sediments, as these have been shown to be the major controls on physical transport and deposition [16,37]. Topographic features, such as the river channel shape and the proximity of valley sides and slope calculations, were extrapolated using ArcGIS v.10 from the LIDAR-Digital Terrain Model (DTM) of the area (Welsh 1 m LIDAR, Natural Resources Wales information ©). The ArcGIS Spatial Analysis tools' 'height contour' and 'slope' tools were used to produce the area's height contour and slope maps. The stream profile was extrapolated using the 'stack profile' tool from the functional surface in the 3D analyst tool from the DTM layer. The attribute tables exported the data for calculating the slope degree using Excel functions. The first slope percentage was calculated as the ratio of the elevation change to the length of the relevant stream segment. Then, the slope percentage was converted into a slope degree for consistency with the elaborated ArcGIS map. Finally, landscape survey notes were collected, including nearby vegetation presence, channel and overbank characteristics, and land use.

2.3. Sediment Sampling

Along the Nant Cwmnewyddion and the Nant Magwr, five sediment sampling areas were identified based on the location of mining and variations in fluvial geomorphological features such as slope and topography (Figure 1). The areas were (Figure 1c) (i) upstream of the mine tips, designated as the 0 m point (T0); (ii) the mine tips around the Wemyss Mine (171 m) both at the headwater (CW) and further downstream of the Mill Race Stream (WM); (iii) along the Graig Goch Mine (1110–2620 m) stream segment (GG); (iv) around a middle-reach (3000–4070 m) depositional area (DC); and (v) the floodplain at approximately 5800 m and 6740 m (RB up/s (up-stream) and RB d/s (down-stream), respectively). All the meter indications (e.g., 171 m) refer to the distance from the mine upstream area T0 (0 m, Figure 1a).

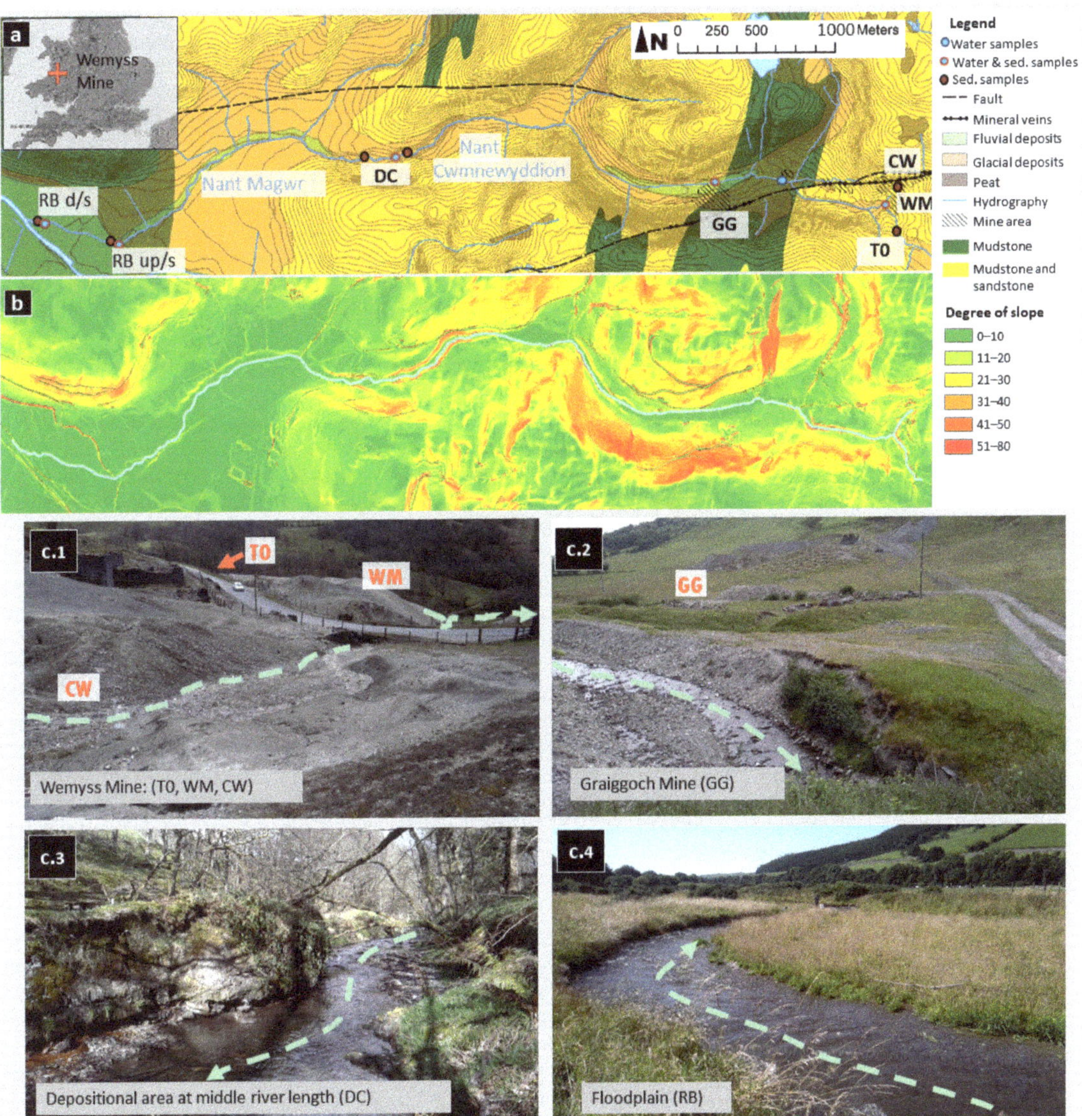

Figure 1. (**a**) Map of the water and sediment (sed.) sampling areas along the Nant Cwmnewyddion and the Nant Magwr. WM: Wemyss Mine; CW: tributary at Wemyss mine; T0: upstream mine areas; GG: Graig Goch Mine; DC: depositional area at middle river length; RB: floodplain upstream (up/s) and downstream (d/s) area. (**b**) Topography and slope map of the research site elaborated with ArcGIS from 1 m resolution LIDAR images (Natural Resources Wales information ©). (**c**) Sampling area views and river paths indicated with the blue dashed line. (**c.1**) Wemyss Mine area and waste heaps with tributary (dashed lines) and sampling areas (T0, WM, and CW). (**c.2**) Graig Goch Mine with GG-labelled samples. (**c.3**) Depositional area at middle river length (DC) with 1 to 2 m riverbank and percolating red water visible on the bottom left. (**c.4**) Floodplain (RB) view of a meander corner.

In each area, sediment samples were collected along transects perpendicular to the stream corridor on 18, 19, 20, and 26 July of 2016. The transect positions depended on fluvial morphological observations (such as depositional or erosional areas), geological and sedimentological features (sediment texture and rocky outcrop), and vegetation. A total of 68 samples distributed over 17 transects were collected with a stainless-steel trowel and hand augers to a maximum depth of 40 cm. The samples were transported to the laboratory and left to dry at 21 °C for 5 days, and their pH (paper test) values and grain size distributions were determined. The obtained pH values were indicative only and used to describe the relative pH differences among the samples. Grain sizes were separated on a vibration platform (Fritsch) using Endcotts sieves following the Wentworth size classes [38]: very coarse sand (2–1 mm), coarse sand (1–0.5 mm), medium sand (0.5–0.25 mm), fine sand (0.25–0.125 mm), very fine sand (0.125 mm–63 μm), and silt and clay (less than 63 μm).

2.4. Sediment Geochemical Analysis

The geochemical characterisation of the sediment samples was carried out using a portable X-ray Fluorescence (pXRF) Genius 9000XRF model Skyray [39]. The p-XRF analysis was carried out for each size fraction to reduce physical matrix effects. Element peak positions were determined with a standard of silver, and data quality was assessed with the Fresh Water Sediment certified reference material CRM016 (Sigma Aldrich, © 2023 Merck KGaA, Darmstadt, Germany). Samples were run six times to increase data precision and to enable the standard deviation to be calculated. The relative standard deviation (the ratio of the standard deviation and average) for Zn and Pb was below 5%. The limit of detection was calculated as 10 times the standard deviation and equal to 30 mg/kg for Zn and 10 mg/kg for Pb. The arithmetic averages for Pb and Zn concentrations were calculated for each sampling transect to homogenise the values and obtain a representative value of the sampled area [40]. A subset of 10 samples was selected for mineralogical characteristics and metal content quantification. The selection criteria were based on the proximity of the sample to the stream water (e.g., samples on the stream corridor) and the highest metal concentrations obtained with pXRF analysis. The finest granulometric fraction (<63 μm) was selected for X-ray diffraction (XRD) and scanning electron microscope (SEM) analysis due to their observed highest metal concentrations. The mineralogy of the finest was analysed with a Rigaku Miniflex XRD. The resultant data were processed with SIeve for Powder Diffraction File 2 (PDF) software (Release 2012) and the Powder Diffraction File-International Centre for Diffraction Data (PDF-ICDD) database. The XRD results were used for a qualitative description of the sediment mineralogy. SEM observations were performed with an FEI-Quanta 200 SEM and an EDS INCA-X-act-Oxford Instruments. Images were first acquired at large scales (100 to 10 μm) for sample investigations, and then X-ray spectra were acquired for single points or as chemical maps at 20 KeV [41]. Analysis of the whole portion sieved at 2 mm was carried out with a Jeol8100 Superprobe (electron microprobe) with wavelength-dispersive spectroscopy (WDS) and an Oxford Instrument INCA microanalytical system (EDS). Samples were mounted in epoxy-impregnated polished mounts [42]. Metal-bearing phases were detected using backscatter electron images (BSEM) and energy-dispersive detection (EDS). Energy data were collected between 0 and 20 eV using a 15 Kv accelerating voltage and a spot size of 1 μm.

2.5. Zinc and Lead Water Concentrations, Loads, and Suspended Sediments

Stream waters were sampled at six sites along the stream to verify Zn and Pb mobility. The 6 areas, indicated in Figure 1a, were located to capture diffuse and point sources (1) at Wemyss Mine (WM, 171 m); (2) downstream of the tributary Frongoch Adit (d/s FA, 880 m); (3) at Graig Goch Mine (GG, 1645 m); (4) at the middle stream length depositional area (DC, 3210 m); and (5) downstream and upstream of the floodplain (RB up/s, 5930 m and RB d/s, 6780 m). Samples were collected twice in two separate campaigns to measure temporal variability due to streamflow conditions. Streamflows at each site were monitored under low- (28 July 2017) and high- (17 June 2016) flow conditions with gulp injections of NaCl salt [40,43]. The high and low streamflow definitions were determined through analysis of the data from the historical gauge station 63001 located downstream of the

Nant Magwar (Ystwyth at Pont Lllowyn; data are available at https://nrfa.ceh.ac.uk/data/station/meanflow, consulted on the 1 August 2019) [40]. pH, conductivity, and temperature were measured directly in the field using a multiprobe Aqua-reader, which was calibrated daily. A portion of each water sample was filtered at 0.45 μm with regenerated cellulose filters (Whatman SPARTAN). Filtered Pb and Zn (0.45 μm, filtered and acidified with 1% v/v of HNO_3, FA) were measured with ICP-MS (ICP-Mass Spectrometer Varian 720), and unfiltered Pb and Zn (unfiltered and acidified with 1% v/v of HNO_3 1%, RA) were measured with ICP-AES (Inductively Coupled Plasma-Atomic Emission Spectrometer-Varian 810). For precision and accuracy, the water analysis included TMDA 70 (Certified Reference Waters for Trace Elements), EP-H (Matrix Material Environ MAT Drinking Water), the repetition of standards, and sample duplicates. The limits of detection (LODs) verified with field and laboratory blanks were 10 μg/L for Zn and 0.33 μg/L for Pb. Zinc- and Pb-bearing suspended sediment concentrations were calculated via the difference between their unfiltered and filtered water concentrations. The Zn and Pb loads in suspended sediment were estimated by multiplying their concentrations by the streamflow. The method and results from these experiments were presented in full in the study by Onnis, et al. [40]. Low- and high-flow datasets are presented herein to discuss the role of fluvial morphology on sediment geochemistry.

3. Results and Discussion

3.1. Fluvial Geomorphological Zones along the Stream

The topography and slope map for the study area is presented in Figure 1a,b. The morphological variabilities of the Nant Cwmnewyddion and the Nant Magwr are typical of mining-impacted rivers, for which active morphological processes restore geomorphological equilibrium disturbed by ore extraction and spoil material [6]. The observed morphological processes include aggradation and degradation events such as the erosion of banks and terraces and sediment deposition. Therefore, areas along the stream were classified as erosional, transport, or depositional (Figure 2).

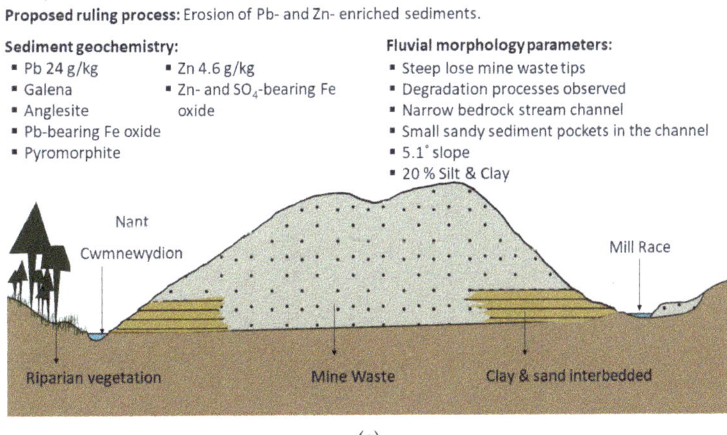

(**a**)

Figure 2. *Cont.*

Graig Goch Mine area (stream segment 1110–2620 m)

Proposed ruling process: Sediment transport area with hyporheic zone trapping Pb in Mn-oxide minerals.

Sediment geochemistry:

- Pb 5 g/kg
- Pb- Mn Oxide
- Pyromorphite
- Galena
- Pb-bearing phyllosilicates
- Zn 1 g/kg
- Zn- bearing Fe oxide
- Zn- bearing phyllosicate

Fluvial morphology parameters:

- Wide valley
- Wide straightened stream channel
- Glacial and fluvial channel bed
- Sand and gravel lens
- 1.2° slope
- 24% Silt & Clay

(b)

Middle stream length depositional area (stream segment 3000–4070 m)

Proposed ruling process: Pb and Zn enriched sediment are trapped and diluted by valley side sediment deposition. Hyporheic and redox conditions are induced in the stream banks.

Sediment geochemistry:

- Pb 20 g/kg, Zn 2 g/kg
- Anglesite
- Sphalerite
- Galena
- Pb- and Zn-bearing Fe oxide
- Plumbojarosite
- Pb 4 g/kg, Zn 1 g/kg
- Pb- Mn oxide
- Pyromorphite
- Sphalerite
- Pb- and Zn- bearing Fe oxide

Fluvial morphology parameters:

- Narrow valley controlled by bedrock
- Deposits in bedrock alcoves
- Meander stream path
- Erosional bank with vertical accretional sediments
- Depositional bank with coarse material
- 1.2° slope
- 24% Silt & Clay

(c)

Floodplain area (stream segment 5800–6740 m)

Proposed ruling process: Pb is stored in the sediment and likely easily released by phyllosicate.

Sediment geochemistry:

- Pb 2–11 g/kg
- Pb-bearing phyllosilicate
- Pb-bearing Fe oxide
- Zn 1-3 g/kg
- Zn-bearing phyllosilicate

Fluvial morphology parameters:

- Wide alluvial valley
- Meanders with later and vertical accretions
- stream channel occasionally channelled by wooden piles
- 0.6° slope
- 17-36 % Silt & Clay

(d)

Figure 2. Summary of sediment geochemical and fluvial morphological parameters that are related to Pb and Zn dispersion. The river sessions are an interpretation of the undergoing processes and scale of sediments and basement are approximate. In solid brown is reported the mudstone geological basement. (**a**) Wemyss Mine (WM, CW); (**b**) Graig Goch Mine; (**c**) Middle river length (DC); (**d**) floodplain (RB).

3.1.1. Erosion from the Mining Heaps (0–180 m Stream Segment)

Over the mine's history, the stream headwater (0–180 m) was continuously fed with mine waste coming from both Wemyss and the adjacent Frongoch Mine sites. The steep valley sides and the highest slope within the catchment (5.1°) limited the headwater capacity to store mine wastes (Figure 1b). At the mine area (WM), the right stream bank (facing downstream) was 0.5–1 m high and constituted clay and loose sand deposits (Figures 1c.1 and 2a). Occasionally, sand was deposited in pools and alcoves of the bedrock streambed. Numerous surveys at the field site showed lateral accretions in the channel and material loss from the tips and stream banks, which caused changes in the riverbed shape and size. Erosion conveyed sediments from the Wemyss Mine tips into the downstream fluvial system, similar to the erosional processes observed by Kincey, et al. [11]. Due to this continuous degradation, the 0–180 m segment was classified as erosional (Figure 2a).

3.1.2. Sediment Transport (1110–2620 m Stream Segment)

Around the Graig Goch Mine (1110–2620 m, GG), the channel crossed a wide valley and had a slope of 1.2° (Figure 1b,c.2). The river channel was straight at this position, which was likely due to the channelisation imposed by the nearby mine workings, road, and farm. The wide stream channel (1–2 m wide) with glacial sand and gravel lens may have enhanced the circulation of the water [40,44]. The fine-grained waste eroded in the upstream segment (0–180 m) was not deposited in this area (Figure 1c), which was likely due to downstream transport across streamflow conditions [42]. Therefore, the 1100–2620 m stream segment, which was mainly characterised by downstream passive sediment transport, was classified as a transport area (Figure 2b).

3.1.3. Middle Stream Length Depositional Area (3000–4070 m Stream Segment)

At the middle river segment (3000–4070 m, DC), the topography indicated a narrow valley, and the river channel was bedrock-controlled with a slope of 1.2° (Figure 1b,c.3). The channel bed was formed alternatively with bedrock and sand lenses, and the banks were covered with dense wooded and shrub-like vegetation (ferns, nettles, and rushes). Due to seasonal recreational land use, the vegetation and the stream bank stability were occasionally managed. The river channel here meandered, creating 0 to 2 m high riverbanks. Interlayered clays and sands characterised the highest banks, whereas the lowest banks contained coarse sand and gravel fluvial deposits. The fine-grained sediments were similar to those observed at the Wemyss Mine site. This segment potentially acted as a 'sediment trap' that was likely filled up during the most active mining years [12]. Depositional structures such as vertical accretions of sand and clay layers and depositional silty–sandy lenses likely control the river bank permeability and interaction with the river water [3,14]. Therefore, the stream segment was classified as a depositional area (Figure 2c).

3.1.4. Floodplain Area (5800–6740 m Stream Segment)

The floodplain stream segment (5810–6740 m, RB) had the lowest slope (0.6°) across the alluvial valley (Figure 1b). The stream channel meandered, except for the segment crossed by the main road, which was channelised using wooden piles. The channel bed comprised largely coarse fluvial deposits (Figure 1c.4). The area was used predominantly as farmland, and channel bed animal crossings occurred in the most downstream segment. The heights of the riverbanks varied from 0 to 2 m and hosted sedimentation structures such as lateral vertical aggradation. Similar features were observed in Welsh rivers (UK) [12] and in a New Zealand mining-impacted catchment [2]. This segment can be considered a depositional area (Figure 2d).

3.2. Sediment Geochemistry and Mineralogy

The sediment pH values varied between 4.5 and 5.0, with the highest readings recorded in the floodplain samples (Table 1). Lead and Zn concentrations ranged from 20 to 24,700 mg/kg and 250 to 4600 mg/kg, respectively. The highest metal values occurred in upstream sediments: 24,700 mg/kg for Pb and 1500 mg/kg for Zn at CW and 23,800 mg/kg for Pb and 4600 mg/kg for Zn at WM. Sediments upstream of the mines (T0) had high Pb and Zn concentrations (7400 mg/kg for Pb and 1300 mg/kg for Zn), which was potentially due to the proximity to the mine waste from the Wemyss and Frongoch Mine. Graig Goch Mine sediments (GG) had lower Pb concentrations than those in the upstream mine heaps. Along the depositional areas (DC and RB), the sediments had some of the lowest concentrations. Sediments from DC had an average of 10,400 mg/kg for Pb and 1000 mg/kg for Zn, and those from RB had average concentrations of 4400 mg/kg for Pb and 1600 mg/kg for Zn (Table 1). Zinc and Pb concentrations throughout the stream were higher than the guideline values reported as the threshold effect level (TEL) and the probable effect level (PEL) from the Canadian soil guidelines adopted by the Environmental Agency [44]. The only values below the TEL (threshold effect level), 123 mg/kg for Zn and 35 mg/kg for Pb [45], were the concentrations measured in a riverbank on the downstream side of the floodplain (Table 1). Therefore, according to the PEL guidelines, the channel and overbank sediments are potentially harmful to the surrounding ecosystem.

Table 1. Sediment geochemistry results presented for each sampled transect. Lead and Zn concentrations (g/kg) were measured with pXRF. The distances are reported in meters from the first sample upstream of the mining area. The potential mineral suite and trace metal associations are presented as a sum of the XRD, SEM, and Jeol8100 Superprobe (WDS) results. Finally, minimums (min), maximums (max), and the averages for the whole area and suggested metal concentration limits are reported as TEL (threshold effect level) and PEL (predicted effect level) derived from the Environment Agency [45]. WM: Wemyss Mine; CW: tributary at Wemyss Mine; T0: upstream of the mine areas; GG: Graig Goch Mine; DC: depositional area at middle river length; RB: floodplain upstream (up/s) and downstream (d/s) areas.

River Area	Transect Distance	Transect	Pb	Zn	pH	Metal-Bearing Mineral Suite
	m		g/kg	g/kg		
Up/s Mines						
T0	0	T0 M	7.4	1.3	4.7	Fe-hydroxide; pyromorphite; anglesite; rutile
Mining Areas						
CW	160	CW J	24.7	1.5	4.4	Monazite; galena; Fe-oxide
WM	171	WM K	23.8	4.6	4.6	Anglesite; galena; Fe-oxide; pyromorphite
GG	1650	GG L	4.5	1.0	4.7	Pb, Mn-oxide; Fe, Mn-oxide; pyromorphite; anglesite
Middle length						
DC	3000	DC D	5.3	0.9	4.8	Anglesite; sphalerite; galena; Pb- and Zn-Fe-oxide; plumbojarosite; pyromorphite; Pb, Mn-oxide
	3400	DC E	3.8	0.8	4.7	
	3600	DC F	20.3	1.5	4.5	
	3800	DC G	11.3	1.1	4.6	
	4070	DC H	11.1	0.9	4.8	
Floodplain						
RB up/s	5800	RB B	2.6	1.3	5	Pb- and Zn-bearing phyllosilicate (illite and chlorite); Pb-Fe oxide
	5900	RB C	11.0	2.7	5	
		RB DEP C	2.2	2.0	5	
	6000	RB D	7.4	1.4	5	

Table 1. *Cont.*

River Area	Transect Distance	Transect	Pb	Zn	pH	Metal-Bearing Mineral Suite
	m		g/kg	g/kg		
RB d/s	6600	RB I	8.3	1.5	4.7	
		RB DEP B	1.7	1.6	5	
	6700	RB DEP A	2.2	1.6	5	
		RB A	0	0.3	4.9	
Tot. min.			0	0.3	4.4	Pb- and Zn-bearing phyllosilicate (illite and chlorite); Pb-Fe oxide
Tot. max.			24.7	4.6	5.0	
Tot. average			8.7	1.5	4.8	
TEL			0.035	0.123		
PEL			0.091	0.315		

Mineralogical analysis showed spatial variability with a widespread presence of bulk minerals such as quartz, feldspar, and phyllosilicate (such as illite and chlorite) associated with zircon, monazite, rutile, and ilmenite. The Zinc and Pb primary minerals were galena and sphalerite, localised predominantly in the upstream part. The secondary mineral geochemistry reflected the presence of phosphate, silicate, sulphate, Pb, and Zn (Table 2). The minerals were anglesite, plumbojarosite, pyromorphite, and Pb- and Zn-bearing phyllosilicates. Iron- and Mn-hydroxides and secondary sulphates were present as secondary weathering products, which were likely both directly precipitated or formed as a replacement for other minerals [9]. The weathering products occurred either at the headwaters with the primary minerals or further downstream, where they were more abundant and varied. The Zn and Pb minerals in the Cwmnewyddion and Magwr river systems are typical of Zn and Pb mining areas with low amounts of pyrite and carbonate [37,46]. The mineralogical suite observed at the Wemyss and Graig Goch Mines corroborates previous findings described by Palumbo-Roe and Dearden [35].

Table 2. River water concentrations and sediment loads for Zn and Pb (F, filtered; UF, unfiltered samples) and streamflows (Q) at sampling sites. WM: Wemyss Mine; CW: tributary at Wemyss Mine; T0: upstream of the mine areas; d/s FA: downstream Frongoch Adit; GG: Graig Goch Mine; DC: depositional area at middle river length; RB: floodplain upstream (up/s) and downstream (d/s) areas.

Sample Area	Date	Distance	Q	Zn	Zn	Pb	Pb	Zn	Pb
		m	L/s	mg/L	mg/L	µg/L	µg/L	mg/s	mg/s
				F	UF	F	UF	Suspended Sediment	
High streamflow									
WM	15 June 2016	171	210	1.4	1.4	390	430	0	8
d/s FA	16 June 2016	880	430	3.0	3.0	120	160	20	19
GG	16 June 2016	1645	600	2.3	2.3	50	90	0	27
DC	17 June 2016	3210	950	1.4	1.4	20	30	39	3
RB up/s	17 June 2016	5930	1500	1.0	1.0	10	30	36	21
RB d/s	15 June 2016	6780	880	0.9	0.9	10	50	50	36
Low streamflow									
WM	29 July 2017	171	40	1.1	1.3	130	180	7	2
d/s FA	29 July 2017	880	130	1.8	2.0	80	140	26	8
GG	28 July 2017	1645	160	0.9	1.3	50	250	52	32

Table 2. *Cont.*

Sample	Date	Distance	Q	Zn	Zn	Pb	Pb	Zn	Pb
Area		m	L/s	mg/L	mg/L	µg/L	µg/L	mg/s	mg/s
				F	UF	F	UF	Suspended Sediment	
DC	28 July 2017	3210	130	1.3	1.6	30	70	36	5
RB up/s	30 July 2017	5930	320	0.8	0.9	20	30	36	2
RB d/s	30 July 2017	6780	310	0.8	1.0	20	30	50	8

In the upstream part of the catchment, sub-angular aggregates of well-shaped minerals (quartz, rutile, zircon, and phyllosilicate) cemented with Fe-oxide amorphous phases characterised the sediment. The iron-oxides had low concentrations of Pb, S, Cl, and P and showed O–Fe ratios between one to two, indicating a potential goethite or hematite mineralogy. Furthermore, sulphides, including galena and sphalerite, were observed with Fe-oxide minerals occurring on weathered grain edges (Figure 3a). Other secondary minerals were sulphates (anglesite, hydrate Pb-sulphates, and Cu-bearing jarosite). The Graig Goch Mine sample GG L1 was characterised by mudstone detritus and bulk minerals agglomerated in 0.1–0.5 mm clusters or crystals. Occasionally, galena grains and Zn- and Pb-bearing Fe-hydroxides were observed in the clusters (Figure 3b). Other Pb-bearing minerals were Mn-Pb oxides, with a three to one Mn-Pb ratio [40], and pyromorphite spotted on weathered cluster edges. The SEM images showed significant quantities of Zn (of up to 420 mg/kg) and Pb (of up to 1300 mg/kg) associated with phyllosilicate phases (Figure 3b).

Stream erosional banks from the meandered depositional area (3000–4070 m) showed different mineral suites. In the middle length of the river (DC), Pb was observed with sulphides, sulphates, Fe–Mn hydroxides, and phyllosilicate minerals. Samples collected from the depositional side of a meander (DC E) had Pb-bearing Fe or Mn oxide phases and pyromorphite associated with galena (Figure 2c). Samples collected from the 2 m high riverbank (DC F) showed higher Pb and Zn concentrations than sample DC E (Table 1). Lead was observed as a primary and galena (Figure 3c) as a secondary mineral, with chemical compositions that were suggestive of anglesite and plumbojarosite phases. The presence of Zn was noted only with electron microprobe analysis, in accordance with the Zn average of the pXRF results, which indicated lower concentrations of Zn in this area (Table 1). Zinc was found in Fe-oxides and minerals that were chemically similar to sulphates (Figure 3c). In the floodplain, samples showed an abundant Fe-hydroxide presence, bearing a variety of metals such as Pb, Zn, and As (Figure 3d). These minerals were often found in a cementing phase around phyllosilicate layers. Zinc, Pb, S, and P were often associated with the phyllosilicate composition (Mg, Fe, Si, Al, and O).

3.3. Zinc and Lead Water Concentrations, Loads, and Suspended Sediments

Aqueous Zn and Pb concentrations and suspended sediment loads bearing Pb and Zn varied along the river system (Table 2 and Figure 4) (see [40] for a full dataset). Filtered and unfiltered Zn concentrations showed the same spatial patterns, with slightly higher concentrations in the latter. During low- and high-streamflow conditions, Zn concentrations increased downstream of the Mill Race and the Wemyss Mine tips (171 m downstream). A second increase was noted downstream of Frongoch Adit (880 m), followed by a decrease to the floodplain (6780 m), where the lowest concentrations were observed. Generally, Zn loads increased along WM and FA (Figure 4). Under low-streamflow conditions, filtered and unfiltered Zn loads decreased at GG and gradually increased at DC and the floodplain (RB). An increase in Zn load was observed at GG and, in contrast with low-flow conditions, Zn loads decreased in the lower part of the floodplain (RB d/s). The lowest load values were recorded at 0 m (T0, 6 mg/s) and the highest at 5930 m (RB u/s, 1488 mg/s) during high-flow conditions. Loads under high-streamflow conditions were approximately seven times higher than low-flow loads. Suspended sediment load values were similar across streamflow conditions (Table 2).

a
O —— 10 µm
Zn —— 10 µm
Fe —— 10 µm
Pb —— 10 µm
CP —— 10 µm
S —— 10 µm
b
Wt % Spectra 1 2 3 4
O 60 51 57 63
Al 0 3 13 0
Si 1 1 16 36
K 0 0 4 0
Fe 36 40 7 0
Zn 0.7 0.3 0.5 0.0
Pb 0.4 1.5 0.6 0.0
50 µm
c
Wt % Spectra 5 6 7 8 9
O 12 7 21 15 45
Al 1 1 12 2 0.4
Si 1 0 10 0.2 50
S 16 24 0 11 0
Fe 0.3 0.3 22 0 0.1
Zn 33 52 0.7 3 0.1
Pb 2.6 6.3 0.0 60 0.0
50 µm
d
O —— 10 µm
Zn —— 10 µm
Pb —— 10 µm
S —— 10 µm
Fe —— 10 µm
CP —— 10 µm

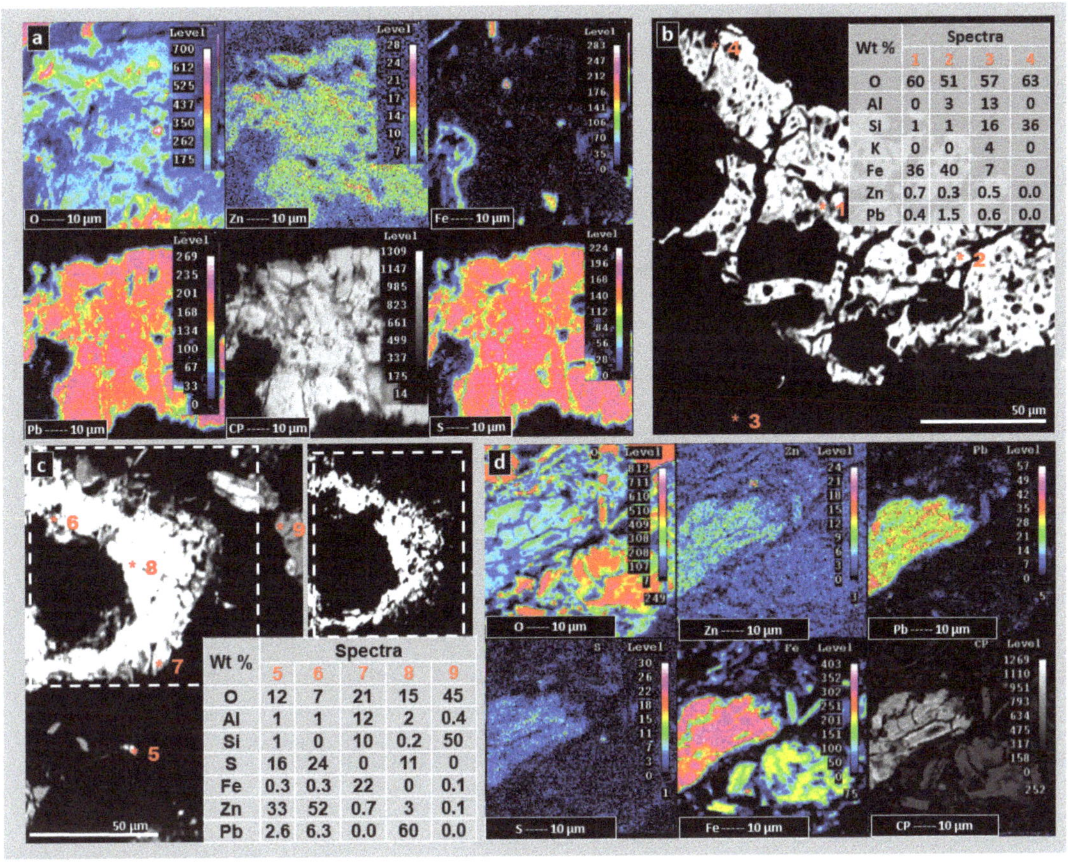

Figure 3. Electronic microprobe images, chemical maps of selected Pb- and Zn-bearing minerals, chemical spectra (position indicated by the asterisk), and backscattered electron images (CP): (**a**) Sample from the headwater mine heaps (WM) showing a galena grain with cubical fractures and weathered edges coated with Fe-oxide. (**b**) Sample from Graig Goch mine area (GG) showing a Fe-hydroxide (spectra 1 and 2) and bearing Pb and Zn illite or muscovite (spectra 3). (**c**) Sample from the middle river length (DC) showing Pb-, Zn-, and S-bearing grains, Pb-bearing Fe-hydroxide, and silicate minerals (likely quartz and illite-like minerals). (**d**) Sample from the upper part of the floodplain (RB), a Pb-bearing Fe-hydroxide.

Filtered and unfiltered Pb concentrations generally had the same spatial pattern, but unfiltered values showed more variability between low- and high-flow conditions (Table 2). The first Pb concentration increase was recorded at 171 m, which was particularly important under high-flow conditions for unfiltered Pb. Across streamflow conditions, unfiltered Pb ranged from 95 µg/L to 248 µg/L at the Graig Goch Mine, whereas filtered Pb had a constant concentration of approximately 50 µg/L. At 3210 m (DC), Pb concentrations varied among streamflows, especially those of unfiltered Pb (Table 2). Although the filtered Pb concentration decreased in the downstream part of the river, unfiltered Pb exhibited a final increase along the floodplain under low- and high-flow conditions. Lead loads showed spatial variability across streamflow conditions (Figure 3). Under high-flow conditions, filtered and unfiltered Pb loads had the highest values at WM and then decreased up to DC. Under low-flow conditions, the Pb load had the highest spike at FA (880 m) and then decreased along GG and DC. Across streamflow conditions, Pb loads increased in the floodplain area, with unfiltered Pb increasing along the entire floodplain stretch and the filtered Pb load, instead, decreasing in the most downstream part. The variations between filtered and unfiltered Pb loads highlighted the spatial variability of the Pb-bearing sediment load along the stream (Table 2 and Figure 3).

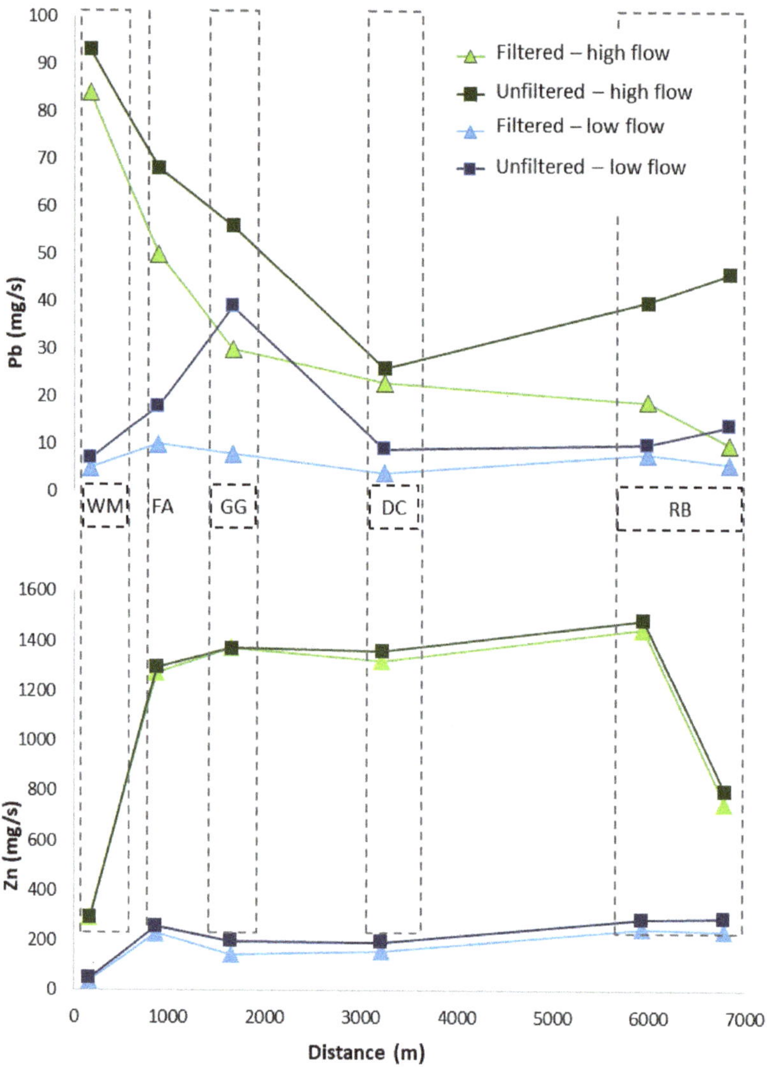

Figure 4. Lead and Zn loads (mg/s) along the stream, with distance (m) from upstream of the mining areas. Loads are reported under high- (green) and low- (blue) streamflow conditions. WM: Wemyss Mine; FA: Frongoch Adit; GG: Graig Goch Mine; DC: depositional area at middle river length; RB: floodplain upstream (up/s) and downstream (d/s) areas.

3.4. Fluvial Geomorphological Controls on Zinc and Lead Geochemistry

The geomorphological features (channels, slopes, meanders, etc.) observed along the stream can be associated with different geochemical settings [47]. Together with hydrological conditions (such as streamflow variations), accumulation, erosion, and sediment permeability can significantly control pH–redox reactions and metal residence times [48]. Examples of geochemical processes that are likely to be controlled by fluvial morphological zones are reported in Figure 2.

In the erosional area (0–180 m), Zn and Pb primary minerals or secondary phases, such as anglesite, can be washed downstream via degradation processes (Figure 2a). Anglesite minerals, being stable in circum-neutral streamwater [9], may contribute to the suspended sediment, particularly during high-flow conditions, as reflected in the unfiltered Pb load increase during high-flow conditions (Table 2). Similarly, Kincey, et al. [11] observed the erosion of the waste from a spoil tip with the exposure of fresh waste to weathering conditions and the transport of metal-enriched sediments.

Sediment deposition was low in the transport stream segment (1110–2620 m), and the gravel and sandy streambed likely promoted oxygenated streamwater circulation (Figure 2b). The manganese hydroxides found in this segment (Table 1) may indicate the mixing of oxygenated stream water with anaerobic water from more reducing groundwater environments [49]. Ground- and hyporheic-water-promoting Mn-hydroxide formations may enhance Pb attenuation processes [40,50]. The filtered Pb load decreased around GG across low- and high-flow conditions (Figure 4), and the suspended load was among the highest calculated values (Table 2), indicating processes promoting Pb(aq) uptake and the mobilisation of sediment-borne Pb. Aqueous Pb can be released from Mn-Pb oxides if redox conditions change [31], for example, under prolonged flooding conditions or sediment deposition, decreased water circulation, and the promotion of anaerobic conditions.

Along the depositional area at 3000–4070 m, the narrow valley and the bedrock-controlled river channel were likely trapping metal-bearing sediments mixed with uncontaminated sediments [6,10]. Figure 4 shows a decrease in Pb loads under both high- and low-streamflow conditions, indicating the deposition of Pb-bearing suspended sediments and the attenuation of aqueous Pb. Lead-bearing sulphate and Fe hydroxide were found in the vertical accretion structures, along with sphalerite grains and Zn associated with Fe, P and S, which were interpreted as plumbojarosite minerals (Figure 3c). Herein, prolonged flooding conditions can generate reducing conditions promoting Fe oxyhydroxide dissolution and plumbojarosite stability [51]. The subsequent drying of the sediments can oxidise sulphide phases (galena and sphalerite) and release associated metals [31], easily reaching the streamwater through the sandy layers or forming secondary Zn sulphate minerals and anglesite [14].

Furthermore, in the sandy pockets that are characteristic of the low riverbanks, the presence of Pb bearing Mn oxides may indicate small areas of water interaction between oxygenated streamwater and sub-surface water (Figure 2c). Secondary Zn sulphate minerals with a high solubility [29] may release Zn in an aqueous form when this sediment is washed out during moderate- to high-streamflow conditions.

In the floodplain (5800–6740 m), the low stream gradient (Figure 1b) likely promotes slow water movement and enhances aqueous Pb and Zn uptake onto mineral surfaces and the deposition of metal-bearing sediments (Figure 4). Lead-bearing phyllosilicates (illite and chlorite) can be easily dispersed along the stream depending on the streamflow conditions, especially under high flow, contributing to the suspended sediment loads (Figure 4); on the contrary, Zn dispersion under high-flow conditions was not observed. Generally, the deposition of metal-bearing sediments as lateral vertical aggradations (Figure 2d) occurring under moderate- to high-streamflow conditions can result in the deposition and prolonged storage of Pb and Zn, which is typical of floodplains [7].

3.5. Implications for Sediment Quality Management

Metal contamination monitoring largely focuses on catchment headwaters wherein mine workings are often located. However, there is ample evidence that centuries of mining activity have produced tonnes of contaminated waste, which high-gradient rivers have remobilised along the rivers [6]. Sediment at the headwater of the Wemyss Mine can release

Zn and Pb from sulphide minerals both as aqueous and solid phases (Table 2 and Figure 4). Metals can be physically (deposited) or chemically (precipitated or adsorbed) trapped along the stream, as with the Graig Goch Mine, or sedimented further downstream where both primary and secondary metal-enriched minerals are observed. Lead dispersion along the catchment seems to be influenced by the fluvial morphological parameters recognised along the classified areas. At a circum-neutral pH, Pb forms more stable secondary minerals, such as anglesite and Pb-Mn oxides. This water–soil interaction controls filtered and unfiltered Pb phase mobility, explaining the Pb-bearing suspended sediment load increase under high-flow conditions and the persistent Pb concentration in downstream sediments. A similar Pb behaviour controlled by the topography was observed in the Le'an River floodplain (Jiangxi Province, China) [52]. Zhu, et al. [37] explained that slope geomorphologically and hydrologically controls sediment movement, to which the Pb fate appears to be linked. On the other hand, the downstream decrease in sediment Zn concentrations is potentially due to the instability of Zn sulphate minerals and the high mobility of Zn in circum-neutral river systems [29,30]. In this study, Zn-sulphates, Zn-bearing Fe-oxides, and phyllosilicate were observed in the stream sediments, although the small differences between filtered and unfiltered Zn loads (Figure 4) indicated that Zn was mostly mobilised as aqueous phases and that the Zn-bearing phases had a high solubility.

The primary and secondary minerals observed along the studied stream represent a storage system of trace metals in solid phases. Although sediment metal concentrations tended to decrease downstream from the former mine, the estimated metal load variations demonstrated that secondary sources in the middle stream length and floodplain can still contribute to the metal load, especially under high-flow conditions. The metals contained in downstream sinks are potentially bioavailable. Byrne, et al. [46], using sequential extractions, found increases in readily extractable Zn and Cd downstream from the Dylife Mine, UK. Chemical metal release or up-take from metal-bearing mineral phases may occur due to changes in pH or redox settings [28,53]; this can occur due to streamflow variations [40], the mixing of water, or the soil formation [54]. Therefore, geochemical and mineralogical investigations of metal-bearing phases are fundamental to anticipating their potential metal release and informing remediation strategies.

Procedures for assessing sediment quality may benefit from the evidence provided herein, where both geomorphological and sedimentary geochemical studies highlighted the metal distribution and suggested the possible mechanisms of metal-enriched sediment mobilisation or aqueous metal release. The geomorphological approach can help identify areas of mine waste exposure and deposition. Fluvial geomorphological description can also identify potential secondary metal sources throughout a river, leading to the first steps in mining-impacted river monitoring. This approach can provide guidance on metal contaminant values for restoration and land use purposes [8]. For instance, despite evidence for the presence of Pb- and Zn-contaminated sediments, the areas are used extensively for grazing (1110–2620 m and 5800–6740 m) and recreational purposes (3000–4070 m). This highlighted the need to consider the ecotoxicity Pb and Zn can pose in the environment along the whole stream, where different lands can be dedicated to different uses (farming, recreational, touristic attractions, access to public paths, etc.), increasing the potential exposure of receptors to metal intake. This approach can also be applied to mining incidents such as dam failures to identify areas of sediment deposition and appraise potential metal behaviours across streamflow conditions.

4. Conclusions

The catchment-scale investigation of metal-bearing sediment distribution aids in understanding the long-term environmental impacts of metal mineral extraction on river systems. Previous studies have evaluated the riverine impact of metal mining waste by estimating metal distribution through either geomorphological or geochemical observations. This novel study combined both aspects with metal stream load estimations to elucidate the geochemical and metal release processes operating in different geomorphological settings.

The semi-quantitative approach that was used to investigate the fluvial geomorphology (topography, slope, grain size distribution, and landscape observations) allowed the identification of fluvial geomorphological zones (erosional, transport, and depositional areas) where metals were released and trapped across different streamflow conditions. Overall, Pb and Zn sediment concentrations were higher than guideline values, and a variety of metal-bearing minerals were observed along the river: (i) Pb and Zn sulphides, anglesite, and Fe-oxide dominated the erosional mining areas; (ii) hyporheic areas found in sediment transport or depositional areas encouraged Pb-bearing Mn oxide formation; and (iii) Pb- and Zn-bearing phyllosilicate and Fe-oxides characterised downstream depositional areas.

Assessing and monitoring sediments in impacted catchments is often complex, costly, and time-consuming. Fluvial morphological parameters provide low-cost monitoring to plan a sampling strategy whereby alluvial or bedrock-controlled depositional areas are targeted. Fluvial depositional parameters coupled with geochemical and mineralogical investigations can provide information on possible Zn and Pb release patterns.

Author Contributions: P.O., P.B. and K.A.H.-E. designed this study. P.O. wrote the manuscript. P.O., P.B., K.A.H.-E., C.O.H. and T.S. assisted with the fieldwork. All authors contributed to and reviewed the manuscript. All authors have read and agreed to the published version of the manuscript.

Funding: John Moores University funded P.O.'s PhD scholarship (2015–2018).

Data Availability Statement: The data are available upon request. Please contact the corresponding author.

Acknowledgments: The authors gratefully thank Liverpool John Moores University for funding P.O.'s PhD scholarship. We gratefully acknowledge Andy Beard for their assistance with the microprobe analysis, Gary Tarbuck for providing the chemical analysis, Dave Williams and Hazel Clarke for their assistance with the fieldwork equipment, Nicola Dempster and Rob Allen for the XRD analysis, Natasha Donn-Arnold for her help with sampling and the pXRF analysis, Paul Gibbson for their assistance with the SEM analysis, and Timothy P. Lane for their help with the ArcGIS map preparation.

Conflicts of Interest: The authors declare no conflict of interest.

References

1. Kossoff, D.; Dubbin, W.E.; Alfredsson, M.; Edwards, S.J.; Macklin, M.G.; Hudson-Edwards, K.A. Mine tailings dams: Characteristics, failure, environmental impacts, and remediation. *Appl. Geochem.* **2014**, *51*, 229–245. [CrossRef]

2. Clement, A.J.H.; Nováková, T.; Hudson-Edwards, K.A.; Fuller, I.C.; Macklin, M.G.; Fox, E.G.; Zapico, I. The environmental and geomorphological impacts of historical gold mining in the Ohinemuri and Waihou river catchments, Coromandel, New Zealand. *Geomorphology* **2017**, *295*, 159–175. [CrossRef]

3. Hudson-Edwards, K.A.; Jamieson, H.E.; Lottermoser, B.G. Mine Wastes: Past, Present, Future. *Elements* **2011**, *7*, 375–380. [CrossRef]

4. Elznicová, J.; Matys Grygar, T.; Popelka, J.; Sikora, M.; Novák, P.; Hošek, M. Threat of Pollution Hotspots Reworking in River Systems: Case Study of the Ploučnice River (Czech Republic). *ISPRS Int. J. Geo-Inf.* **2019**, *8*, 37. [CrossRef]

5. Lewin, J.; Macklin, M.G. Metal mining and floodplain sedimentation in Britain. In *International Geography*; Gardiner, V., Ed.; John Wiley & Sons Ltd: Hoboken, NJ, USA, 1987.

6. Miller, J.R. The role of fluvial geomorphic processes in the dispersal of heavy metals from mine sites. *J. Geochem. Explor.* **1997**, *58*, 101–118. [CrossRef]

7. Dennis, I.A.; Coulthard, T.J.; Brewer, P.; Macklin, M.G. The role of floodplains in attenuating contaminated sediment fluxes in formerly mined drainage basins. *Earth Surf. Process. Landf.* **2009**, *34*, 453–466. [CrossRef]

8. Fornasaro, S.; Morelli, G.; Rimondi, V.; Fagotti, C.; Friani, R.; Lattanzi, P.; Costagliola, P. A GIS-based map of the Hg-impacted area in the Paglia River basin (Monte Amiata Mining District—Italy): An operational instrument for environmental management. *J. Geochem. Explor.* **2022**, *242*, 107074. [CrossRef]

9. Hudson-Edwards, K.A.; Macklin, M.G.; Curtis, C.D.; Vaughan, D.J. Processes of Formation and distribution of Pb-, Zn-, Cd-, and Cu-bearing Minerals in the Tyne Basin, Northeast England: Implication for Metal-Contaminated River Systems. *Environ. Sci. Technol.* **1996**, *30*, 72–80. [CrossRef]

10. Hudson-Edwards, K.A.; Macklin, M.G.; Miller, J.R.; Lechler, P.J. Sources, distribution and storage of heavy metals in the Rio Pilcomayo, Bolivia. *J. Geochem. Explor.* **2001**, *72*, 229–250. [CrossRef]

11. Kincey, M.; Warburton, J.; Brewer, P.A. Contaminated sediment flux from eroding abandoned historical metal mines: Spatial and temporal variability in geomorphological drivers. *Geomorphology* **2018**, *319*, 199–215. [CrossRef]

12. Foulds, S.A.; Brewer, P.A.; Macklin, M.G.; Haresign, W.; Betson, R.E.; Rassner, S.M. Flood-related contamination in catchments affected by historical metal mining: An unexpected and emerging hazard of climate change. *Sci. Total Environ.* **2014**, *476*, 165–180. [CrossRef] [PubMed]

13. Nordstrom, D.K. Hydrogeochemical processes governing the origin, transport and fate of major and trace elements from mine wastes and mineralized rock to surface waters. *Appl. Geochem.* **2011**, *26*, 1777–1791. [CrossRef]

14. Lynch, S.; Batty, L.; Byrne, P. Environmental Risk of Metal Mining Contaminated River Bank Sediment at Redox-Transitional Zones. *Minerals* **2014**, *4*, 52–73. [CrossRef]

15. Palumbo-Roe, B.; Wragg, J.; Banks, V.J. Lead mobilisation in the hyporheic zone and river bank sediments of a contaminated stream: Contribution to diffuse pollution. *J. Soils Sediments* **2012**, *12*, 1633–1640. [CrossRef]

16. Macklin, M.G.; Brewer, P.A.; Hudson-Edwards, K.A.; Bird, G.; Coulthard, T.J.; Dennis, I.A.; Lechler, P.J.; Miller, J.R.; Turner, J.N. A geomorphological approach to the management of rivers contaminated by metal mining. *Geomorphology* **2006**, *79*, 423–447. [CrossRef]

17. Sracek, O.; Kříbek, B.; Mihaljevič, M.; Majer, V.; Veselovský, F.; Vencelides, Z.; Nyambe, I. Mining-related contamination of surface water and sediments of the Kafue River drainage system in the Copperbelt district, Zambia: An example of a high neutralization capacity system. *J. Geochem. Explor.* **2012**, *112*, 174–188. [CrossRef]

18. Andrew Marcus, W. Copper dispersion in ephemeral stream sediments. *Earth Surf. Process. Landf.* **1987**, *12*, 217–228. [CrossRef]

19. Obiora, S.C.; Chukwu, A.; Chibuike, G.; Nwegbu, A.N. Potentially harmful elements and their health implications in cultivable soils and food crops around lead-zinc mines in Ishiagu, Southeastern Nigeria. *J. Geochem. Explor.* **2019**, *204*, 289–296. [CrossRef]

20. Abdel-Tawwab, M. Effect of feed availability on susceptibility of Nile tilapia, *Oreochromis niloticus* (L.) to environmental zinc toxicity: Growth performance, biochemical response, and zinc bioaccumulation. *Aquaculture* **2016**, *464*, 309–315. [CrossRef]

21. Zhang, X.; Yang, L.; Li, Y.; Li, H.; Wang, W.; Ye, B. Impacts of lead/zinc mining and smelting on the environment and human health in China. *Environ. Monit. Assess* **2012**, *184*, 2261–2273. [CrossRef]

22. Smith, K.M.; Abrahams, P.W.; Dagleish, M.P.; Steigmajer, J. The intake of lead and associated metals by sheep grazing mining-contaminated floodplain pastures in mid-Wales, UK: I. Soil ingestion, soil-metal partitioning and potential availability to pasture herbage and livestock. *Sci. Total Environ.* **2009**, *407*, 3731–3739. [CrossRef]

23. Aghili, S.; Vaezihir, A.; Hosseinzadeh, M. Distribution and modeling of heavy metal pollution in the sediment and water mediums of Pakhir River, at the downstream of Sungun mine tailing dump, Iran. *Environ. Earth Sci.* **2018**, *77*, 128. [CrossRef]

24. Cook, N.; Ehrig, K.; Rollog, M.; Ciobanu, C.; Lane, D.; Schmandt, D.; Owen, N.; Hamilton, T.; Grano, S. 210Pb and 210Po in Geological and Related Anthropogenic Materials: Implications for Their Mineralogical Distribution in Base Metal Ores. *Minerals* **2018**, *8*, 211. [CrossRef]

25. De Giudici, G.; Medas, D.; Cidu, R.; Lattanzi, P.; Podda, F.; Frau, F.; Rigonat, N.; Pusceddu, C.; Da Pelo, S.; Onnis, P.; et al. Application of hydrologic-tracer techniques to the Casargiu adit and Rio Irvi (SW-Sardinia, Italy): Using enhanced natural attenuation to reduce extreme metal loads. *Appl. Geochem.* **2018**, *96*, 42–54. [CrossRef]

26. Elmayel, I.; Esbri, J.M.; Efren, G.O.; Garcia-Noguero, E.M.; Elouear, Z.; Jalel, B.; Farieri, A.; Roqueni, N.; Cienfuegos, P.; Higueras, P. Evolution of the Speciation and Mobility of Pb, Zn and Cd in Relation to Transport Processes in a Mining Environment. *Int. J. Environ. Res. Public Health* **2020**, *17*, 4912. [CrossRef]

27. Pavlowsky, R.T.; Lecce, S.A.; Owen, M.R.; Martin, D.J. Legacy sediment, lead, and zinc storage in channel and floodplain deposits of the Big River, Old Lead Belt Mining District, Missouri, USA. *Geomorphology* **2017**, *299*, 54–75. [CrossRef]

28. Jarvis, A.P.; Davis, J.E.; Orme, P.H.A.; Potter, H.A.B.; Gandy, C.J. Predicting the Benefits of Mine Water Treatment under Varying Hydrological Conditions using a Synoptic Mass Balance Approach. *Environ. Sci Technol* **2019**, *53*, 702–709. [CrossRef] [PubMed]

29. Lee, G.; Bigham, J.M.; Faurea, G. Removal of trace metals by coprecipitation with Fe, Al and Mn from natural waters contaminated with acid mine drainage in the Ducktown Mining District. *Appl. Geochem.* **2002**, *17*, 569–581. [CrossRef]

30. Smith, K.S.; DeGraff, J.V. Strategies to predict metal mobility in surficial mining environments. In *Understanding and Responding to Hazardous Substances at Mine Sites in the Western United States*; Geological Society of America: Boulder, CL, USA, 2007; Volume 17.

31. Lynch, S.F.L.; Batty, L.C.; Byrne, P. Environmental risk of severely Pb-contaminated riverbank sediment as a consequence of hydrometeorological perturbation. *Sci. Total Environ.* **2018**, *636*, 1428–1441. [CrossRef]

32. Natural Resource of Wales. Abandoned Mine Case Study, Wemyss Lead and Zinc Mine. 2016. Available online: https://naturalresources.wales/media/679806/wemyss-mine-case-study_2016_06.pdf (accessed on 27 January 2023).

33. Dyfed, A.T. *Metal Mines Remediation Project Part 3: Wemyss Archaeological Assessment*; Report No 2016/05–Part 3; DAT Archaeological Services: Llandeilo, UK, 2016; pp. 1–51.

34. Sheppard, T.H. *The Geology of the Area around Lampeter, Llangybi and Llanfair Clydogau*; Integrated Geoscience Surveys South Programme, Internal Report IR/04/064; British Geological Survey: Nottingham, UK, 2004. Available online: http://nora.nerc.ac.uk/id/eprint/509306/1/IR04064.pdf (accessed on 27 January 2023).

35. Palumbo-Roe, B.; Dearden, R. The hyporheic zone composition of a mining-impacted stream: Evidence by multilevel sampling and DGT measurements. *Appl. Geochem.* **2013**, *33*, 330–345. [CrossRef]

36. Watts, G.; Battarbee, R.W.; Bloomfield, J.P.; Crossman, J.; Daccache, A.; Durance, I.; Elliott, J.A.; Garner, G.; Hannaford, J.; Hannah, D.M.; et al. Climate change and water in the UK—Past changes and future prospects. *Prog. Phys. Geogr. Earth Environ.* **2015**, *39*, 6–28. [CrossRef]
37. Zhu, Y.-M.; Lu, X.X.; Zhou, Y. Suspended sediment flux modeling with artificial neural network: An example of the Longchuanjiang River in the Upper Yangtze Catchment, China. *Geomorphology* **2007**, *84*, 111–125. [CrossRef]
38. Wentworth, C.K. A Scale of Grade and Class Terms for Clastic Sediments. *J. Geol.* **1922**, *30*, 377–392. [CrossRef]
39. Radu, T.; Diamond, D. Comparison of soil pollution concentrations determined using AAS and portable XRF techniques. *J. Hazard. Mater.* **2009**, *171*, 1168–1171. [CrossRef] [PubMed]
40. Onnis, P.; Byrne, P.; Hudson-Edwards, K.A.; Frau, I.; Stott, T.; Williams, T.; Edwards, P.; Hunt, C.O. Source apportionment of mine contamination across streamflows. *Appl. Geochem.* **2023**, *151*, 105623. [CrossRef]
41. *INCA Energy Applications Training Notes*; Oxford Instrument: Abingdon-on-Thames, UK, 2008.
42. Davis, A.; Drexler, J.W.; Ruby, M.V.; Nicholson, A. Micromineralogy of mine wastes in relation to lead bioavailability, Butte, Montana. *Environ. Sci. Technol.* **1993**, *27*, 1415–1425. [CrossRef]
43. Richardson, M.; Moore, R.D.; Zimmermann, A. Variability of tracer breakthrough curves in mountain streams: Implications for streamflow measurement by slug injection. *Can. Water Resour. J. Rev. Can. Des. Ressour. Hydr.* **2017**, *42*, 21–37. [CrossRef]
44. Mayes, W.M.; Gozzard, E.; Potter, H.A.; Jarvis, A.P. Quantifying the importance of diffuse minewater pollution in a historically heavily coal mined catchment. *Environ. Pollut.* **2008**, *151*, 165–175. [CrossRef]
45. Environment Agency. *Assessment of Metal Mining-Contaminated River Sediments in England and Wales*; Environment Agency: Bristol, UK, 2008.
46. Byrne, P.; Reid, I.; Wood, P.J. Sediment geochemistry of streams draining abandoned lead/zinc mines in central Wales: The Afon Twymyn. *J. Soils Sediments* **2010**, *10*, 683–697. [CrossRef]
47. Ciszewski, D.; Grygar, T.M. A Review of Flood-Related Storage and Remobilization of Heavy Metal Pollutants in River Systems. *Water Air Soil Pollut.* **2016**, *227*, 239. [CrossRef]
48. Hudson-Edwards, K.A. Sources, mineralogy, chemistry and fate of heavy metal-bearing particles in mining-affected river systems. *Mineral. Mag.* **2003**, *67*, 205–217. [CrossRef]
49. Hudson-Edwards, K.A.; Macklin, M.G.; Curtis, C.D.; Vaughan, D.J. Chemical remobilization of contaminant metals within floodplain sediments in an incising river system: Implications for dating and chemostratigraphy. *Earth Surf. Process. Landf.* **1998**, *23*, 671–684. [CrossRef]
50. Harvey, J.W.; Fuller, C.C. Effect of enhanced manganese oxidation in the hyporheic zone on basin-scale geochemical mass balance. *Water Resour. Res.* **1998**, *34*, 623–636. [CrossRef]
51. Fuller, C.C.; Harvey, J.W. Reactive Uptake of Trace Metals in the Hyporheic Zone of a Mining-Contaminated Stream, Pinal Creek, Arizona. *Environ. Sci. Technol.* **2000**, *34*, 1150–1155. [CrossRef]
52. Forray, F.L.; Smith, A.M.L.; Drouet, C.; Navrotsky, A.; Wright, K.; Hudson-Edwards, K.A.; Dubbin, W.E. Synthesis, characterization and thermochemistry of a Pb-jarosite. *Geochim. Cosmochim. Acta* **2010**, *74*, 215–224. [CrossRef]
53. Chen, Y.; Liu, Y.; Liu, Y.; Lin, A.; Kong, X.; Liu, D.; Li, X.; Zhang, Y.; Gao, Y.; Wang, D. Mapping of Cu and Pb contaminations in soil using combined geochemistry, topography, and remote sensing: A case study in the Le'an River floodplain, China. *Int. J. Environ. Res. Public Health* **2012**, *9*, 1874–1886. [CrossRef]
54. Wilkin, R.T. Contaminant Attenuation Processes at Mine Sites. *Mine Water Environ.* **2008**, *27*, 251–258. [CrossRef]

 *minerals*

Article

Rapid Removal of Cr(VI) from Aqueous Solution Using Polycationic/Di-Metallic Adsorbent Synthesized Using Fe^{3+}/Al^{3+} Recovered from Real Acid Mine Drainage

Khathutshelo Lilith Muedi [1], Vhahangwele Masindi [2,3], Johannes Philippus Maree [4] and Hendrik Gideon Brink [1,*]

[1] Department of Chemical Engineering, Faculty of Engineering, Built Environment and Information Technology, University of Pretoria, Pretoria 0028, South Africa
[2] Magalies Water, Scientific Services, Research & Development Division, Erf 3475, Stoffberg Street, Brits 0250, South Africa
[3] Department of Environmental Sciences, School of Agriculture and Environmental Sciences, University of South Africa (UNISA), P.O. Box 392, Florida 1710, South Africa
[4] Department of Water and Sanitation, University of Limpopo, University Street, Polokwane 0727, South Africa
* Correspondence: deon.brink@up.ac.za; Tel.: +27-84-206-8338

Abstract: The mining of valuable minerals from wastewater streams is attractive as it promotes a circular economy, wastewater beneficiation, and valorisation. To this end, the current study evaluated the rapid removal of aqueous Cr(VI) by polycationic/di-metallic Fe/Al (PDFe/Al) adsorbent recovered from real acid mine drainage (AMD). Optimal conditions for Cr(VI) removal were 50 mg/L initial Cr(VI), 3 g PDFe/Al, initial pH = 3, 180 min equilibration time and temperature = 45 °C. Optimal conditions resulted in $\geq$95% removal of Cr(VI), and a maximum adsorption capacity of Q = 6.90 mg/g. Adsorption kinetics followed a two-phase pseudo-first-order behaviour, i.e., a fast initial Cr(VI) removal (likely due to fast initial adsorption) followed by a slower secondary Cr(VI) removal (likely from Cr(VI) to Cr(III) reduction on the surface). More than 90% of adsorbed Cr(VI) could be recovered after five adsorption–desorption cycles. A reaction mechanism involving a rapid adsorption onto at least two distinct surfaces followed by slower in situ Cr(VI) reduction, as well as adsorption-induced internal surface strains and consequent internal surface area magnification, was proposed. This study demonstrated a rapid, effective, and economical application of PDFe/Al recovered from *bona fide* AMD to treat Cr(VI)-contaminated wastewater.

Keywords: chromium removal; acid mine drainage; wastewater streams; circular economy; wastewater beneficiation; polycationic/di-metallic adsorbent (nanocomposite)

check for
updates

Citation: Muedi, K.L.; Masindi, V.; Maree, J.P.; Brink, H.G. Rapid Removal of Cr(VI) from Aqueous Solution Using Polycationic/ Di-Metallic Adsorbent Synthesized Using Fe^{3+}/Al^{3+} Recovered from Real Acid Mine Drainage. *Minerals* **2022**, *12*, 1318. https://doi.org/ 10.3390/min12101318

Academic Editor: María Ángeles Martín-Lara

Received: 13 September 2022
Accepted: 13 October 2022
Published: 19 October 2022

Publisher's Note: MDPI stays neutral with regard to jurisdictional claims in published maps and institutional affiliations.

1. Introduction

Despite the socio-economic benefits provided by the gold and coal industries, the mining of these commodities is notorious for inducing significant environmental degradation [1,2]. Mining entails the excavation of large quantities of rock to obtain the targeted minerals or minerals. During the extraction of mineral resources, hard rock (such as gold) or soft rock (such as coal) is exposed to air which typically leads to the oxidation of associated minerals. Generally, coal and gold are associated with sulphide-bearing minerals, e.g., FeS, FeAsS, ZnS, CuS, and NiS amongst others, resulting in the production of metalliferous acidic drainage rich in sulphates [3,4]. Due to the acidic nature of this wastewater stream, minerals in the surrounding geology leach into the surrounding body of water hence increasing the electrical conductivity and total dissolved solids. Acid mine drainage (AMD) or acid rock drainage (ARD) generally contains significant quantities of Al, Fe, Mn, and sulphate as major contaminants. However, the presence of minor, but not insignificant, levels of toxic and hazardous heavy metals, radionuclides, metalloids, oxyanions, and rare earth metals has been reported [4–8]. As such, the AMD matrix needs to be contained and treated prior to discharge to the receiving environments [9,10].

Research by the authors of the study demonstrated the valorisation of acid mine drainage for the recovery of Al/Fe tri-valent metals through the synthesis of an adsorbent that was tested for the removal of arsenic [1] and Congo Red [2] from wastewater. The synthesized adsorbent achieved remarkably high adsorption capacities for both As (102–129 mg·g^{-1}) and Congo Red (411 mg·g^{-1}), illustrating the potential of the recovered adsorbent for the effective treatment of hazardous pollutants from solution.

Chromium is used as an additive in myriad production processes; chromium is a crucial component of chemical production, metallurgical industries, refractories, and foundries [11]. A report by the US Geological Survey indicated that South Africa and Kazakhstan are the world's largest producers of chromium, with approximately 95 percent of global chromium reserves [12]. This toxic and hazardous chemical species emanates from natural (geogenic) and man-made (anthropogenic) sources. Natural sources include rocks, volcanic eruptions, and minerals, while anthropogenic sources comprise mining, tanning, and manufacturing of paints, plastics, ceramics, glass, salts, dyes, and dietary supplements. These sources release chromium rich effluents to different spheres of the environment [13–15].

(Eco)-toxicological studies highlighted that the intake of water with elevated levels of chromium ions, specifically the hexavalent chromium, could lead to direct detrimental impacts on human health through bioaccumulation, animals, aquatic organisms, and the environment at large [16]. According to toxicological study, the maximum allowed limit in drinking water should be ≤ 0.05 ppm [17]. In response to this stringent regulatory framework and standard, various studies devised ways to treat, remove, and recover chromium ions from the water and wastewater matrices. These include electrocoagulation [18], nano-filtration [19], freeze desalination [20], ion exchange [21], photo-catalysis [22], adsorption, precipitation, bio-(phyto)-remediation, and crystallization [11,13,14,23–25]. However, these technologies still pose challenges regarding operational cost and disposal of secondary sludge which reduces their desirability. Consequently, adsorption has emerged as a promising technology due to its effectiveness, affordable costs, and reliance on locally available materials [1,9]. However, the dependence of traditional adsorbent manufacturing on virgin and pristine materials poses notable risks to the environment; hence, there is a demand to find alternative sources of materials to produce adsorbents.

The use of Fe and Al adsorbents has been demonstrated successfully for the treatment of aqueous Cr(VI) [26–30], and dominant mechanisms reported involved ion exchange [26,27,30], surface reduction of Cr(VI) to Cr(III) reduction [28], and electrostatic forces [29,30].

The current study will be the first known to explore the feasibility of using Polycationic/dimetal Al^{3+} and Fe^{3+}-derived adsorbent (PDFe/Al) synthesized using real AMD for the removal of hexavalent chromium from aqueous solution. Chiefly, wastewater treatment technologies need to be feasible, effective, and affordable. Therefore, this study introduces an affordable and effective technique to treat chromium-contaminated water using a material recovered from acid mine drainage—a well-known environmental hazard—thereby providing a potential solution for both waste streams.

2. Materials and Methods

2.1. Sample Collection and Preparation of Working Solution

Crude acid mine effluent (AMD) was collected from a coal mine in Mpumalanga, South Africa. Potassium dichromate ($K_2Cr_2O_7$) was purchased from Sigma Aldrich. Caustic soda (NaOH), sulfuric acid (98.5% H_2SO_4) and hydrochloric acid (37% HCl) were purchased from Merck. All chemicals were used as received, with no further purification. Ultrapure water (18.2 MΩ-cm) was used for the synthesis of all solutions. For the preparation of hexavalent chromium (Cr(VI)) stock solution, $K_2Cr_2O_7$ salt was used. To avoid contamination, experimental glassware was meticulously cleaned before and after each use. To simulate the working solution, 1000 mg·L^{-1} Cr(VI) stock solution was prepared by dissolving 2.835 g of

$K_2Cr_2O_7$ salt in ultrapure water (18.2 MΩ-cm) in a 1000 mL volumetric flask. The container was then filled to the specified level (mark). Thereafter, serial dilutions were made from the prepared working solution.

2.2. Synthesis of PDFe/Al from Authentic AMD

The PDFe/Al was synthesized using the method previous described [1,2]. A known AMD volume was reacted with 30% NaOH at a pH = 4.5 to selectively precipitate trivalent Fe and Al as -OOH compounds and the mixture was stirred for 30 min at room temperature with an overhead stirrer. Subsequently, the mixture was heated to 100 °C while stirring. The precipitates were vacuum filtered (Whatman® Grade 40 ash-less) and dried. The recovered material was vibratory-ball-milled at 700 rpm and calcined at 800 °C. The milled samples were sieved to a maximum size of 32 µm and stored in a plastic "zip-lock" bag until use.

2.3. Optimisation Studies

The synthesised Fe/Al di-metal composite was then used for the removal of Cr(VI) from aqueous solutions. Optimised parameters include initial Cr(VI) concentration (mg/L), adsorbent dose (g), agitation time (g), temperature (°C), and initial solution pH. In all experiments, 250 mL of Cr(VI) rich solutions was added to 500 mL volumetric flasks. All experiments were performed in triplicate for quality control and quality assurance, and the data were reported as mean ± standard deviations. The impacts of various parameters on the adsorption process were investigated, and the results are summarized in Table 1.

Table 1. Parameters tested for Cr(VI) adsorption using PDFe/Al.

Experiment No	Initial Cr(VI) Concentration (mg/L)	Initial pH	Adsorbent Dose (g)	Agitation Time (min)	Temperature (°C)
1	1; 5; 10; 20; 30; 40; 50; 100; 150; 200	4–5	1	180	25
2	10	2, 3, 4, 5, 6, 7, 8, 9, 10 (±0.2)	1	180	25
3	10	4–5	0.1; 0.5; 1; 2; 3; 4; 5 (±0.0005)	180	25
4	10	4–5	1	10; 30; 60; 90; 120; 180; 240; 300	25
5	10	4–5	1	180	25; 35; 45; 55; 65

The various initial concentration were obtained by diluting a concentrated stock solution of 1000 mg·L^{-1} Cr(VI) as shown in Table 1. The effect of agitation time was determined by measuring the Cr(VI) concentrations at various time. The influence of initial solution pH was investigated by changing the pH with 0.1 M NaOH/0.1 HNO$_3$. Batch adsorption studies were carried out in a thermal shaker/incubator at various temperatures.

2.4. Characterisation of the Feedstock and Product Minerals

Inductively Coupled Plasma Mass Spectrometry (ICP-MS) was used to assess the fate of Cr(VI) in aqueous samples (7500ce, Agilent, Alpharetta, GA, USA).

The PDFe/Al and Cr-residue products were analysed as described previously [1,2] using X-ray diffraction (XRD: Panalytical X'PertPRO (Malvern Panalytical, Malvern, UK) equipped with a Cu-K radiation source), Htable igh-resolution scanning electron micrographs (HR-SEM-EDX) were used to examine the surface morphology and composition of solid materials (CarlZeiss Sigma VP FE-SEM with Oxford EDX Sputtering System, Carl Zeiss AG, Oberkochen, Germany). Fourier Transform Infrared Spectrometer—Attenuated Total Reflectance (FTIR-ATR: Perkin-Elmer Spectrum 100), BET (Micromeritics Tri-Star II 3020, Surface area and porosity, Poretech CC, Jefferson, TX, USA), and Thermo Gravimetric Analyzer (TGA: SelectScience TGA Q500, TA instruments, Bath, UK).

2.5. Point of Zero Charge (PZC)

The point of zero charge (PZC) of the PDFe/Al was determined using a method proposed by Smičiklas et al. (2000) [31]. 0.1 g of the PDFe/Al were added to each of nine flasks containing 50 mL 0.1 M KNO_3 solutions adjusted to pH 2–10 (using 0.1 M HNO_3 and/or NaOH) and left to equilibrate for 24 h at $\pm$ 25 °C. Subsequently, the suspensions were filtered, and the pH of the filtrates determined.

2.6. PDFe/Al Adsorption Capacity and Removal Efficiency

2.6.1. Adsorption Capacity

The adsorption capacities of PDFe/Al were determined using Equation (1) [32]:

$$Q = \frac{(C_0 - C) \times V}{m} \tag{1}$$

where Q (mg·g^{-1}) is the adsorption capacity; C_0 (mg·L^{-1}) is the initial concentration of Cr(VI), C (mg·L^{-1}) is the measured concentration of Cr(VI), respectively; V (L) is the volume of the Cr(VI) solution; and m (g) is the dosage of PDFe/Al.

2.6.2. Percentage Removal

The removal efficiency of Cr(VI) by PDFe/Al was determined using Equation (2):

$$\%R_e = \frac{C_0 - C_e}{C_0} \times 100 \tag{2}$$

where $\%R_e$ is the removal efficiency of the PDFe/Al; C_0 is the initial Cr(VI) concentration (mg/L); C_e is the equilibrium Cr(VI) concentration (mg·L^{-1}).

2.7. Regeneration Study

Kumari et al. (2006) [32] described a method for studying the regeneration of PDFe/Als. 250 mL of a 150 mg·L^{-1} Cr(VI) solution was treated with 1 g of PDFe/Al for 90 min in a batch experiment. The adsobent was centrifuged to separate it from the supernatant and the recovered and washed five times with 250 mL ultra-pure water (to remove residual Cr(VI)) and dried. 250 mL of 0.1 M HNO_3 was added to the dried sample at room temperature, the HNO_3 extract was collected and tested for Cr(VI) ions. Equation (3) was used to calculate the regeneration percentage:

$$\%\text{Desorption} = \frac{C_{des}}{C_o} \times 100 \tag{3}$$

where C_{des} (mg·L^{-1}) is the concentration of Cr(VI) ions in the desorption eluent; C_o (mg·L^{-1}) is the initial concentration of Cr(VI) ions.

3. Results

3.1. Characterisation of PDFe/Al before and after Cr(VI) Adsorption

3.1.1. FTIR Analysis

Figure 1 depicts the functional groups of the PDFe/Al before and after Cr(VI) adsorption using a Fourier Transform Infrared Spectrometer (FTIR). Table 2 summarises the peak positions for raw PDFe/Al and Cr-PDFe/Al. For the raw PDFe/Al and Cr- PDFe/Al, significant -OH stretching was measured between 4000 and 3500 cm^{-1}. At *circa* 1630 cm^{-1} and 1100 cm^{-1}, HOH stretching is observed [33]. After Cr(VI) adsorption, a change in the stretching of HOH group is observed as a shift of wave band 1096.3 to 1103.2 cm^{-1} [34]. The characteristic peaks for aluminium oxides at ~535 cm^{-1} [35] ~606 cm^{-1} and ~705 cm^{-1} [36], and iron oxides at ~606 cm^{-1} and ~795 cm^{-1} [36,37] provide evidence for the successful synthesis of Fe/Al oxides. In addition, a new absorption band at, corresponding to O-Cr-O, is observed at 610.9 cm^{-1} [38,39], confirming the successful adsorption of Cr(VI) to the adsorbent surface.

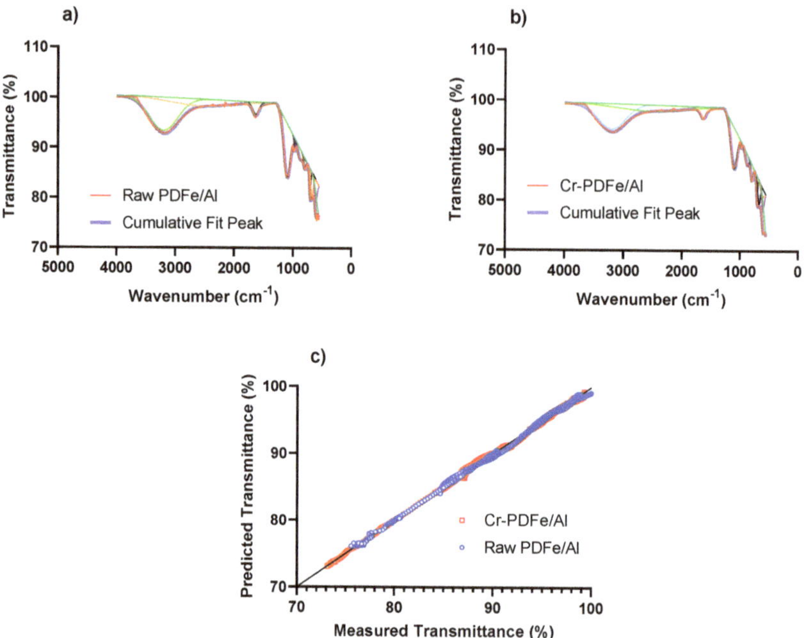

Figure 1. The deconvoluted peaks of the FTIR spectra for PDFe/Al (**a**) before and (**b**) after Cr(VI) adsorption. (**c**) shows the Transmittance for Raw PDFe/Al and Cr-PDFe/Al.

Table 2. A summary of peak positions for raw PDFe/Al and Cr-PDFe/Al.

Raw PDFe/Al	Cr-PDFe/Al	Likely Attributable Source
535.5	532.8	AlOOH [35], Si-O-Al [28]
606.4	604.4	Fe-O [36,37], O-Al-O [36]
-	610.9	O-Cr-O [38,39]
705.2	704.9	AlOOH [36]
795.3	796.2	Fe-O [36], Silica [28]
976.4	976.9	Si-O [40]
1096.3	1103.2	HOH stretching [33], S-O groups [41], Si-O-Al linkages [40]
1629.9	1630.2	HOH stretching [28,33]
3197.3	3200.0	O-H [1,2]

3.1.2. XRD Mineralogical Composition

Figure 2 depicts the XRD analyses of the PDFe/Al before and after Cr(VI) adsorption. Iron (Fe) is observed to be the dominant species in the PDFe/Al in the form of goethite/iron(III) oxide hydroxide (FeO(OH)) with clear diffraction peaks showing the crystallinity of the adsorbent and some amorphousness. In addition, the presence of aluminium in the form of aluminium oxide (Al$_2$O$_3$) is observed, thus confirming the composite nature of the material.

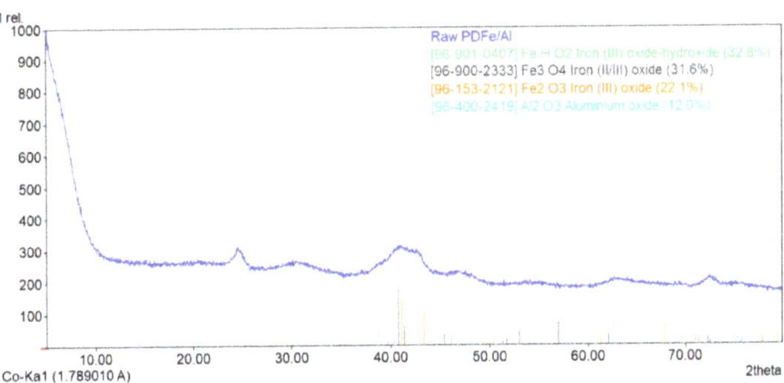

Figure 2. XRD diffractogram of the PDFe/Al before Cr(VI) adsorption. Error between selected phases and experimental profile = 3.3%.

Significant peaks between $2\theta \approx 35°$ to $50°$ are observed to have increased after Cr(VI) adsorption. This is attributed to the diffusion of Cr(VI) ions into the pores of the PDFe/Al adsorbent through chemisorption, with the subsequent reduction to Cr(III) (as confirmed by the presence of CrOOH) [42].

Figure 3 shows the diffractogram of the PDFe/Al after Cr(VI) adsorption.

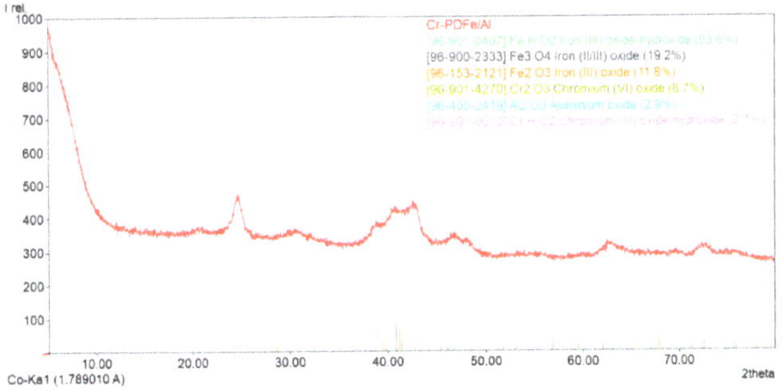

Figure 3. XRD diffractogram of the PDFe/Al after Cr(VI) adsorption. Error between selected phases and experimental profile = 3.4%.

Table 3 shows the EDS and XRF mole percentage results for raw PDFe/Al and Cr-PDFe/Al. Characterisation before Cr(VI) adsorption shows the presence of Fe and Al in the synthesised adsorbent which confirm the dimetallic nature of the material.

The presence of Fe and S shows that AMD was generated from pyrite oxidation. The presence of Si, Ca, Mg, and other trace elements in the material are impurities that resulted from co-precipitation from AMD. After Cr(VI) adsorption, it is observed that chromium is the only element showing a significant difference between the PDFe/Al and Cr-PDFe/Al. Overall, the composition of Fe, Al, O and C was observed to be preserved during adsorption, therefore demonstrating the chemically stability of the material.

Figure 4 shows the relationship between 2θ values for PDFe/Al before and after Cr(VI) adsorption.

As indicated in Figure 4, the plot for 2θ values for PDFe/Al before and after Cr(VI) adsorption shows a clear shift in 2θ values of $0.66°$ thus confirming internal adsorption induced strains as previously reported in Muedi et al. 2022 [2].

3.1.3. BET Surface and Porosity Analysis

Figure 5 and Table 4 depicts the porosity of the PDFe/Al before and after Cr(VI) adsorption using Brunauer-Emmet-Teller (BET). It is clear that a significant increase in both the internal surface area and volume was observed The total surface area of the raw PDFe/Al was reported to be around 37.5841 m^2/g, which then changed to 95.5269 m^2/g after Cr(VI) adsorption, while an eight-fold increase in internal pore volume (0.003 cm^3/g to 0.02 cm^3/g) was observed after Cr(VI) adsorption. These results are consistent with that observed previously for Congo Red adsorption [2].

Table 3. EDS and XRF mole percentage results.

Element	PDFe/Al			Cr-PDFe/Al			Fe/Al vs. Cr-PDFe/Al
	EDS	XRF	*p*-Value [1]	EDS	XRF	*p*-Value [2]	*p*-Value [3]
Fe	21.11 ± 10.42	37.51	0.22	32.38 ± 15.39	38.91	0.72	0.21
O	67.99 ± 2.53	59.83	0.042	59.75 ± 19.51	60.07	0.99	0.38
S	7.34 ± 3.72	0.24	0.16	4.59 ± 2.52	0.07	0.18	0.21
Cr [4]	0 ± 0	0.00	-	1.46 ± 0.57	0.08	0.091	0.00043
Al	0.83 ± 0.38	1.15	0.49	1.02 ± 0.47	0.62	0.48	0.51
Si	0.19 ± 0.32	0.14	0.90	0.07 ± 0.11	0.13	0.65	0.46
K	0 ± 0	0.00	-	0.15 ± 0.1	0.00	0.24	0.01
Ca	1.75 ± 3.74	0.15	0.72	0 ± 0	0.00	-	0.33
Cl	0.51 ± 1.13	0.00	0.70	0 ± 0	0.00	-	0.35
Na	0.18 ± 0.4	0.62	0.37	0.3 ± 0.31	0.00	0.43	0.61
Mg	0.11 ± 0.25	0.29	0.56	0.2 ± 0.32	0.00	0.59	0.63
Ti	0 ± 0	0.00	0.22	0.06 ± 0.13	0.00	0.70	0.35
Total	100.00 ± 0.00	99.94	-	100.00 ± 0.00	99.89	-	-

[1] *p*-value for two tailed *t*-test comparison between EDS and XRF for PDFe/Al; [2] *p*-value for two tailed *t*-test comparison between EDS and XRF for Cr-PDFe/Al; [3] *p*-value for two-tailed *t*-test comparison between EDS data sets for PDFe/Al and Cr-PDFe/Al (5 repeat measurements each); [4] Only element showing a significant difference between the PDFe/Al and Cr-PDFe/Al when considering the adjusted *p*-value significance level as calculated using the Holm-Šídák method.

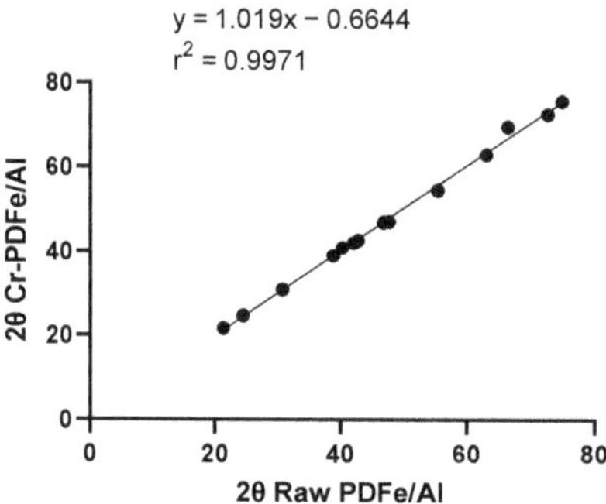

Figure 4. 2θ values for PDFe/Al before and after Cr(VI) adsorption.

The increased total surface area implies that the adsorbed chemical has the potential to enhance the adsorbent's surface area, indicating the material's potential for subsequent usage after Cr(VI) ion adsorption.

Figure 6 depicts the nitrogen adsorption–desorption isotherms of the produced PDFe/Al before and after Cr(VI) adsorption. The adsorbed quantities increased as relative pressure increased, which could indicate that the adsorbent's non-rigid nature and the placement of the distinctive shoulder are compatible with condensate destabilization at the P/P0 ratio, which is limiting the entire process. Furthermore, type IV adsorption isotherm behaviour was observed for both the raw PDFe/Al and the Cr-adsorbed PDFe/Al.

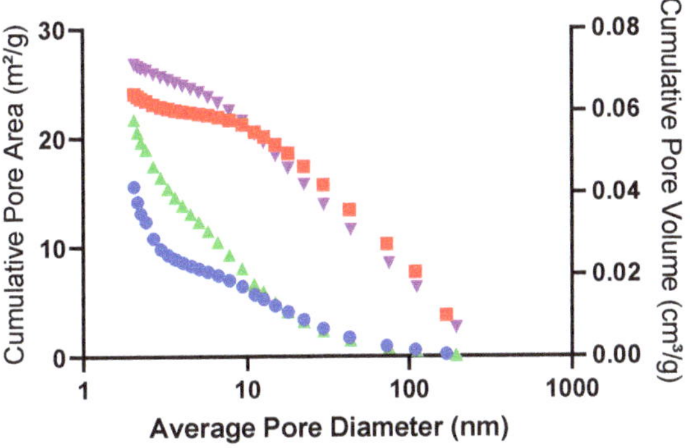

Figure 5. Porosity of the PDFe/Al before and after Cr(VI) adsorption.

Table 4. Shows a summary of the porosity of PDFe/Al before and after Cr(VI) adsorption.

Material	Parameter	Value
PDFe/Al	Total Surface Area (BET):	37.5841 m^2/g
	Micropore Volume:	0.003066 cm^3/g
	Micropore Area:	13.6686 m^2/g
	External Surface Area:	23.9156 m^2/g
Cr-PDFe/Al	Total Surface Area (BET):	95.5269 m^2/g
	Micropore Volume:	0.020797 cm^3/g
	Micropore Area:	52.9926 m^2/g
	External Surface Area:	42.5343 m^2/g

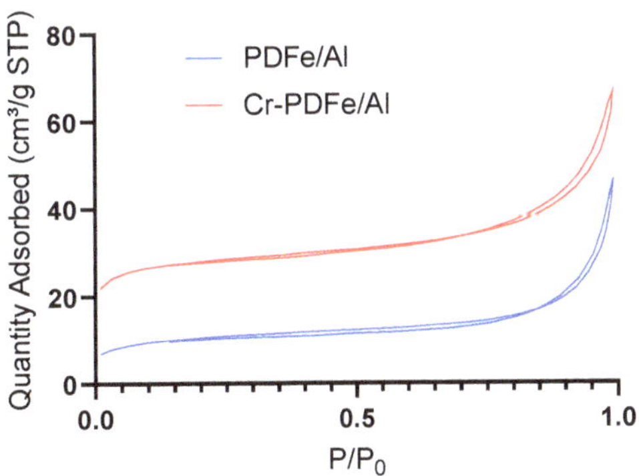

Figure 6. The nitrogen adsorption–desorption isotherms of the synthesized PDFe/Al before and after Cr(VI) adsorption.

3.1.4. SEM Morphology

Figure 7 shows the morphology of the PDFe/Al before and after Cr(VI) adsorption using scanning electron microscopy (SEM).

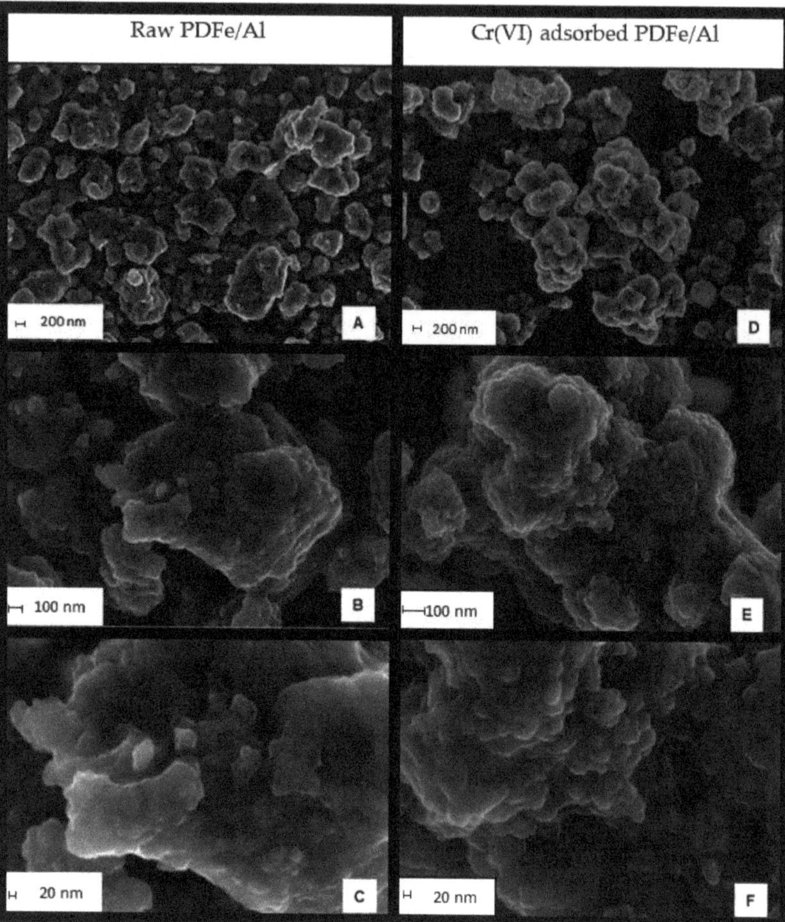

Figure 7. Morphology of the PDFe/Al before adsorption (**A–C**) and after Cr(VI) (**D–F**) adsorption.

The morphological features of the synthesized PDFe/Al before and after Cr(VI) adsorption are presented in Figure 7. Figure 7A–C show the morphology of raw PDFe/Als of various sizes, with non-uniform pressed-like structures with irregular agglomerates scattered unevenly. Figure 7D–F show the morphology of PDFe/Als following Cr(VI) adsorption in various sizes, with blood cell-like formations that are irregularly dispersed and lumped together. The shift in structural forms could indicate the presence of chromium heavy metal in the material.

3.1.5. EDX Elemental Mapping

Figure 8 shows the mapping of the elemental composition of the PDFe/Al before and after Cr(VI) adsorption using energy dispersive X-ray (EDX). The raw PDFe/Al confirms the co-precipitation of Fe and Al from AMD, (Figure 8A–D). Figure 8E–H illustrates the elemental composition of the material after Cr(VI) adsorption, where chromium is observed to be present.

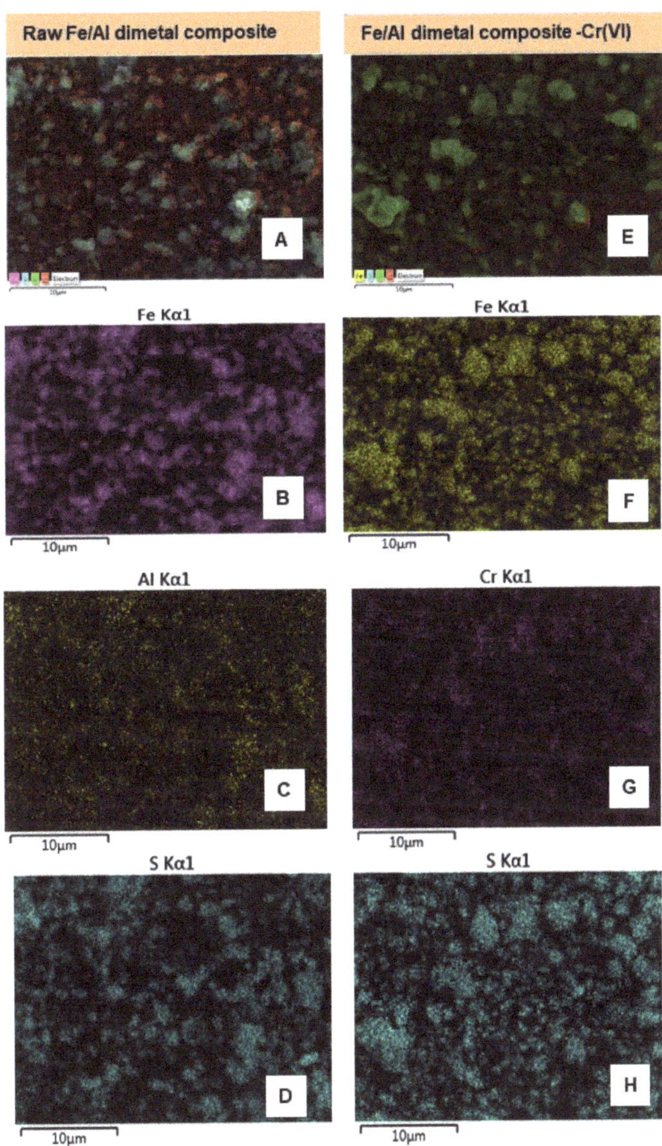

Figure 8. Mapping of elemental composition of the PDFe/Al before adsorption (**A–D**) and after Cr(VI) (**E–H**) adsorption.

3.1.6. TGA Thermal Stability

The thermal stability of PDFe/Al before and after Cr(VI) adsorption is demonstrated in Figure 9. The calcination process was broken down into three stages, with the first involving the loss of moisture from the material at temperatures ranging from 100 to 350 °C. At temperatures between 400 and 550 °C, the second step involves the loss of chemically bonded HOH, while the third stage involves the loss of the hydroxyl group (-OH) at temperatures over 550 °C.

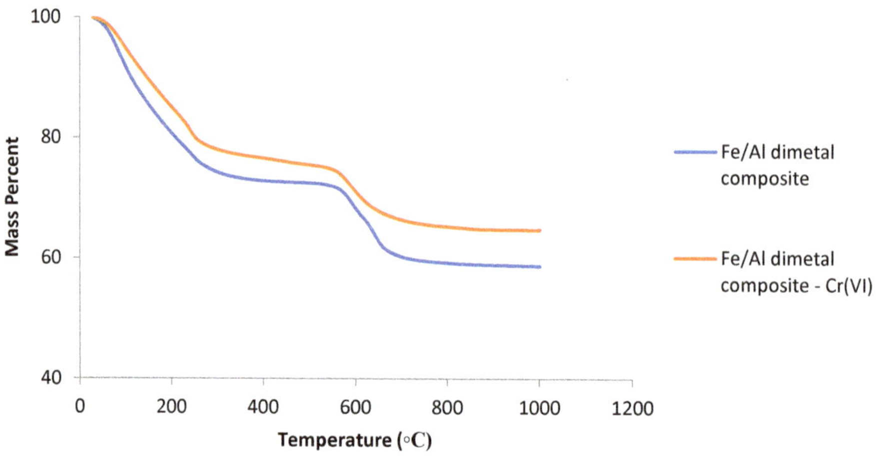

Figure 9. Thermal stability of the PDFe/Al before and after Cr(VI) adsorption.

3.2. Batch Adsorption Experiments

The effects of operational parameters on the removal of Cr(VI), as summarised in Table 1, are illustrated in Figure 10.

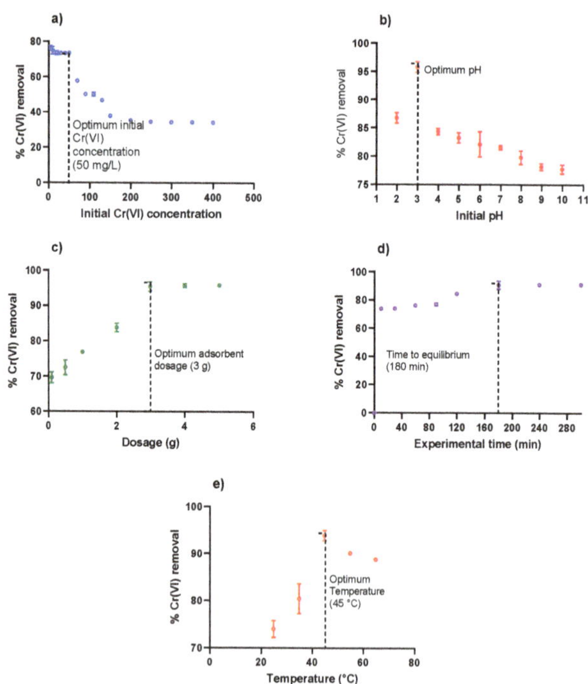

Figure 10. Effects of different operational parameters on the adsorption of Cr(VI), where (**a**) shows % Cr(VI) removal versus initial Cr(IV) concentration; (**b**) shows % Cr(VI) versus initial pH; (**c**) shows % Cr(VI) removal versus dosage; (**d**) shows % Cr(VI) removal versus agitation time; and (**e**) shows % Cr(VI) removal versus temperature.

3.2.1. Effect of Concentration

As illustrated in Figure 10a, a steep increase in the residual concentration of Cr(VI) is observed after 50 mg/L, thus indicating that the material became oversaturated with Cr(VI) oxyanions. After 50 mg/L, an increase in the residual concentration is observed, thus indicating over-saturation of the adsorbent matrices with Cr(VI) oxyanions. This indicates that the material could not adsorb Cr(IV) oxyanions of concentration >50 mg/L. Therefore, it can be concluded that 50 mg/L is the optimum concentration for Cr(VI) adsorption using PDFe/Al.

3.2.2. Effect of Initial Solution pH

As illustrated in Figure 10b, a significant amount of Cr(VI) oxyanions were removed from the aqueous system at a pH of 3. This corroborates the point of zero charge (PZC) of the adsorbent which was found to be $pH_{PZC} = 3.02$. Moreover, [43] also reported that the adsorption of Cr(VI) on magnetite (Fe_3O_4) decreases with an increase in pH after level 3. An optimal pH = 3 was also obtained by [44]. In addition, it was observed that the final pH in all experimental runs were between 2.07 and 2.79 indicating that in all cases, except for the run in which the initial pH was 2, a decrease in pH was observed.

3.2.3. Effect of Adsorbent Dosage

As illustrated in Figure 10c, a steep increase is observed from 0.1 to 3 g during Cr(VI) removal, after which the trend takes a gentle slope. This could indicate the depletion of available sites in the adsorbent; hence, the material could not adsorb more Cr(VI) oxyanions. This behaviour is also observed in Figure 10c, where the maximum adsorption capacity of 1 g PDFe/Al in 250 mL of 10 mg/L Cr(VI) solution was observed to be approximately 1 mg/g, hence 3g of PDFe/Al nanocomposite is optimum dosage for Cr(VI) removal. Gürü et al. 2008 [45] also reported that the highest removal of chromium was observed at 3 g using diatomite.

3.2.4. Effect of Agitation Time

As illustrated in Figure 10d, it was observed that the removal of Cr(VI) from an aqueous system increases with time. After 180 min of agitation, the graph takes a gentle slope, hence no significant removal. Most studies reported 4 h as the optimal time for removal of Cr(VI) from water [46,47]. From the results, 180 min is the optimal agitation time for Cr(VI) removal.

3.2.5. Effect of Temperature

As illustrated in Figure 10e, it was observed that the removal of Cr(VI) from an aqueous system increased with an increase in temperature, particularly between 25–35 °C. Temperatures higher than 35 °C make Cr(VI) to go back into solution. Additionally, [48] also reported that maximum removal of Cr(VI) was achieved at 40 °C. Therefore, 35 °C is an ideal temperature for maximum removal of Cr(VI) from an aqueous system

3.3. Adsorption Kinetics

Adsorption kinetics for the adsorption of Cr(VI) by PDFe/Al was studied to demonstrate the mechanisms and rates of adsorption, as shown in Table 5 and Figure 11.

Figure 11 shows different kinetic models fitted to the kinetic data for the adsorption of hexavalent chromium (Cr(VI)) onto the adsorbent. As shown in Figure 11a, a good trend was obtained from the pseudo-first order (PFO) model ($R^2 = 0.945$). However, there seems to be a lot of uncertainty depicted by the model the future observations in respect of Cr(VI) adsorption application, thereby making it risky to apply PFO kinetic model.

In Figure 11b, a good trend was obtained from the pseudo-second order (PSO) model, where results were obtained in the same manner as PFO. However, the PSO depends on the adsorbed amount of the adsorbate. The results obtained show a better trend than PFO ($R^2 = 0.962$) with great prediction of the adsorption rate of Cr(VI).

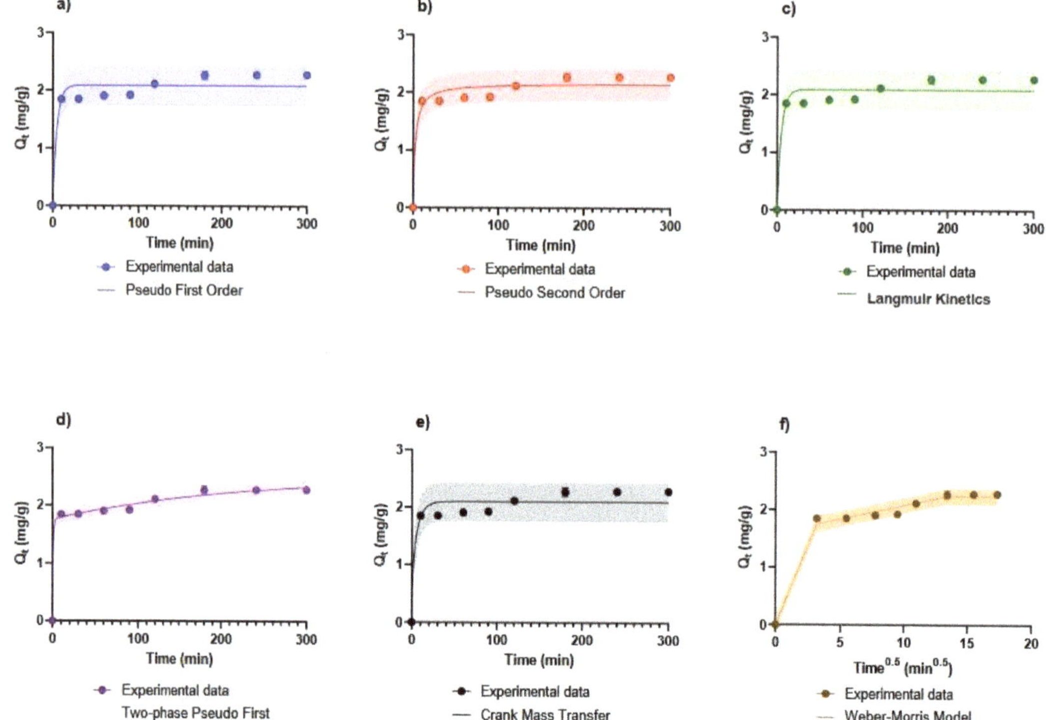

Figure 11. Different kinetic models fitted to the kinetic data for the adsorption of Cr(VI) to the adsorbent. The models were: (**a**) Pseudo First Order, (**b**) Pseudo Second Order, (**c**) Langmuir Kinetics, (**d**) Two-phase Pseudo First Order, (**e**) Crank Mass Transfer Model, (**f**) Weber-Morris Model. The shaded areas represent the 95% prediction intervals for the respective model fits. The optimised model parameters are reported in Table 5.

In Figure 11c, the Langmuir kinetics show a trend with good results ($R^2 = 0.929$) for the adsorption of Cr(VI); however, there is uncertainty in the diffusion and extrapolation of the model.

In Figure 11d, the two-phase pseudo-first order adsorption (TPA) model, which is based on two parallel adsorption processes: rapid and slow adsorption mechanisms, outperforms the PFO and PSO kinetic models. In comparison to previous models, the findings obtained demonstrate the best trend for Cr(VI) adsorption ($R^2 = 0.993$). As a result, it demonstrates how Cr(VI) is rapidly adsorbed and gradually slows down as saturation approaches.

In Figure 11e, the Crank diffusion model was investigated to determine pore diffusion since the diffusion coefficient (D_e) remains constant under the condition that the diffusion is uniform within the sphere. The results obtained show that an effective diffusion coefficient $D_e = 2.61 \times 10^{-13}$ m^2·s^{-1} provides a good prediction of the Cr(VI) adsorption process. This indicates that the system is significantly limited by mass transport; the molecular diffusion coefficient of chromate is 1.4494×10^{-9} m^2·s^{-1} [49] which is four orders of magnitude greater that the effective diffusivity.

Weber Morris' intra-particle diffusion model was used to study the effects of inter-particle and intra-particle diffusion on Cr(VI) adsorption in Figure 11f. The many phases of adsorption visible in the multilinear fit of the data are depicted in this model. The first phase depicts in-particle diffusion, the second phase depicts in-particle diffusion, and the last phase depicts adsorption to the adsorbent. A good fit of the data was obtained ($R^2 = 0.991$) and the results showed a D_e value between $D_{e1} = 7.02 \times 10^{-12}$ m$^2 \cdot$s^{-1} and $D_{e2} = 5.11 \times 10^{-13}$ m$^2 \cdot$s^{-1}. These results correlate well with the predicted effective diffusivity from the Crank diffusion model.

Table 5. Kinetic models for the adsorption of Cr(VI) by PDFe/Al.

Kinetic Law	Differential Form	Analytical Form	Fitted Parameters	R^2/RMSE
Pseudo first order [50]	$\frac{dQ}{dt} = k_1(Q_e - Q)$	$Q = Q_e(1 - e^{-k_1 t})$	$Q_e = 2.092$ mg$\cdot$g^{-1} $k_1 = 0.211$ min^{-1}	0.945/ 0.1604 mg$\cdot$g^{-1}
Pseudo second order [50]	$\frac{dQ}{dt} = k_2(Q_e - Q)^2$	$Q = \frac{(k_2 Q_e^2)t}{1 + k_1 Q_e t}$	$Q_e = 2.165$ mg$\cdot$g^{-1} $k_2 = 0.188$ L$^2 \cdot$mg$^{-2} \cdot$min^{-1}	0.962/ 0.1336 mg$\cdot$g^{-1}
Langmuir adsorption [51]	$\frac{dQ}{dt} = k_{ad}C(Q_{max} - Q) - \frac{k_{ad}}{K_L}Q$ $K_L = \exp\left(\frac{\Delta S^o}{R} - \frac{\Delta H^o}{RT}\right)$		$k_{ad} = 0.00491$ L$\cdot$mg$^{-1} \cdot$min^{-1} $Q_{max}, \Delta S^o, \Delta H^o$ obtained from Langmuir isotherm	0.945/ 0.1603 mg$\cdot$g^{-1}
Two phase adsorption [52–54]	$\frac{dQ_{slow}}{dt} = k_{fast}Q_{fast} - k_{slow}Q_{slow}$ $Q_e = Q_{fast} + Q_{slow}$	$Q = Q_{fast}\left(1 - e^{-k_{fast}t}\right)$ $+ Q_{slow}\left(1 - e^{-k_{slow}t}\right)$ $Q_e = Q_{fast} + Q_{slow}$	$k_{fast} = 1.11$ min^{-1} $Q_{fast} = 1.77$ mg/g $k_{slow} = 0.00444$min^{-1} $Q_{slow} = 0.77$ mg/g	0.993/ 0.0563 mg$\cdot$g^{-1}
Crank internal mass transfer model [50]	$\frac{\partial Q}{\partial t} = k_{CR}\frac{\partial}{\partial r}\left(\frac{r^2 \partial Q}{\partial r}\right)$ $k_{CR} = \frac{D_e}{r^2}$	$Q = Q_e\frac{6}{\pi^2}\sum_{n=1}^{\infty}\frac{1}{n^2}\exp(-k_{CR}n^2\pi^2 t)$ $k_{CR} = \frac{D_e}{r^2}$	$D_e = 2.61 \times 10^{-13}$ m$^2 \cdot$s^{-1} $r = 32$ μm	0.946/ 0.159 mg$\cdot$g^{-1}
Weber and Morris [50,55]		$Q = k_{WM}t^{\frac{1}{2}} + C,$ $k_{WM} = \frac{6}{\pi^{\frac{1}{2}}}\sqrt{\frac{D_e}{r^2}}$	$D_{e1} = 4.59 \times 10^{-13}$ m$^2 \cdot$s^{-1} $r = 32$ μm $D_{e2} = 3.30 \times 10^{-15}$ m$^2 \cdot$s^{-1} $D_{e3} = 0$ m$^2 \cdot$s^{-1}	0.991/ 0.0665 mg$\cdot$g^{-1}

3.4. Adsorption Isotherms

The adsorption isotherms of Cr(VI) adsorption by PDFe/Al dimetal composite were studied on various mechanisms that determine the adsorption process. The isotherm models are summarized in Table 6 and illustrated in Figure 12.

Figure 12a depicts the Langmuir adsorption isotherm, with a maximum adsorption capacity (Q_{max}) of 6.67 mg$\cdot$g^{-1}, which is acceptable for a mineral based adsorbent, especially considering that the adsorbent was recovered from authentic industrial waste AMD. To compare, Alemu et al. [56] and Panda et al. [57] tested the removal of Cr(VI) using industrially obtained basalt rock ($Q_{max} = 0.079$ mg$\cdot$g^{-1}) and dolochar ($Q_{max} = 0.904$ mg$\cdot$g^{-1}), respectively. In addition, the maximum adsorption capacities for comparable Fe/Al materials were in the range of 2.3–59.9 mg$\cdot$g^{-1}, indicating that the current study compares well with results from the literature [26–30].

The Freundlich adsorption isotherm is shown in Figure 12b, where the intensity parameter indicates the adsorption favorability, with K_F values of K_F (298 K) = 0.8477; K_F (318 K) = 1.15; K_F (328 K) = 1.13; K_F (338 K) = 1.098. The adsorption of Cr(VI) by PDPDFe/Al from the aqueous system is highly favorable in this investigation, with $n_F = 2.48$.

The two-surface Langmuir adsorption isotherm is depicted in Figure 12c, which posits that the adsorbent's surface contains various surface types with varying adsorption capabilities. The results show that $Q_{max,1} = 1.80$ mg$\cdot$g^{-1} for one surface and $Q_{max,2} = 5.096$ mg$\cdot$g^{-1} for the other, for a total maximum adsorption of $Q = 6.896$ mg$\cdot$g^{-1}.

Table 6 shows a summary of tested adsorption isotherms, fitted parameters, coefficient of determination (R^2), and root-mean-square error (RMSE) as a measure of goodness of fit for various isotherm models.

3.5. Regeneration Study

As shown in Figure 13, a regeneration study was carried out to determine the possibility of recovering and reusing PDFe/Al following Cr(VI) adsorption.

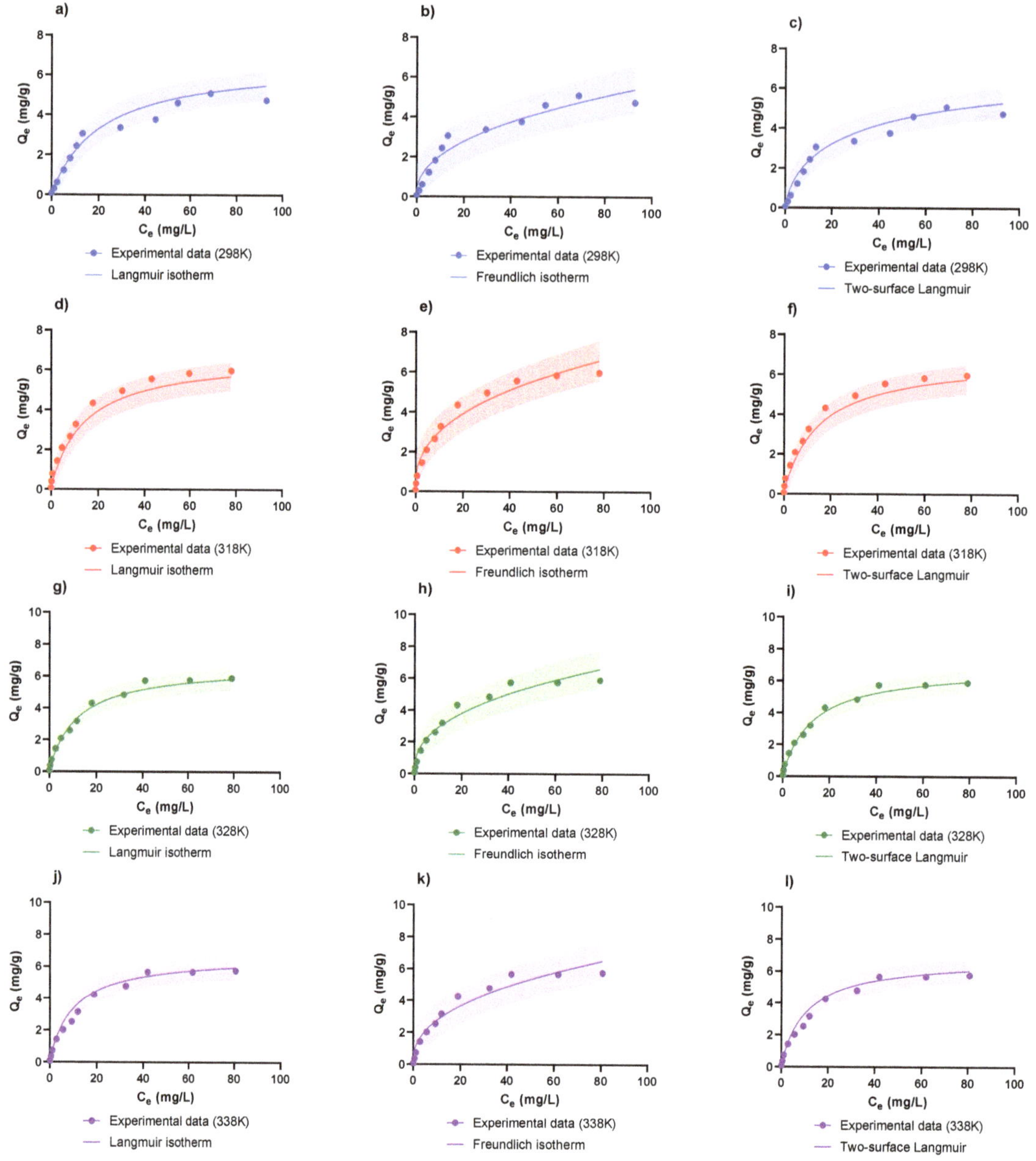

Figure 12. The non-linear fits of the isotherm models from: (**a**) Langmuir (298K), (**b**) Freundlich (298K), (**c**) Two Surface Langmuir (298K), (**d**) Langmuir (318K), (**e**) Freundlich (318K), (**f**) Two-surface Langmuir (318K), (**g**) Langmuir (328K), (**h**) Freundlich (328K), (**i**) Two-surface Langmuir (328K), (**j**) Langmuir (338K), (**k**) Freundlich (338K) and (**l**) Two-surface Langmuir (338K). The shaded areas indicate the 95% prediction intervals.

A desorption investigation was carried out to regenerate the material after Cr(VI) adsorption, as shown in Figure 13. The material had the ability to be utilized for Cr(VI) adsorption more than four times. During the first four cycles, it was observed that the material matrices could still adsorb the oxyanions from an aqueous system. However, after four reuse cycles, it was observed that the material started losing efficacy, probably due to the loosening of the material matrices, which probably lost layers in the process of adsorption–desorption.

Table 6. Summary of tested adsorption isotherms, fitted parameters, coefficient of determination (R^2), and root-mean-square error (RMSE) as a measure of goodness of fit for various isotherm models.

Kinetic Law	Differential Form	Fitted Parameters	R^2/RMSE
Langmuir [50,58,59]	$Q_e = \frac{k_L(T)Q_{max}C_e}{1+k_L(T)C_e}$, $k_L(T) = \exp\left(\frac{\Delta S^o}{R} - \frac{\Delta H^o}{RT}\right)$	$\Delta S^o = 26.25 \; J.(mol\cdot K)^{-1}$ $\Delta H^o = 15.3 \; kJ\cdot mol^{-1}$ $Q_{max} = 6.67 \; mg\cdot g^{-1}$ $0.0616 < R_L < 0.953$	$0.9717/0.3412 \; mg\cdot g^{-1}$
Freundlich [58]	$Q_e = K_F(T)C_e^{\frac{1}{n}}$	$K_F(298\;K) = 0.8477 \; mg\cdot g^{-1}(L\cdot mg^{-1})^{n_F^{-1}}$ $K_F(318\;K) = 1.15 \; mg\cdot g^{-1}(L\cdot mg^{-1})^{n_F^{-1}}$ $K_F(328\;K) = 1.13 \; mg\cdot g^{-1}(L\cdot mg^{-1})^{n_F^{-1}}$ $K_F(338\;K) = 1.098 \; mg\cdot g^{-1}(L\cdot mg^{-1})^{n_F^{-1}}$ $n_F = 2.48$	$0.959/0.410 \; mg\cdot g^{-1}$
Two-surface Langmuir [59]	$Q_e = \sum_{i=1}^{2} \frac{k_{L,i}(T)Q_{max,i}C_e}{1+k_{L,i}(T)C_e}$, $k_{L,i}(T) = \exp\left(\frac{\Delta S_i^o}{R} - \frac{\Delta H_i^o}{RT}\right)$	$\Delta S_1^o = -134.5 \; J\cdot(mol\cdot K)^{-1}$ $\Delta H_1^o = -37.1 \; kJ\cdot mol^{-1}, Q_{max,1} = 1.80 \; mg\cdot g^{-1}$ $\Delta S_2^o = 76.7 \; J\cdot(mol\cdot K)^{-1}$ $\Delta H_1^o = 32.0 \; kJ\cdot mol^{-1}, Q_{max,2} = 5.096 \; mg\cdot g^{-1}$ $0.0221 < R_{L,1} < 0.952, \; 0.0553 < R_{L,2} < 0.976$	$0.975/0.320 \; mg\cdot g^{-1}$

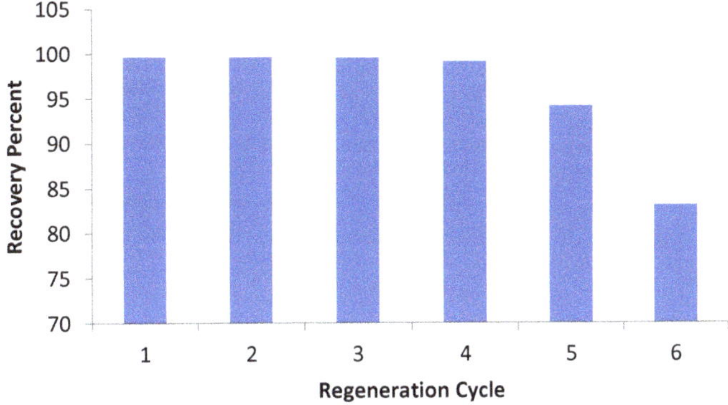

Figure 13. Recovery efficiency of PDFe/Al after Cr(VI) adsorption.

4. Discussion

4.1. Summary of Results

The results obtained in the study provide clear evidence for the potential of PDFe/Al synthesized from authentic AMD for the adsorption of Cr(VI) from aqueous solution. The surface characterization results demonstrated that the adsorption of Cr(VI) to the adsorbent involved several surface interactions with the presence of Cr(VI) and Cr(III) observed after adsorption. In addition, the adsorption of Cr(VI) resulting in adsorption induced strains within the adsorbent matrix which resulted in an increase in both the internal surface area and surface volumes. The surface analyses of the adsorbent showed the presence of Fe, O, Al, S, Cr (only after adsorption), and other minor constituents, confirming the heterogenous nature of the AMD used for the synthesis.

The batch adsorption results demonstrated that the adsorption followed a two-phase adsorption process with an initially fast process followed by a significantly slower adsorption.

The isotherm results were best described by a two-surface Langmuir kinetic model with distinct thermodynamic and saturation properties. The material further demonstrated good reusability with Cr(VI) in excess of 90% possible after five adsorption/desorption cycles.

4.2. Comparison of Fe/Al Dimetal Nanocomposite for Removal of Cr(VI) from an Aqueous System

Comparisons of rate constants, times to reach 99% of equilibrium [36], and maximum Cr(VI) adsorption capacities of different studies employing Fe- and Al-based adsorbents from the literature to the current material are made in Table 7.

Table 7. Comparison of Fe/Al dimetal nanocomposite with other Fe and Al based adsorbents in the removal of Cr(VI).

Adsorbent	Q_{max} (mg·g^{-1})	Q_e (mg·g^{-1})	k_2	$t_{2,99}$ (min) [36]	References
PDFe/Al nanocomposite	6.67	2.165	0.188 g·mg^{-2} min^{-1}	279	This study
Aluminium Oxide	78.1	72.5	6.9×10^{-4} g·mg^{-2} min^{-1}	1978	[26]
Mesoporous iron–zirconium bimetal oxide	59.88	52.46	0.02 g·mg^{-2} min^{-1}	2619	[27]
Zirconium oxide intercalated sodium montmorillonite scaffold	52.46	44.46	0.0023 g·mg^{-2} min^{-1}	2277	[28]
Hematite	2.299	0.5	0.0088 g·mg^{-2} min^{-1}	5952	[29]
Hydrochloric acid-modified kaolinite	18.15	0.34	0.089 g·mg^{-2} min^{-1}	2631	[30]
Acetic acid-modified kaolinite	10.42	0.38	0.66 g·mg^{-2} min^{-1}	513	[30]

From Table 7, it can be concluded that the synthesized Fe/Al dimetal nanocomposite demonstrated a maximum adsorption capacity comparable to other Fe and Al based adsorbents (within the same order of magnitude). In addition, it was observed that the PDFe/Al achieved equilibrium nearly an order of magnitude faster (279 min) than most of the studies (1978–5952 min), except for the acetic acid-modified kaolinite that achieved equilibrium within 513 min. These results are significant, as they demonstrate the rapidity of the adsorbent in removing the Cr(VI), a requirement for the application of an adsorption system industrially. To illustrate the point, the mesoporous iron–zirconium bimetal oxide [27] demonstrated a maximum adsorption capacity of *circa* 60 mg·g^{-1}; however, the time to 99% equilibrium took 2619 min (~44 h). This slower adsorption rate would invariably affect the residence times required for successful water treatment which negates the higher adsorption capacity displayed by the adsorbent.

In addition, the fact that the PDFe/Al dimetal nanocomposite was recovered and synthesised from authentic acid mine drainage, as opposed to synthetic chemicals, further supports the potential impact of the adsorbent as it provides an avenue for AMD valorisation through PDFe/Al synthesis and subsequent application for water treatment.

4.3. Proposed Mechanism

The results from the study indicate that the adsorption of Cr(VI) by PDFe/Al is significantly diffusion controlled with a predicted effective diffusivity of between 7.02×10^{-12} m^2·s^{-1} and 4.79×10^{-13} m^2·s^{-1} which is between 3 and 4 orders of magnitude less than the molecular diffusion coefficient of chromate (1.4494×10^{-9} m^2·s^{-1} [49]). The adsorption isotherms indicated that the surface likely consisted of at least two surfaces with distinct adsorption properties. The one surface had exothermic adsorption properties with a corresponding decrease in entropy while the other supported endothermic adsorption with an increase in entropy predicted. The decrease in entropy on the first surface likely demonstrates the increased organisation resulting from adsorption of chromium from solution, while the increased entropy on the second surface result from the displacement of protons from the surface resulting in the observed decrease in pH [1]. It is interesting to note that the standard enthalpy and entropy values for the single surface Langmuir model were between the corresponding values from the two surface Langmuir model, indicating the single surface Langmuir model represented a net result of the two-surface model.

The results further showed that the adsorption of Cr(VI) induced significant internal strain and corresponding deformation within the support matrix which result in the formation of cracks therefore increased surface area and volume. Finally, it was observed that the adsorbed Cr(VI) was reduced to Cr(III) on the surface of the adsorbent. This likely corresponds to the oxidation of the Iron(II/III) oxide observed in the XRD profile after Cr(VI) adsorption [60]. The reduction of Cr(VI) to Cr(III) is known to have a standard Gibbs free energy of -390 kJ/mol [61] while the oxidation of Iron(II/III) oxide has a Gibbs free energy of -42 kJ/mol [61]. The resulting oxidation-reduction reaction has a standard Gibbs free energy of -450 kJ/mol and is therefore highly favourable and spontaneous.

5. Conclusions

The synthesis of polycationic di-metal iron/aluminium nanocomposite was executed through selective precipitation followed by calcination and vibratory ball milling. The material was successfully recovered and synthesised from industrial acid mine drainage and applied in the removal of hexavalent chromium (Cr(VI)) from an aqueous system. Adsorption parameters were optimized using the one-factor-at-a-time (OFAAT) approach on a batch experimental set-up. The fate of Cr(VI) in the aqueous solution, as well as the surface of the material, was determined; the results demonstrated not only the successful adsorption of Cr(VI) from solution, but also confirmed the presence and in situ reduction of Cr(VI) to Cr(III) on the surface.

Recorded optimised conditions for Cr(VI) adsorption are: 50 mg/L initial Cr(VI) concentration; 3 g of PDFe/Al; pH 3; 180 min agitation time; and temperature of 35 °C. The synthesized PDFe/Al was observed to have a Langmuir adsorption capacity of $Q_{max} = 6.67$ mg·g^{-1} for Cr(VI) with >95% removal.

Adsorption kinetics followed a two-phase pseudo-first-order as opposed to pseudo-first or pseudo-second-order behaviour. The adsorption process followed two-surface Langmuir adsorption model. The PDFe/Al achieved more than 90% Cr(VI) recovery after five adsorption/desorption cycles, thereby demonstrating the potential reusability.

It was proposed that a diffusion limited adsorption mechanism involving two surfaces with distinct adsorption characteristics were responsible for the adsorption. The mechanism further involved the reduction of Cr(VI) to Cr(III) while simultaneously significant internal adsorption induced strains resulted in the formation of internal cracks within the adsorbent matrix resulting in increased pore surface area and pore volume.

Comparison of the maximum adsorption capacities and time required to reach equilibrium of the PDFe/Al to literature studies demonstrated that the PDFe/Al exhibited a comparable adsorption capacity while a superior adsorption rate was measured.

This study demonstrates the industrial potential of the synthesized PDFe/Al for the simultaneous valorisation of AMD and the treatment of Cr(VI), thereby directly contributing towards sustainable mining.

Author Contributions: Conceptualization, K.L.M., V.M. and J.P.M.; methodology, K.L.M.; software, H.G.B.; validation, V.M. and H.G.B.; formal analysis, K.L.M. and H.G.B.; investigation, K.L.M.; resources, V.M., J.P.M. and H.G.B.; data curation, K.L.M., H.G.B. and V.M.; writing—original draft preparation, K.L.M., V.M. and H.G.B.; writing—review and editing, K.L.M., V.M. and H.G.B.; visualization, K.L.M. and H.G.B.; supervision, H.G.B., V.M., J.P.M.; project administration, H.G.B.; funding acquisition H.G.B. All authors have read and agreed to the published version of the manuscript.

Funding: This research was funded by the National Research Foundation (NRF) South Africa, grant number MND200517522430 and 145848. Austrian Agency for International Cooperation in Education and Research (OeAD): Africa UniNet P056.

Data Availability Statement: The data presented in this study are openly available in the University of Pretoria Research Data Repository at https://doi.org/10.25403/UPresearchdata.21342765.

Conflicts of Interest: The authors declare no conflict of interest.

References

1. Muedi, K.L.; Brink, H.G.; Masindi, V.; Maree, J.P. Effective removal of arsenate from wastewater using aluminium enriched ferric oxide-hydroxide recovered from authentic acid mine drainage. *J. Hazard. Mater.* **2021**, *414*, 125491. [CrossRef] [PubMed]
2. Muedi, K.L.; Masindi, V.; Maree, J.P.; Haneklaus, N.; Brink, H.G. Effective Adsorption of Congo Red from Aqueous Solution Using Fe/Al Di-Metal Nanostructured Composite Synthesised from Fe(III) and Al(III) Recovered from Real Acid Mine Drainage. *Nanomaterials* **2022**, *12*, 776. [CrossRef] [PubMed]
3. Baker, B.J.; Banfield, J.F. Microbial communities in acid mine drainage. *FEMS Microbiol. Ecol.* **2003**, *44*, 139–152. [CrossRef]
4. Amos, R.T.; Blowes, D.W.; Bailey, B.L.; Sego, D.C.; Smith, L.; Ritchie, A.I.M. Waste-rock hydrogeology and geochemistry. *Appl. Geochem.* **2015**, *57*, 140–156. [CrossRef]
5. Sheoran, V.; Sheoran, A.; Choudhary, R.P. Biogeochemistry of acid mine drainage formation: A review. In *Mine Drainage and Related Problems*; Nova Science Publishers Inc.: Hauppauge, NY, USA, 2011; pp. 119–154.
6. Sheoran, A.; Sheoran, V.; Choudhary, R.P. Geochemistry of acid mine drainage: A review. *Perspect. Environ. Res.* **2011**, *4*, 217–243.
7. Naidu, G.; Ryu, S.; Thiruvenkatachari, R.; Choi, Y.; Jeong, S.; Vigneswaran, S. A critical review on remediation, reuse, and resource recovery from acid mine drainage. *Environ. Pollut.* **2019**, *247*, 1110–1124. [CrossRef]
8. Cheng, H.; Hu, Y.; Luo, J.; Xu, B.; Zhao, J. Geochemical processes controlling fate and transport of arsenic in acid mine drainage (AMD) and natural systems. *J. Haz. Mat* **2009**, *165*, 13–26. [CrossRef]
9. Masindi, V.; Foteinis, S.; Chatzisymeon, E. Co-treatment of acid mine drainage and municipal wastewater effluents: Emphasis on the fate and partitioning of chemical contaminants. *J. Haz. Mat* **2022**, *421*, 126677. [CrossRef]
10. Masindi, V.; Fosso-Kankeu, E.; Mamakoa, E.; Nkambule, T.T.I.; Mamba, B.B.; Naushad, M.; Pandey, S. Emerging remediation potentiality of struvite developed from municipal wastewater for the treatment of acid mine drainage. *Environ. Res.* **2022**, *210*, 112944. [CrossRef]
11. Dhal, B.; Thatoi, H.N.; Das, N.N.; Pandey, B.D. Chemical and microbial remediation of hexavalent chromium from contaminated soil and mining/metallurgical solid waste: A review. *J. Hazard. Mater.* **2013**, *250–251*, 272–291. [CrossRef]
12. Papp, J.F. *Mineral Commodity Summaries*; U.S. Geological Survey: Reston, VA, USA, 2017.
13. Zhang, Y.K.; Qi, S.; Chen, H.H. A review of remediation of chromium contaminated soil by washing with chelants. *Adv. Mater. Res.* **2014**, *838–841*, 2625–2629. [CrossRef]
14. Sultana, M.Y.; Akratos, C.S.; Pavlou, S.; Vayenas, D.V. Chromium removal in constructed wetlands: A review. *Int. Biodeterior. Biodegrad.* **2014**, *96*, 181–190. [CrossRef]
15. Jobby, R.; Jha, P.; Yadav, A.K.; Desai, N. Biosorption and biotransformation of hexavalent chromium [Cr(VI)]: A comprehensive review. *Chemosphere* **2018**, *207*, 255–266. [CrossRef]
16. Shahid, M.; Shamshad, S.; Rafiq, M.; Khalid, S.; Bibi, I.; Niazi, N.K.; Dumat, C.; Rashid, M.I. Chromium speciation, bioavailability, uptake, toxicity and detoxification in soil-plant system: A review. *Chemosphere* **2017**, *178*, 513–533. [CrossRef] [PubMed]
17. WHO. Guidelines for drinking-water quality. In *Fourth Edition Incorporating the First Addendum*; WHO: Geneva, Switzerland, 2017.
18. Wang, Z.; Shen, Q.; Xue, J.; Guan, R.; Li, Q.; Liu, X.; Jia, H.; Wu, Y. 3D hierarchically porous NiO/NF electrode for the removal of chromium(VI) from wastewater by electrocoagulation. *Chem. Eng. J.* **2020**, *402*, 126151. [CrossRef]
19. Giagnorio, M.; Steffenino, S.; Meucci, L.; Zanetti, M.C.; Tiraferri, A. Design and performance of a nanofiltration plant for the removal of chromium aimed at the production of safe potable water. *J. Environ. Chem. Eng.* **2018**, *6*, 4467–4475. [CrossRef]
20. Melak, F.; Du Laing, G.; Ambelu, A.; Alemayehu, E. Application of freeze desalination for chromium (VI) removal from water. *Desalination* **2016**, *377*, 23–27. [CrossRef]
21. Shao, Z.; Huang, C.; Wu, Q.; Zhao, Y.; Xu, W.; Liu, Y.; Dang, J.; Hou, H. Ion exchange collaborating coordination substitution: More efficient Cr(VI) removal performance of a water-stable CuII-MOF material. *J. Hazard. Mater.* **2019**, *378*, 120719. [CrossRef]
22. Kretschmer, I.; Senn, A.M.; Meichtry, J.M.; Custo, G.; Halac, E.B.; Dillert, R.; Bahnemann, D.W.; Litter, M.I. Photocatalytic reduction of Cr(VI) on hematite nanoparticles in the presence of oxalate and citrate. *Appl. Catal. B Environ.* **2019**, *242*, 218–226. [CrossRef]
23. Kalita, E.; Baruah, J. 16—Environmental remediation. In *Colloidal Metal Oxide Nanoparticles*; Thomas, S., Tresa Sunny, A., Velayudhan, P., Eds.; Elsevier: Amsterdam, The Netherlands, 2020; pp. 525–576. [CrossRef]
24. Islam, M.A.; Angove, M.J.; Morton, D.W. Recent innovative research on chromium (VI) adsorption mechanism. *Environ. Nanotechnol. Monit. Manag.* **2019**, *12*, 100267. [CrossRef]
25. Dimos, V.; Haralambous, K.J.; Malamis, S. A review on the recent studies for chromium species adsorption on raw and modified natural minerals. *Crit. Rev. Environ. Sci. Technol.* **2012**, *42*, 1977–2016. [CrossRef]
26. Alvarez-Ayuso, E.; Garcia-Sanchez, A.; Querol, X. Adsorption of Cr(VI) from synthetic solutions and electroplating wastewaters on amorphous aluminium oxide. *J. Hazard. Mater.* **2007**, *142*, 191–198. [CrossRef] [PubMed]
27. Wang, Y.; Liu, D.; Lu, J.; Huang, J. Enhanced adsorption of hexavalent chromium from aqueous solutions on facilely synthesized mesoporous iron-zirconium bimetal oxide. *Colloids Surf. A-Physicochem. Eng. Asp.* **2015**, *481*, 133–142. [CrossRef]

28. Rathinam, K.; Atchudan, R.; Edison, T. Zirconium oxide intercalated sodium montmorillonite scaffold as an effective adsorbent for the elimination of phosphate and hexavalent chromium ions. *J. Environ. Chem. Eng.* **2021**, *9*, 106053. [CrossRef]

29. Adegoke, H.; Adekola, F. Equilibrium sorption of hexavalent chromium from aqueous solution using synthetic hematite. *Colloid J.* **2012**, *74*, 420–426. [CrossRef]

30. Dim, P.; Mustapha, L.; Termtanun, M.; Okafor, J. Adsorption of chromium (VI) and iron (III) ions onto acid-modified kaolinite: Isotherm, kinetics and thermodynamics studies. *Arab. J. Chem.* **2021**, *14*, 103064. [CrossRef]

31. Smičiklas, I.; Milonjić, S.K.; Pfendt, P.; Raičević, S. The point of zero charge and sorption of cadmium (II) and strontium (II) ions on synthetic hydroxyapatite. *Sep. Purif. Technol.* **2000**, *18*, 185–194. [CrossRef]

32. Kumari, P.; Sharma, P.; Srivastava, S.; Srivastava, M.M. Biosorption studies on shelled Moringa oleifera Lamarck seed powder: Removal and recovery of arsenic from aqueous system. *Int. J. Miner. Process.* **2006**, *78*, 131–139. [CrossRef]

33. Bordoloi, S.; Nath, S.K.; Gogoi, S.; Dutta, R.K. Arsenic and iron removal from groundwater by oxidation–coagulation at optimized pH: Laboratory and field studies. *J. Hazard. Mater.* **2013**, *260*, 618–626. [CrossRef]

34. Naiya, T.K.; Singha, B.; Das, S.K. FTIR Study for the Cr(VI) Removal from Aqueous Solution Using Rice Waste. In Proceedings of the International Conference on Chemistry and Chemical Process, Singapore, 25–28 February 2019.

35. Tang, B.; Ge, J.; Zhuo, L.; Wang, G.; Niu, J.; Shi, Z.; Dong, Y. A facile and controllable synthesis of gamma-Al_2O_3 nanostructures without a surfactant. *Eur. J. Inorg. Chem.* **2005**, *2005*, 4366–4369. [CrossRef]

36. Prabakaran, E.; Pillay, K.; Brink, H. Hydrothermal synthesis of magnetic-biochar nanocomposite derived from avocado peel and its performance as an adsorbent for the removal of methylene blue from wastewater. *Mater. Today Sustain.* **2022**, *18*, 100123. [CrossRef]

37. Hwang, S.; Umar, A.; Dar, G.; Kim, S.; RI, B. Synthesis and characterization of iron oxide nanoparticles for phenyl hydrazine sensor applications. *Sens. Lett.* **2014**, *12*, 97–101. [CrossRef]

38. Bhatt, R.; Sreedhar, B.; Padmaja, P. Chitosan supramolecularly cross linked with trimesic acid—Facile synthesis, characterization and evaluation of adsorption potential for chromium(VI). *Int. J. Biol. Macromol.* **2017**, *104*, 1254–1266. [CrossRef] [PubMed]

39. Ablouh, E.; Hanani, Z.; Eladlani, N.; Rhazi, M.; Taourirte, M. Chitosan microspheres/sodium alginate hybrid beads: An efficient green adsorbent for heavy metals removal from aqueous solutions. *Sustain. Environ. Res.* **2019**, *29*, 1–11. [CrossRef]

40. Xia, Z.; Baird, L.; Zimmerman, N.; Yeager, M. Heavy metal ion removal by thiol functionalized aluminum oxide hydroxide nanowhiskers. *Appl. Surf. Sci.* **2017**, *416*, 565–573. [CrossRef]

41. Peak, D.; Ford, R.G.; Sparks, D.L. An in situ ATR-FTIR investigation of sulfate bonding mechanisms on goethite. *J. Colloid Interface Sci.* **1999**, *218*, 289–299. [CrossRef]

42. Ou, J.; Sheu, Y.; Tsang, D.; Sun, Y.; Kao, C. Application of iron/aluminum bimetallic nanoparticle system for chromium-contaminated groundwater remediation. *Chemosphere* **2020**, *256*, 127158. [CrossRef]

43. Gallios, G.P.; Vaclavikova, M. Removal of chromium (VI) from water streams: A thermodynamic study. *Environ. Chem. Lett.* **2008**, *6*, 235–240. [CrossRef]

44. Yang, J.; Yu, M.; Chen, W. Adsorption of hexavalent chromium from aqueous solution by activated carbon prepared from longan seed: Kinetics, equilibrium and thermodynamics. *J. Ind. Eng. Chem.* **2015**, *21*, 414–422. [CrossRef]

45. Gürü, M.; Venedik, D.; Murathan, A. Removal of trivalent chromium from water using low-cost natural diatomite. *J. Hazard. Mater.* **2008**, *160*, 318–323. [CrossRef]

46. Hu, C.-Y.; Lo, S.-L.; Liou, Y.-H.; Hsu, Y.-W.; Shih, K.; Lin, C.-J. Hexavalent chromium removal from near natural water by copper–iron bimetallic particles. *Water Res.* **2010**, *44*, 3101–3108. [CrossRef] [PubMed]

47. Gaffer, A.; Al Kahlawy, A.A.; Aman, D. Magnetic zeolite-natural polymer composite for adsorption of chromium (VI). *Egypt. J. Pet.* **2017**, *26*, 995–999. [CrossRef]

48. Enniya, I.; Rghioui, L.; Jourani, A. Adsorption of hexavalent chromium in aqueous solution on activated carbon prepared from apple peels. *Sustain. Chem. Pharm.* **2018**, *7*, 9–16. [CrossRef]

49. Iadicicco, N.; Paduano, L.; Vitagliano, V. Diffusion coefficients for the system potassium chromate water at 25 degrees C. *J. Chem. Eng. Data* **1996**, *41*, 529–533. [CrossRef]

50. Largitte, L.; Pasquier, R. A review of the kinetics adsorption models and their application to the adsorption of lead by an activated carbon. *Chem. Eng. Res. Des.* **2016**, *109*, 495–504. [CrossRef]

51. Chu, K.H. Fixed bed sorption: Setting the record straight on the Bohart-Adams and Thomas models. *J. Hazard. Mater.* **2010**, *177*, 1006–1012. [CrossRef]

52. Brusseau, M.L.; Jessup, R.E.; Suresh, P.; Rao, C. Nonequilibrium Sorption of Organic Chemicals: Elucidation of Rate-Limiting Processes. *Environ. Sci. Technol.* **1991**, *25*, 134–142. [CrossRef]

53. Cornelissen, G.; VanNoort, P.; Parsons, J.; Govers, H. Temperature dependence of slow adsorption and desorption kinetics of organic compounds in sediments. *Environ. Sci. Technol.* **1997**, *31*, 454–460. [CrossRef]

54. Wang, Z.; Zhao, J.; Song, L.; Mashayekhi, H.; Chefetz, B.; Xing, B. Adsorption and desorption of phenanthrene on carbon nanotubes in simulated gastrointestinal fluids. *Environ. Sci. Technol.* **2011**, *45*, 6018–6024. [CrossRef]

55. Weber, W.J.; Morris, J.C. Kinetics of Adsorption on Carbon from Solution. *J. Sanit. Eng. Div.* **1963**, *89*, 31–60. [CrossRef]

56. Alemu, A.; Lemma, B.; Gabbiye, N.; Alula, M.; Desta, M. Removal of chromium (VI) from aqueous solution using vesicular basalt: A potential low cost wastewater treatment system. *Heliyon* **2018**, *4*, e00682. [CrossRef] [PubMed]

57. Panda, H.; Tiadi, N.; Mohanty, M.; Mohanty, C.R. Studies on adsorption behavior of an industrial waste for removal of chromium from aqueous solution. *S. Afr. J. Chem. Eng.* **2017**, *23*, 132–138. [CrossRef]
58. Girods, P.; Dufour, A.; Fierro, V.; Rogaume, Y.; Rogaume, C.; Zoulalian, A.; Celzard, A. Activated carbons prepared from wood particleboard wastes: Characterisation and phenol adsorption capacities. *J. Hazard. Mater.* **2009**, *166*, 491–501. [CrossRef] [PubMed]
59. Langmuir, I. The adsorption of gases on plane surfaces of glass, mica and platinum. *J. Am. Chem. Soc.* **1918**, *40*, 1361–1403. [CrossRef]
60. Chang, J.; Wang, H.; Zhang, J.; Xue, Q.; Chen, H. New insight into adsorption and reduction of hexavalent chromium by magnetite: Multi-step reaction mechanism and kinetic model developing. *Colloids Surf. A-Physicochem. Eng. Asp.* **2021**, *611*, 125784. [CrossRef]
61. Vansek, P. *Corrosion: Materials*; Electrochemical Series; CRC Press: London, UK, 2018; pp. 665–671.

minerals

Article

Ferric Hydroxide Recovery from Iron-Rich Acid Mine Water with Calcium Carbonate and a Gypsum Scale Inhibitor

Tumelo Monty Mogashane [1,2,*], Johannes Philippus Maree [1,3,*], Munyaradzi Mujuru [1], Mamasegare Mabel Mphahlele-Makgwane [1] and Kwena Desmond Modibane [2]

1 Department of Water and Sanitation, University of Limpopo, Private Bag X1106, Sovenga 0727, South Africa
2 Department of Chemistry, University of Limpopo, Private Bag X1106, Sovenga 0727, South Africa
3 ROC Water Technologies, P.O. Box 70075, Die Wilgers, Pretoria 0041, South Africa
* Correspondence: monty.tumelo@gmail.com (T.M.M.); maree.jannie@gmail.com (J.P.M.)

Abstract: The focus of this study was to improve the Reverse Osmosis Cooling (ROC) process by using $CaCO_3$ for neutralization and selective recovery of $Fe(OH)_3$ at pH 3.5. By using a specific inhibitor, ferric hydroxide was recovered separately from gypsum and other metals present in mine water. Ferric hydroxide was processed to pigment, a product that is imported and used as colorant in paints and tiles. In addition to pigment recovery, aluminum hydroxide and calcium carbonate can also be recovered from mine water. The following conclusions were made: (i) the rate of gypsum crystallization, in the absence of Fe^{3+}, is influenced by the over saturation concentration in solution, the seed crystal concentration and temperature; (ii) gypsum crystallization from an over-saturated solution, in the presence of $Fe(OH)_3$ sludge, required an inhibitor dosage of 100 mg/L to keep gypsum in solution for a period of 30 min; (iii) gypsum crystallization from an over-saturated solution, in the presence of both $Fe(OH)_3$ sludge and $CaCO_3$ reactant, required a higher inhibitor dosage than 100 mg/L to keep gypsum in solution for a period of 30 min. A dosage of 200 mg/L kept gypsum in solution for the total reaction period; (iv) when only $Fe(OH)_3$ is present in the slurry, gypsum inhibition is more effective when $Fe(OH)_3$ sludge is allowed to settle after the initial mixing; (v) when both $Fe(OH)_3$ and $CaCO_3$ are present in the slurry, gypsum inhibition is more effective when the inhibitor is added over a period of time (10 min) rather than applying the total dosage at time zero; (vi) $Fe(OH)_3$ can be changed to yellow pigment (Goethite) by heating to 150 °C and to red pigment (Hematite) by heating to 800 °C. Pigment of nano particle size was produced; (vii) in the case of Na_2CO_3, the TDS increased from 12,660 mg/L in the feed to 13,684 mg/L due to the replacement of metal ions (Fe^{3+}, Al^{3+}, Fe^{2+}, Mn^{2+} and Ca^{2+}) with Na^+ in solution. In the case where $CaCO_3$ was used for the removal of Fe^{3+} and Al^{3+}, $Ca(OH)_2$ for the removal of Fe^{2+}, Mn^{2+}, and Na_2CO_3 for the removal of Ca^{2+}, the TDS dropped from 12,661 mg/L to 2288 mg/L, due to gypsum precipitation. The alkali cost in the case of calcium alkalis amounted to ZAR29.43/m^3 versus ZAR48.46/m^3 in the case of Na_2CO_3.

Keywords: acid mine drainage; neutralization; crystallization; inhibition; calcium sulfate; pigment

Citation: Mogashane, T.M.; Maree, J.P.; Mujuru, M.; Mphahlele-Makgwane, M.M.; Modibane, K.D. Ferric Hydroxide Recovery from Iron-Rich Acid Mine Water with Calcium Carbonate and a Gypsum Scale Inhibitor. *Minerals* **2023**, *13*, 167. https://doi.org/10.3390/min13020167

Academic Editor: Jean-Francois Blais

Received: 29 November 2022
Revised: 13 January 2023
Accepted: 22 January 2023
Published: 24 January 2023

1. Introduction

Mining processes produce huge volumes of solid and liquid waste that must be properly managed to reduce environmental risk. Mine tailings, the by-product of mineral processing, are typically dumped as slurry in sizable impoundments or storage facilities [1–3]. Acid Mine Drainage (AMD) is difficult and expensive to treat because it contains many hazardous metals. If AMD is not adequately managed it causes significant environmental degradation, water contamination, aquatic ecosystem, and has a severe health impact on nearby communities [1,4,5]. According to South African legislation, mine water must be treated to a level suitable for drinking water, and ideally there should be zero-waste [6]. The physical and chemical stability of the acid mine waste and reservoirs is the primary issue with regard to the management of AMD [5,7]. Furthermore, the next problem is the storage of sludge before it is collected, and the sludge must be treated to easily regulate and stabilize pollutants in it; however, this process is costly. Before treatment plants can be built, Environmental Impact Assessment (EIA) studies are required if waste is produced. By stopping the production of any waste such lengthy investigations and approval procedures can be avoided. Therefore, operations must be able to produce products that can be sold, rather than mixed sludge that must be dumped on the ground or brines in evaporation ponds. For instance, it has been reported that AMD sludge can be recycled to create adsorbents that can remove pollutants from wastewater such as rare earth elements and heavy metals [7].

Gypsum is crucial to the pre-treatment and desalination phases of the mine water treatment process. Gypsum precipitates during the pre-treatment phase when calcium alkalis are used to neutralize the metal hydroxides. Gypsum scaling of the membranes during the desalination process must be prevented by regularly washing the membranes with chemical solutions and dosing anti-scalants [8]. Calcium sulfate occurs in three different crystalline forms: calcium sulfate dihydrate (gypsum), $CaSO_4 \cdot 2H_2O$; calcium sulfate hemihydrate (plaster of Paris), $CaSO_4 \cdot \frac{1}{2}H_2O$; and calcium sulfate anhydrite ($CaSO_4$). Gypsum is the most common calcium sulfate scale found in cooling water and reverse osmosis (RO) based desalination systems, whereas $CaSO_4 \cdot \frac{1}{2}H_2O$ and $CaSO_4$ are the most frequently formed salts in high temperature processes such as multi-stage distillation and geothermal [9,10]. The solubility of all forms of calcium sulfate changes with increasing temperature. The supersaturation level of the solution affects the force that causes crystallization [11]. Additionally, the presence of metal ions like Fe and Al and contaminants complicates the scaling problem and could render the conventional antiscalant ineffective [12].

Several experiments were performed to assess calcium sulfate scale formation and inhibition in the presence of metals at low pH [13,14]. Genesys International Limited has formulated several antiscalants specific to mining including Genmine AS34, Genmine AS45, Genmine AS26 and Genmine AS65. Genmine AS26 was developed particularly for acidic mine waters and is mainly effective at inhibiting the formation of $CaSO_4$ scale at low pH [14]. Investigations into the dissociation of sulfate and bi-sulfate ions at low pH levels were followed by threshold jar tests to screen water chemistries under various conditions, and finally evaluations of membrane performance and scaling inhibition using actual membrane coupons with a Flat Sheet Test rig [13]. Researchers have suggested several strategies over the past three decades for preventing the formation of scale, including the use of acids, chelants, ion exchangers, and inhibitors [15–17]. Early researchers on gypsum scaling mainly focused on the kinetics of scale formation, while later studies put the emphasis on the effects of external factors such as hydrodynamics [8,18,19].

ROC Water Technologies has developed the ROC process for treatment of mine water and continues to identify further improvements [20]. In the ROC process, acidic or neutral mine water is treated with Na_2CO_3 and/or NaOH and/or MgO in the pre-treatment stage to allow selective precipitation of metals ($Fe(OH)_3$, $Al(OH)_3$, $CaCO_3$, MnO_2 and $Mg(OH)_2$ [20–22]. After pre-treatment, the sodium-rich water is passed through a membrane stage to produce drinking water and brine. The brine has a sufficient TDS concentration to allow Na_2SO_4 crystallization upon cooling. Figure 1 shows the modified process configuration of the ROC process that made provision for various improvements. These improvements could result in reduced treatment costs and recovery of valuable products from mine water.

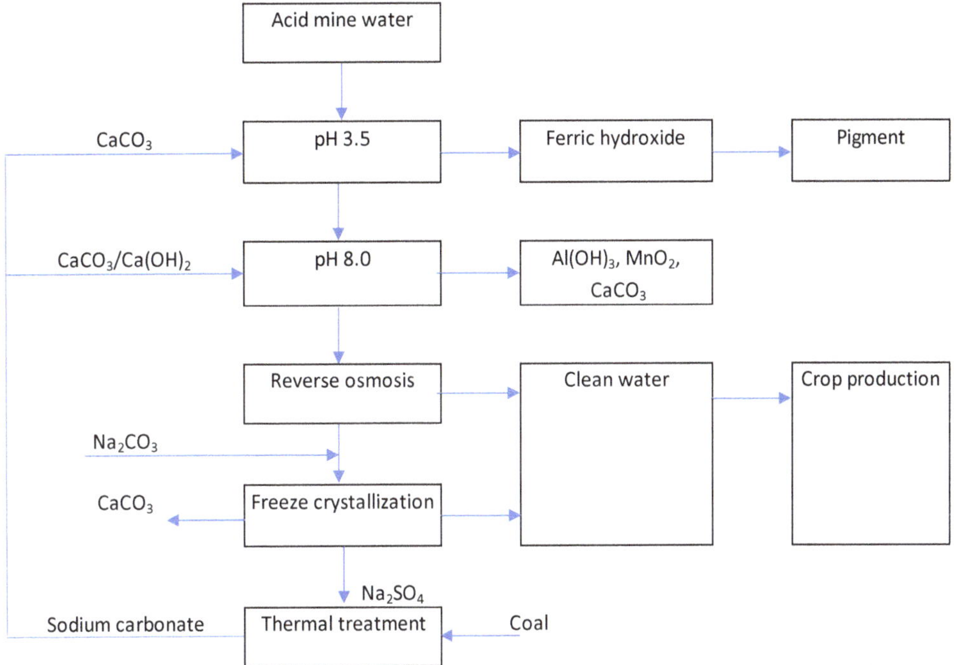

Figure 1. Ideal solution for mine water treatment.

$CaCO_3$ neutralization

If Na_2CO_3 can be replaced with $CaCO_3$ or $Ca(OH)_2$ for selective recovery, the alkali cost will be reduced due to its different prices: Na_2CO_3 (ZAR5 000/t) compared to the conventional alkalis: $CaCO_3$ (ZAR750/t) and $Ca(OH)_2$ (ZAR2 500/t). In order to prevent simultaneous precipitation of gypsum and metal hydroxides, an anti-scalant is used to keep gypsum in solution for the period needed for metal removal.

Pigment formation

Pigment can be recovered from the $Fe(OH)_3$-sludge precipitated at pH 3.5. In a wide range of applications, including energy generation and storage, catalytic transformations, and water treatment, metal oxide nanoparticles are becoming increasingly important [23,24]. The nanotechnology community is very interested in controlling size and shape since these factors are crucial in deciding how well nanoparticles work, affecting qualities like reactivity, conductivity, and magnetic behavior. Supercritical fluid nanoparticle synthesis is a reliable and simple way to meet the need to control size and form for a variety of metal oxide nanoparticles, in addition to using environmentally friendly solvents [25,26]. As reported, AMD contains higher concentrations of Fe ions that are worth recovering. Few researchers have recovered Fe from mine water for the synthesis of pigments and other industrial products [2]. The possibility of recovering iron compounds from AMD is very essential, feasible and doable given the amount of AMD produced annually [6,27,28]. Iron oxide nanoparticles (NPs) have a great adsorption capacity, are inexpensive, have improved stability, and are simple to separate, giving them potential for industrial scale wastewater treatment [29,30].

The following objectives were set for this investigation: (i) kinetics of gypsum crystallization; (ii) identify an inhibitor that will prevent gypsum crystallization in the presence of $Fe(OH)_3$; (iii) recovery of $Fe(OH)_3$ and processing to pigment ($FeOOH$ (goethite) and

Fe_2O_3 (hematite)); and (iv) determine the feasibility when Na_2CO_3 is replaced with $CaCO_3$ for pigment recovery.

2. Materials and Methods

2.1. Feedstock

Neutralization ($CaCO_3$ (Kulu Lime, South Africa)). Acid mine water samples were collected from Emalahleni, Mpumalanga Province, South Africa. The AMD collected from the Top Dam water was rich in Fe^{3+} due to extensive exposure to oxygen over a long period in a shallow pond. Reagents of high purity were used in this study. Nitric acid, hydrochloric acid, Hydrophilic polypropylene membrane filter (0.45 μm), Anhydrous sodium sulfate, Na_2SO_4, $CaCl_2$, $KMnO_4$, $NaOH$ and NH_4 were all supplied by Sigma-Aldrich (Chemie GmbH, Germany). Table 1 shows the chemical compositions of the Top Dam water. The Top Dam had a high concentration (4500 mg/L Fe^{3+}), which made it the preferred water to treat. The Fe^{3+} had a high concentration due to evaporation and would provide a higher yield of pigment compared to other waters.

Table 1. Chemical compositions of water in the Top Dam.

	Top Dam
pH	2.3
Acidity (mg/L $CaCO_3$)	14,981.0
H^+ (mg/L H)	40.0
Na^+ (mg/L)	50.0
K^+ (mg/L)	30.0
Mg^{2+} (mg/L)	300.0
Ca^{2+} (mg/L)	500.0
Mn^{2+} (mg/L)	200.0
Fe^{2+} (mg/L)	400.0
Fe^{3+} (mg/L)	4500.0
t-Fe (mg/L)	5000.0
Al^{3+} (mg/L)	300.0
Si^{4+} (mg/L)	60.0
Sr^{2+} (mg/L)	0.0
Ba^{2+} (mg/L)	0.0
SO_4^{2-} (mg/L)	19,095.2
Cl^- (mg/L)	200.0
TDS (mg/L)	25,475.2
Cations (meq/L)	397.8
Anions (meq/L)	397.8

Inhibition. Solutions, over-saturated with respect to $CaSO_4$, were prepared by mixing Na_2SO_4, $CaCl_2$ and the inhibitor solutions. The inhibitors tested were commercial materials. The desired concentrations of the inhibitor were obtained by dilution. AMD from the Top Dam at the Khwezela Colliery site (Mpumalanga, South Africa) was the feed water.

Inhibitor B is a fully neutralized and low molecular weight polyacrylic acid. Inhibitor A is an aqueous solution of polymeric phosphates with multifunctional additives and C is an aqueous solution of phosphonates and carboxylates with multifunctional additives. $Fe(OH)_3$ produced during neutralization of AMD to pH 3.5 was used for pigment studies.

2.2. Equipment

Neutralization and Inhibition. Batch studies were performed in 1000 mL beakers to determine the rate of neutralization. A portable pH/Electrical Conductivity (EC) meter (HACH HQ4OD, Aqualytic, South Africa) was used to measure pH and EC readings of the samples during the experiments. A high temperature muffle furnace (Carbolite, type s30 fitted with 2AU ESF Eurotherm, England) was used to heat recovered Fe(III) sludge. A 4-paddle stirrer (Model 1924, Electronics, India) was used for stirring solutions.

2.3. Procedure

Neutralization. Acid mine water from the Top Dam at the Navigation mine site (Mpumalanga, South Africa) was used as feed water. Beaker studies were carried out to measure the rate of neutralization with $CaCO_3$. Acid water (500 mL) was poured into beakers (1000 mL) and stirred at 250 rpm. Alkali was added at the beginning of time. Samples were taken at regular intervals (0, 10, 30, 60 and 180 min), filtered, pH recorded and analyzed for acidity, conductivity, Fe and Ca. Sludge was added to serve as a promoter of crystal growth.

Inhibition. Calcium sulfate was produced by mixing equal volumes of a 0.25 M $CaCl_2 \cdot 2H_2O$ solution and a 0.25 M Na_2SO_4 solution at room temperature. The inhibitor dosage varied between 0 and 400 mg/L. Gypsum seed crystals were added to catalyze the gypsum crystallization. The kinetics of the reactions in the absence and presence of inhibitors were monitored.

Pigment formation. A mass of 5 g $Fe(OH)_3$ was placed in porcelain crucibles and subjected to various temperatures for different time periods.

2.4. Experimental

Neutralization. The effects of the following parameters were investigated over the given reaction time: Alkali dosage and temperature (25–55 °C).

Inhibition. Inhibitors A, B and C were assessed to prevent or slow down the rate of gypsum crystallization.

Pigment formation. The effects of the following parameters were investigated:

- Temperature (150 °C and 800 °C);
- Heating period (60 min).

2.5. Analytical

Standard procedures were used to collect samples at various phases, filter them through Whatman No. 1 paper, and measure their contents of Fe(II), Fe(III), pH, Ca, and alkalinity [31]. Metals were analyzed using Inductively coupled plasma-atomic emission spectroscopy (ICP-OES) (iCAP 7000 Series, ANATECH, South Africa). Fe(II) concentrations were determined by adding filtered sample (5 mL), 0.1 N H_2SO_4 (5 mL) and Zimmerman-Reinhardt (ZR) reagent (5 mL) to an Erlenmeyer flask and titrating the solution with 0.05 N $KMnO_4$ until pale pink [31]. Because magnesium was absent, calcium was chosen as the measure of overall total hardness. Filtered sample (5 mL), deionized water (45 mL), dilute NH_4 (5 mL) and two drops of Eriochrome Black T indicator were added to an Erlenmeyer flask (100 mL). The solution was titrated with 0.01 M EDTA to a blue endpoint. Acidity was determined by titration of sample (5 mL) to pH 8.3 using 0.1/1 N NaOH [31]. Fe, Al, Na, Ca, Ni, Mg, K and Mn were analyzed using ICP-OES. The pH/EC meters were calibrated before the start of each set of experiments and during the experiment using calibration buffers.

2.6. Characterisation of the Sludge

Morphological and elemental properties of the synthesized pigments were determined using High Resolution Field Emission Scanning Electron Microscopy (HR-FESEM) equipped with the means to perform Energy-dispersive X-ray spectroscopy (EDS). Specifically, the Auriga Cobra FIB-FESEM (Model: Sigma VP FE-SEM with Oxford EDS Sputtering System, Make: Carl Zeiss, Supplier: Carl Zeiss, USA). Particle size distribution was determined from SEM images of the particles by manually counting and delineating the particles displayed on the image. The ferrite crystallinity was determined using X-ray diffraction on a Bruker D2, 30 kV, 10 mA utilizing monochromatic CuKα radiation (k = 1.54184 Å) from 5–90° (XRD Analytical Solutions and Consulting). Al_2O_3 was used in the form of corundum as a reference material for the applied standard reflection, which was pretreated at high temperature to obtain a highly crystalline with a zero full width at half maximum (FWHM). The patterns were recorded from 10° to 90° (2θ) with a scanning speed of 4°/min at 30 kV and 10 mA. The width of the standard incident and receiving Soller slits were 2.5° and 0.5°, respectively.

2.7. OLI

In this work the behavior of metals dissolved in water under the influence of alkalis like Na_2CO_3 (Botash, Botswana) and $CaCO_3$ was predicted using the OLI ESP software program. The reactions were modelled using the OLI Analyzer System by running a simulated AMD sample with fictitious settings for temperature, pressure, and pH [32]. Base titrants Na_2CO_3, $CaCO_3$ and MgO (Chamotte Mining, South Africa) were employed. When MgO was used to neutralize the pH to 3.5 and then Na_2CO_3 to elevate the pH to 8.6, the influence of temperature on the solubility of $MgSO_4$ and Na_2SO_4 was identified. The OLI Systems Chem Analyzer would display a calculated summary of the simulated outcomes once the input values were run. According to the precise qualities, this might be utilized to forecast the actual reactions to incorporate into the treatment techniques. As a result, it was applied to improve an AMD neutralization–precipitation–desalination process.

3. Results and Discussion

3.1. Kinetics of Gypsum Crystallization—No Inhibitor

Limestone ($CaCO_3$) and/or lime ($Ca(OH)_2$) are currently used by the mining industry for neutralization of acid mine drainage. $CaCO_3$ has the lowest cost (typically ZAR700/t) and $Ca(OH)_2$ has the second lowest cost (typically ZAR2 200/t). Calcium alkalis can only be utilized for selective recovery of metals if co-precipitation of gypsum can be avoided through the dosing of inhibitors. Liu and Nancollas determined the rate of gypsum crystallization and found it to be a second order reaction with respect to the over-saturation concentration of gypsum in solution. It is also related to the gypsum seed crystal concentration (Equation (1)) [15,16]. This finding was confirmed, as shown in Figure 2. Figure 3 confirmed that the reaction order is 2 with respect to the over-saturation concentration of gypsum in solution.

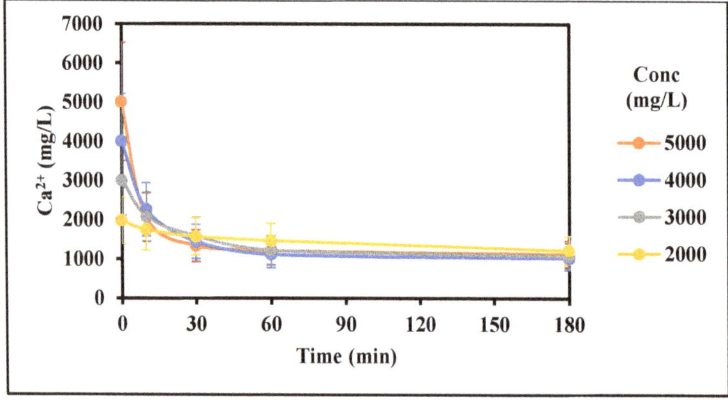

Figure 2. Effect of concentration on the rate of gypsum crystallization (2000–5000 mg/L $CaSO_4$ (as Ca), 5 g/L gypsum seed, 0 g/L Fe^{3+}, 0 mg/L Inh, 25 °C, stirring rate: 200 rpm; Stirring time (min): 180/180).

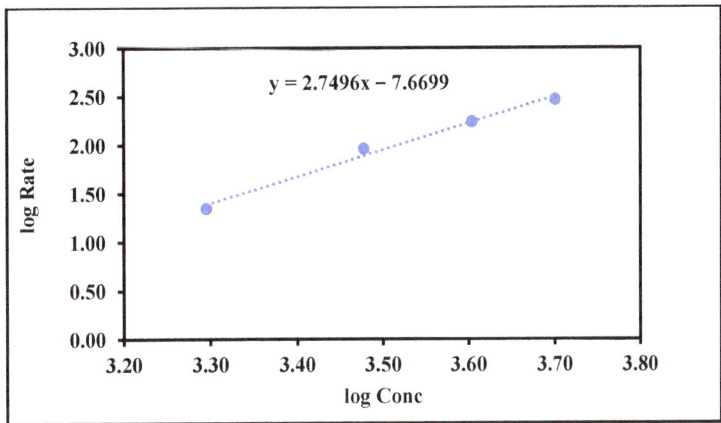

Figure 3. Determination of the reaction order.

The effects of temperature on the formation of gypsum crystals were also investigated. Figure 4 shows the effect of temperature on the reaction rate, the higher the temperature, the faster the rate of crystallization. The Arrhenius equation log k = log A − E/(2.303RT) was used to estimate the value of the reaction rate k at other temperatures. The amounts E, R, and log A have the values 4.80 kcal/mole (activation energy), 1.987 (a constant) and 5.52 cal mole^{-1} degree^{-1} (gas constant), respectively (Figure 5).

$$R = k \cdot S \cdot (C - C_0)^2 \tag{1}$$

where R—rate; k—; S—Surface area; C—$CaSO_4$ concentration; C_0—Equilibrium $CaSO_4$ concentration.

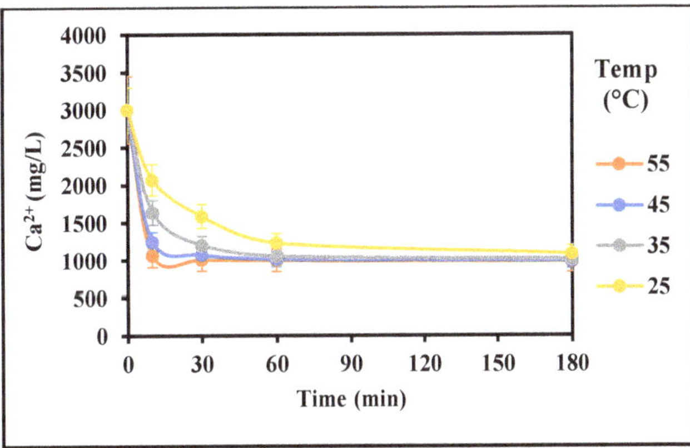

Figure 4. Effect of temperature on the rate of gypsum crystallization (3000 mg/L $CaSO_4$ (as Ca), 5 g/L gypsum seed, 0 g/L Fe^{3+}, 0 mg/L Inh, 25–55 °C, stirring rate: 200 rpm; Stirring time (min): 180/180).

3.2. Inhibition of Gypsum Crystallization in Artificial Mine Water

The aim of this study was to recover $Fe(OH)_3$ in the presence of a solution that is over-saturated with respect to $CaSO_4$ in solution prepared artificially. A solution that is over-saturated with respect to $CaSO_4$ was prepared by mixing the inhibitor with solutions of Na_2SO_4 and $CaCl_2$. The effect of $Fe(OH)_3$ on the inhibition of $CaSO_4$ was determined by dosing $FeCl_3$ and the equivalent amount of NaOH. Furthermore, the inhibition of gypsum crystallization when real mine water was neutralized with $CaCO_3$ was also investigated. A study by Rabizadeh (2016) [18] showed that when the pH in the reacting solution was switched from 4 to 7, the efficiency of the low molecular weight poly(acrylic acid) in preventing gypsum formation increased, while it resulted in an adverse effect on the performance of poly(acrylic acid) with higher molecular weight by forming a "net-structure" in the solution [18].

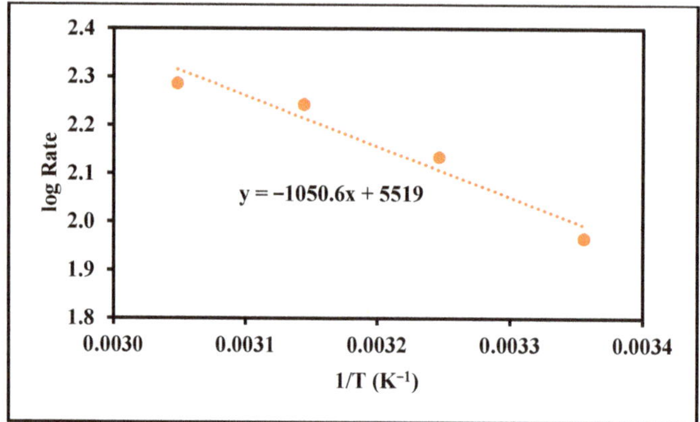

Figure 5. Determination of the activation energy.

3.2.1. Artificial Solutions

Figure 6 compares three commercially available inhibitors that were identified by Mogashane as the most promising [28]. A dosage of 100 mg/L inhibitor A kept 4000 mg/L $CaSO_4$ (as Ca) in solution for a period of 30 min in the presence of 4 g/L Fe^{3+} at pH 3.5 when stirred at a rate of 200 rpm for the total period. All three inhibitors were poli acrylate. The results reported by Rabizadeh (2016) [18] demonstrated the effects of 20 ppm poly(aspartic acid) (PASP) and poly(acrylic acid) (PAA) compounds on the crystallization of gypsum. The comparison showed that PAA worked better than the other anti-scalants, and it inhibited the formation of gypsum [18].

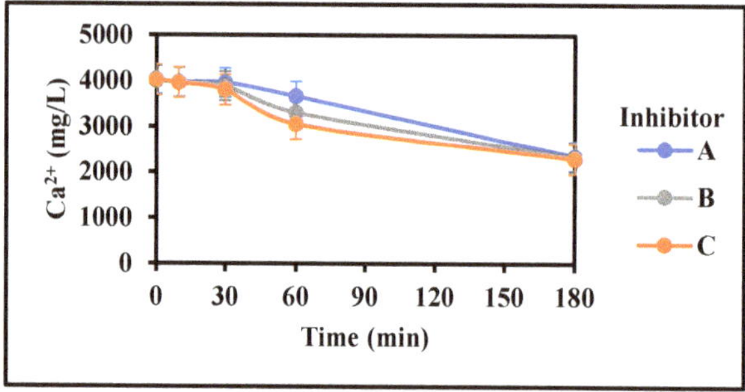

Figure 6. Comparison of various inhibitors on the rate of gypsum crystallization with initial mixing (4000 mg/L $CaSO_4$ (as Ca), 0 g/L gypsum seed, 4 g/L Fe^{3+}, 100 mg/L Inh, 25 °C, stirring rate: 200 rpm; Stirring time (min): 180/180; pH 3.5).

Figure 7 shows the effect of mixing time in the inhibition of gypsum. Gypsum inhibition was most effective when the solution was only stirred for the first 5 min of the total reaction period of 180 min. This can be ascribed to the settling of the $Fe(OH)_3$ sludge to the bottom after mixing was stopped. The inhibitor remained in the over-saturated gypsum solution and was more effective than when in contact with suspended material on which it could absorb.

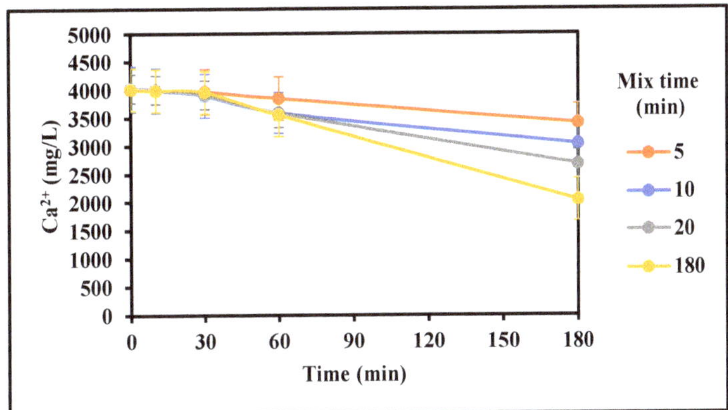

Figure 7. Effect of mixing time on the rate of gypsum crystallization with initial mixing (4000 mg/L $CaSO_4$ (as Ca), 0 g/L gypsum seed, 4 g/L Fe^{3+}, 0–100 mg/L Inh. A, 25 °C, stirring rate: 200 rpm; Stirring time (min): 5 to 180/180).

Figures 8 and 9 compare the effect of inhibitor A doses when stirred for the full 180 min of the reaction and only for 5 min of the reaction, respectively. As shown in Figure 7, inhibition was much more effective when the solution was stirred for only 5 min than when stirred for the full 180 min.

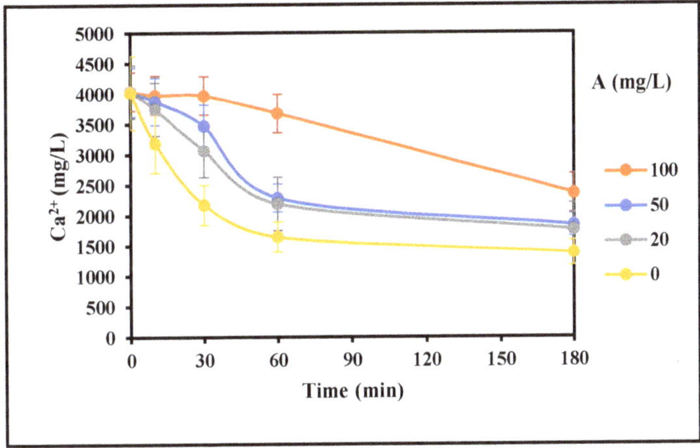

Figure 8. Effect of inhibitor A concentration on the rate of gypsum crystallization with mixing (4000 mg/L $CaSO_4$ (as Ca), 0 g/L gypsum seed, 4 g/L Fe^{3+}, 0–100 mg/L Inh. A, 25 °C, stirring rate: 200 rpm; Stirring time (min): 180/180).

3.2.2. Acid Mine Water

The oversaturated gypsum solution was prepared by mixing $CaCl_2$ and Na_2SO_4 solutions in the presence of an inhibitor. $Fe(OH)_3$ sludge was produced by adding an $FeCl_3$ solution and NaOH to adjust the pH to 3.5. It was realized that this condition differs from the situation that will be applied in practice. In the real situation, a $CaCO_3$ (limestone) slurry will be dosed to an $Fe_2(SO_4)_3$ solution to raise the pH to 3.5. This differs in the sense that the inhibitor needs to inhibit gypsum crystallization, not only in the absence of any solids, or in the presence of only $Fe(OH)_3$, but in the presence of both $Fe(OH)_3$ and $CaCO_3$ solids. Figure 10 shows that when Ca^{2+} came from a solution of $CaSO_4$, it remained completely in solution for 30 min. In the case where Ca^{2+} came from $CaCO_3(s)$, gypsum crystallization took place at a slow rate during the first 30 min. It was concluded that higher inhibitor doses may be needed when $CaCO_3$ is used for gypsum inhibition to allow selective recovery of pigment. Figures 10–12 show that gypsum crystallization could still be suppressed when $CaCO_3$ is dosed, but that higher inhibitor concentrations were needed.

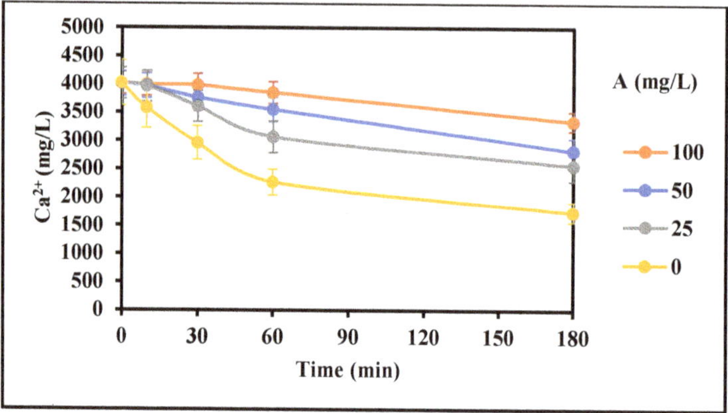

Figure 9. Effect of inhibitor A concentration on the rate of gypsum crystallization with initial mixing (4000 mg/L $CaSO_4$ (as Ca), 0 g/L gypsum seed, 4 g/L Fe^{3+}, 0–100 mg/L Inh. A, 25 °C, stirring rate: 200 rpm; Stirring time (min): 5/180).

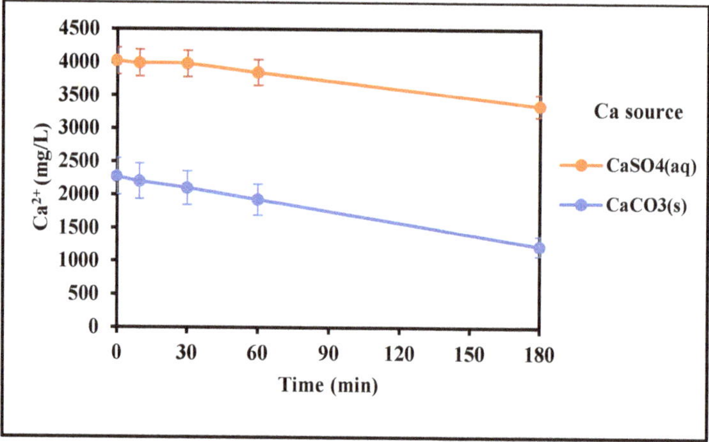

Figure 10. Effect of $CaCO_3$ addition period on gypsum inhibition (Top Dam water, 15 g/L Acidity, 4 g/L Fe^{3+}, 22 g/L $CaCO_3$, 0 g/L gypsum seed, 100 mg/L Inh. A, 25 °C, stirring rate: 200 rpm; Stirring time (min): 180/180; $CaCO_3$ addition: Time 0 to 10 min).

Figure 10 compares the inhibition of gypsum crystallization when Ca originated from solution (when $CaSO_4$ and $CaCl_2$ was mixed) and when added as a solid to neutralize Fe^{3+} and H^+ in solution. The Ca^{2+} in the artificial mine water was higher than in the actual mine water due to different Fe^{3+} concentrations, namely 3220 mg/L Fe^{3+} and 503 mg/L Fe^{2+} in the artificial mine water and 2560 mg/L Fe^{3+} and 251 mg/L Fe^{2+} in the actual mine water. In the artificial mine water no gypsum crystallization took place during the first 30 min, while in the presence of $CaCO_3$(s) some crystallization was noticed. This indicated that $CaCO_3$ as a solid absorbs a portion of the inhibitor and requires a higher inhibitor dosage.

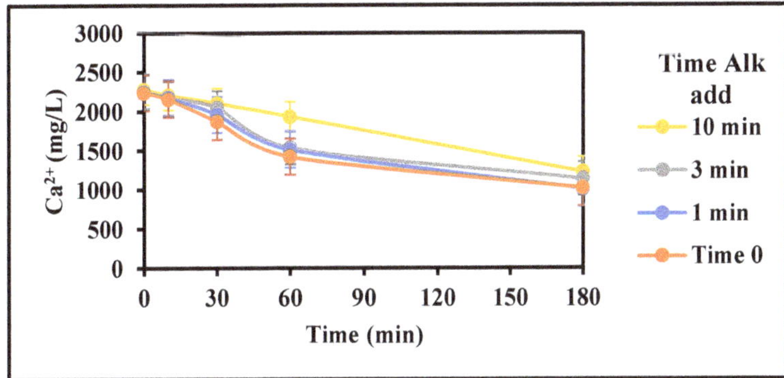

Figure 11. Effect of inhibitor concentration gypsum inhibition (Top Dam water, 15 g/L Acidity, 4 g/L Fe^{3+}, 22 g/L $CaCO_3$, 0 g/L gypsum seed, 100, 200, 400 mg/L Inh A, 25 °C, stirring rate: 200 rpm; Stirring time (min): 180/180; $CaCO_3$ addition period: 10 min).

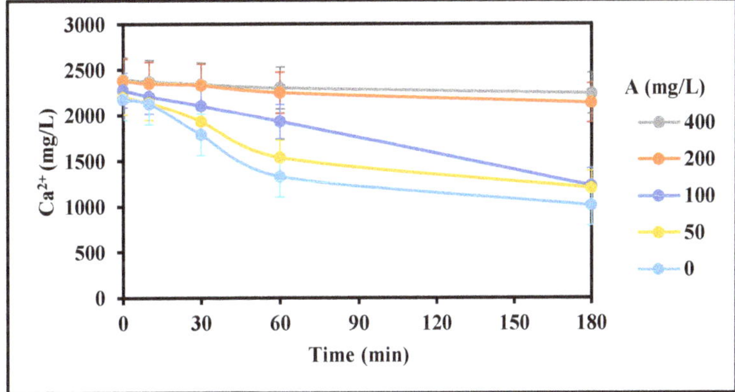

Figure 12. Effect of $CaCO_3$ solids on the efficiency of inhibition of gypsum crystallization (see Table 2 for metal concentrations).

Figure 11 shows the effect of time (immediately, or over 1, 3 and 10 min) in which $CaCO_3$ is added to actual acid mine water. When $CaCO_3$ was added over a 10 min period it was noticed that gypsum remained for a longer time in solution than when added over shorter periods. This can be ascribed to the rapid dissolution of $CaCO_3$ at low pH values, leaving less solid $CaCO_3$ in suspension that could absorb the inhibitor. This conclusion needs to be confirmed with more experiments.

Figure 12 shows that 200 mg/L Inhibitor A keeps gypsum in solution for more than 180 min when actual mine water containing 4 g/L Fe^{3+} is neutralized with $CaCO_3$. This is in line with the study of Fazel et al. (2019) who investigated calcium sulfate scale formation in acidic pH and in the presence of a variety of soluble metals. The study revealed that Inhibitor A was able to prevent scale formation at low pH, especially for AMD waters, and gave near 100% inhibition [13].

3.3. Pigment Formation

$Fe(OH)_3$ produced during neutralization of mine water with $CaCO_3$ can be changed to yellow pigment (Goethite) by heating to 150 °C, and to red pigment (Hematite) by heating to 800 °C (Figure 13). Mogashane et al. (2022) investigated the effect of temperature on the color when $Fe(OH)_3$ produced during neutralization of AMD with Na_2CO_3 is heated. Their study showed that temperature has a dominant influence on the color [28]. Figure 13 shows the effect of temperature on the color when $Fe(OH)_3$ is heated. It indicates that temperature has a dominant influence on the color. Figure 13 shows examples of goethite and hematite that were produced from acid mine-water.

Table 2. Metals concentrations in Figure 3.

Parameter	Unit	Artificial Mine Water	Actual Mine Water
Fe^{3+}	mg/L	3220	2569
Fe^{2+}	mg/L	503	251
Ca^{2+} after Fe^{3+} removal	mg/L	5000	3460
AS26 inhibitor	mg/L	100	100
Prepared from		$CaSO_4$ and $FeCl_3$	Mine water and $CaCO_3$
Stirring rate	rpm	200	200
Stirring time	min	180/180	180/180
Temperature	°C	25	25

Figure 13. Pigments produced from iron-rich mine water using $CaCO_3$ for neutralization.

The XRD analysis was employed to assess the crystallite sizes and phase structures of Goethite Nanoparticles (GNPs) and Hematite Nanoparticles (HNPs). Particles are classified as nanoparticles if their size is smaller than 100 nm. The $Fe(OH)_3$ was recovered by adjusting the coal leachate with $CaCO_3$ to pH 3.5. As shown in Figure 14, after separation of the $Fe(OH)_3$ from the water the $Fe(OH)_3$-sludge was dried and heated to different temperatures. Goethite (yellow) was produced at 150 °C and hematite (red, 4B) at 800 °C. The XRD pattern showed that the synthesized product contained iron oxide nanoparticles both in crystalline and in amorphous state. The black line (not marked) represents the XRD for Sample 1B at 25 °C possessing the intense reflection around 11°, 22° and 30° 2theta due to the presence of the diffractogram shows the reflections of $Al(OH)_3$ (JCPDS#70-2038), $Ca(OH)_2$ (JCPDS#01-073-5492) and $CaCO_3$ (JCPDS#47-1743) [19], respectively. This advocates that the Sample 1B contains layered double hydroxides that may have formed with interlayer carbonate and sulfate anions even in this slightly acidic environment, which was in agreement with the observation from EDS. The sample was not converted to goethite. Furthermore, the XRD pattern (Figure 14) obtained after heating samples at different temperatures revealed that goethite (2B) is completely transformed to hematite (4B). The blue line (300 °C) corresponding with goethite is shown by the red line in Figure 14. The diffraction peaks of HNPs (4B) are well defined, indicating that the crystalline nature of the prepared hematite products represents a practical route to prepare α-Fe_2O_3 of high purity. The formed XRD patterns correspond to the characteristic α-Fe_2O_3 pattern reported for hematite in literature [33]. However, the XRD pattern of hematite showed lower intensity as compared to the one reported in the literature [33], due to the presence of impurities. This shows that HNPs were successfully recovered in a crystalline state and these results were found to be in good agreement with the previously reported values [33,34].

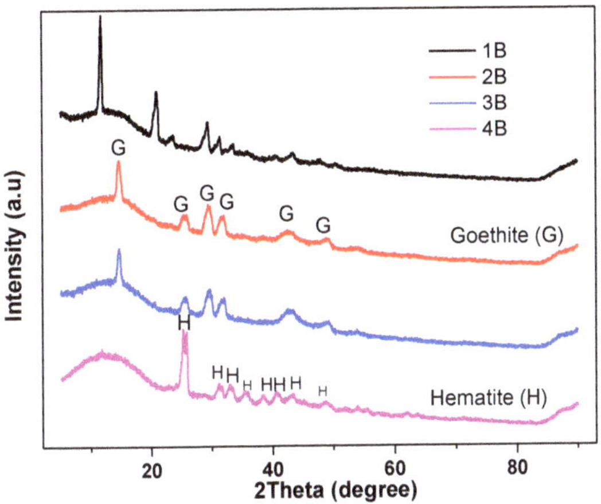

Figure 14. XRD patterns of Nanoparticles (NPs) samples produced from AMD. Notes: 1B = (25 °C-CaCO$_3$ as alkali); 2B = Goethite (150 °C-CaCO$_3$ as alkali); 3B = Goethite (300 °C-CaCO$_3$ as alkali); 4B = Hematite (800 °C-CaCO$_3$ as alkali).

The XRD pattern of the goethite nanoparticles (GNPs) (2B) sample was also evaluated as shown in Figure 14. The diffraction peaks are slightly broad, indicating a smaller crystal size. Its XRD patterns contain all the major peaks referring to JCPDS card No. 29-0713 [35], thus representing the formation of α-FeOOH. The diffraction peaks and lattice parameters of the GNPs in this study were found to be in good agreement with those reported in the literature [33,35]. The XRD results showing low in the α-FeOOH were contaminated with Ca, coming from the CaCO$_3$, S and Al^{3+}, coming from the SO$_4{}^{2-}$ and Al^{3+} in the mine water.

The average crystallite sizes of the samples (1B, 3B), Hematite (4B) and Goethite (2B) were calculated using the Debye–Scherrer formula (Equation (2)) [36]:

$$D = 0.9\lambda/\beta Cos\theta \tag{2}$$

where β is the full-width at half-maximum (FWHM) measured in radians, λ is the X-ray wavelength of Cu-Kα radiation (1.5406 Å), and θ is Bragg's angle. The crystallite sizes (Table 3) were found to be 11.9 nm, 7.02 nm, 7.27 nm and 8.07 nm for samples 1B, 2B, 3B and 4B, respectively. The HNPs showed to have large particle size as compared to Goethite nanoparticles owing to the improved density of active centers for nucleation in the prepared nano-product. The findings revealed that nano-pigments can be produced from Fe(OH)$_3$ that is recovered in the pre-treatment stage of the ROC process. With CaCO$_3$ as alkali, Fe(OH)$_3$ crystals with a smaller crystallite size are recovered due to a fast reaction rate [28].

Table 3. Particle size parameters for Goethite and Hematite nanoparticles.

	Temperature (°C)	FWHM (2θ)	B = FWHM*PI/180°	Average Crystallite Size D (nm) = 0.9λ/βCosθ	d-Spacing (Å) = λ/2sin(θ)
1B	25	0.74	0.00654	11.9	4.25
2B	150	0.73	0.00637	7.02	3.02
3B	300	4.50	0.0393	7.27	2.81
4B	800	7.47	0.0652	8.07	3.51

In addition, using XRD data analysis, Bragg's law (Equation (3)), the interplanar spacing d_{hkl} was also calculated for the as-synthesized HNPs and GNPs

$$\Lambda = 2d\sin(\theta) \tag{3}$$

where λ is the wavelength of the X-ray beam (1.5406 Å), d is the interplanar d spacing, and θ is the diffraction angle. The d spacing values for $Fe(OH)_3$ (1B, at 2θ angle of 20.9°), Goethite (2B, at 2θ angle of 36.8°), Goethite (3B, at 2θ angle of 31.8°) and Hematite (4B, at 2θ angle of 25.4°) are given in Table 3 and found to be 2.81 Å and 3.02 Å for Goethite nanoparticles and 3.51 Å for Hematite nanoparticles.

Figure 15 shows the SEM images and their average particle size distribution histogram of Nanoparticles samples produced from AMD. It also reveals the presence of spherical particles, which are distributed across the surface of the material. The morphological properties further demonstrated that the surface properties of the samples are homogeneous, hence confirming that a high-grade material was synthesized. A particle size distribution histogram of the material indicated that Goethite (2B) and Hematite 4B had the average particle size of 1.96 μm and 1.45 μm, respectively. Moreover, it was clear that the morphological properties of the samples were the same at different magnifications. The results obtained from this study substantiated the EDS results in terms of purity and homogeneity.

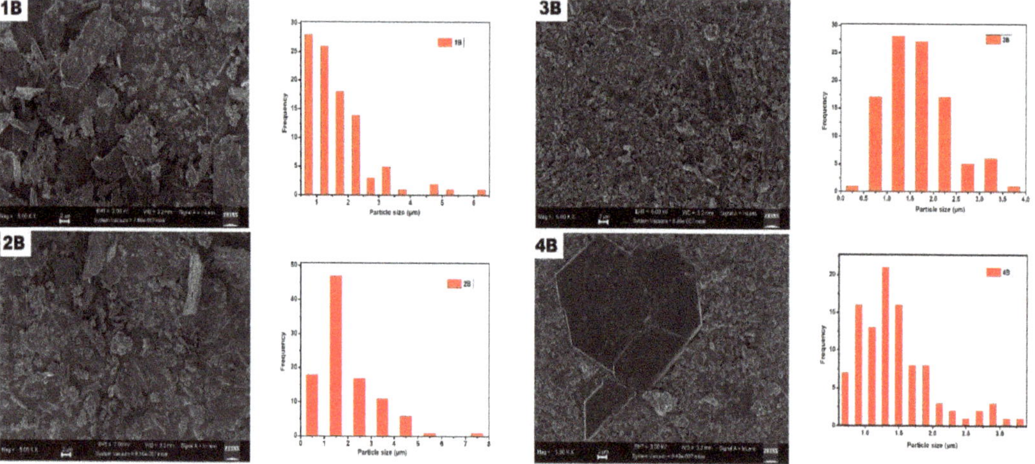

Figure 15. Scanning electron microscope (SEM) pictures and their particle size distribution histogram of Nanoparticles samples produced from AMD.

Figure 16 shows the mapping of the elemental composition of the samples (1B, 3B), Hematite (4B) and Goethite (2B). The synthesized pigments were found to be enriched with Fe and O as principal elements. Moreover, traces of other elements (Ca, C, Al, S, Si and Mn) were present. The results obtained show that the synthesized samples 1B, 2B, 3B and 4B consisted of 55.8%, 53.1%, 49.8% and 43.4%, respectively, of Fe-O mineral. This demonstrated that goethite and hematite nanoparticles were produced. However, significant levels of sulfate were found in all the samples; this was likely the result of oxyhydrosulfates that formed during the precipitation of Fe-O minerals. $CaCO_3$ alkali was added to AMD when preparing these samples. The results attained confirmed that the synthesized pigments were rich in Fe-O constituents. The XRD results showed that the a-FeOOH was contaminated with Ca, coming from the $CaCO_3$, S and Al^{3+}, coming from the SO_4^{2-} and Al^{3+} in the mine water. Further studies will be carried out to determine: (i) the value of the contaminated pigment; and (ii) if the pigment impurity can be improved through an acid wash.

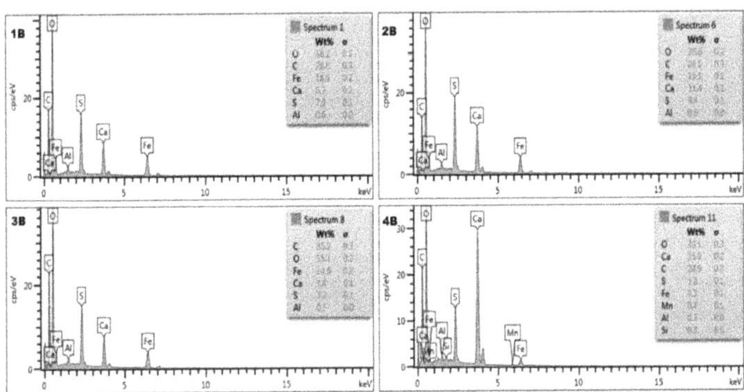

Figure 16. Energy dispersive spectroscopy (EDS) plots of nanoparticles samples produced from AMD. 1B = (25 °C-CaCO$_3$ as alkali); 2B = Goethite (150 °C-CaCO$_3$ as alkali); 3B = Goethite (300 °C-CaCO$_3$ as alkali); 4B = Hematite (800 °C-CaCO$_3$ as alkali).

3.4. Alkali Selection

3.4.1. CaCO$_3$

OLI software was used to identify which alkali (CaCO$_3$, Na$_2$CO$_3$ and Ca(OH)$_2$) will be most suitable for removal of the residual metals in solution after removal of Fe^{3+} with CaCO$_3$ (Tables 4–7). Na$_2$CO$_3$ will only be attractive if the pre-treatment stage is combined with reverse osmosis to achieve complete desalination. If only the pre-treatment stage is used for water treatment, sodium carbonate is disqualified due to the negative impact of sodium on the environment. Ca(OH)$_2$ would be an attractive option as it can remove HCO$_3^-$ as CaCO$_3$ and the residual metals as hydroxides, including Mg^{2+}.

Table 4. Removal of Fe^{3+} and other metals with only CaCO$_3$ at 1 atm (OLI simulation).

	pH	Pressure [atm] (Y2)	Fe(+3) Aq [mg] (Y2)	Fe(OH)$_3$ (Bernalite)—Sol [mg] (Y2)	CO$_2$—Vap [mg] (Y2)	Al(+3) Aq [mg] (Y2)	MnCO$_3$ (Rhodochrosite)—Sol [mg] (Y2)	Na(+1) Aq [mg] (Y2)	Mg(+2) Aq [mg] (Y2)	Ca(+2) Aq [mg] (Y2)	CaSO$_4$.2H$_2$O (Gypsum) [mg] (Y2)	C(+4) Aq [mg] (Y2)	CaCO$_3$ (Calcite)—Sol [mg] (Y2)	S(+6) Aq [mg] (Y2)	Cl(−1) Aq [mg] (Y2)
0	2.5	1.0	2000	0	0	300	0	100	200	300	0	0	0	3074	150
1000	2.6	1.0	1690	593	0	300	0	100	200	604	416	120	0	2997	150
2000	2.6	1.0	1314	1313	0	300	0	100	200	590	2194	240	0	2666	150
3000	2.7	1.0	938	2033	0	300	0	100	200	577	3971	360	0	2335	150
4000	2.7	1.0	562	2751	360	300	0	100	200	564	5745	382	0	2004	150
5000	2.8	1.0	192	3460	794	300	0	100	200	554	7509	383	0	1676	150
6000	3.5	1.0	4.0	3820	1228	226	0	100	200	549	9252	385	0	1351	150
7000	3.7	1.0	1.4	3825	1663	50	0	100	200	546	10,983	386	0	1029	150
8000	5.5	1.0	0.0	3827	1645	0	194	100	200	618	12,397	468	0	765	150
9000	5.9	1.0	0.0	3827	1417	0	385	100	200	756	13,228	593	171	611	150
10,000	5.9	1.0	0.0	3827	1417	0	385	100	200	756	13,228	593	1171	611	150

Table 5. Removal of Fe^{3+} and other metals with only $CaCO_3$ at 0.1 atm (OLI simulation).

CaCO₃ [mg]	pH	Pressure [atm] (Y2)	Fe(+3) Aq [mg] (Y2)	Fe(OH)₃ (Bernalite)—Sol [mg] (Y2)	CO₂—Vap [mg] (Y2)	Al(+3) Aq [mg] (Y2)	MnCO₃ (Rhodochrosite)—Sol [mg] (Y2)	Na(+1) Aq [mg] (Y2)	Mg(+2) Aq [mg] (Y2)	Ca(+2) Aq [mg] (Y2)	CaSO₄·2H₂O (Gypsum) [mg] (Y2)	C(+4) Aq [mg] (Y2)	CaCO₃ (Calcite)—Sol [mg] (Y2)	S(+6) Aq [mg] (Y2)	Cl(−1) Aq [mg] (Y2)
0	2.5	0.1	2000	0	0	300	0	100	200	300	0	0.0	0	3074	150
1000	2.6	0.1	1690	593	413	300	0	100	200	603	420	7.3	0	2996	150
2000	2.6	0.1	1314	1313	853	300	0	100	200	589	2200	7.3	0	2665	150
3000	2.7	0.1	938	2033	1292	300	0	100	200	575	3978	7.4	0	2333	150
4000	2.7	0.1	562	2752	1732	300	0	100	200	563	5753	7.4	0	2003	150
5000	2.9	0.1	192	3460	2171	300	0	100	200	552	7518	7.4	0	1674	150
6000	3.5	0.1	4	3820	2611	225	0	100	200	546	9263	7.4	0	1349	150
7000	3.7	0.1	1	3825	3051	49	0	100	200	543	10,996	7.5	0	1026	150
8000	6.6	0.1	0	3827	3141	0	353	100	200	579	12,564	28.7	0	734	150
9000	6.8	0.1	0	3827	3122	0	390	100	200	590	12,693	39.3	896	710	150
10,000	6.8	0.1	0	3827	3122	0	390	100	200	590	12,693	39.3	1896	710	150

Table 6. Removal of remaining metals with Na_2CO_3 after Fe^{3+} removal with $CaCO_3$ (OLI simulation).

Na₂CO₃ [mg/L]	pH	Ca(+2) Aq [mg] (Y2)	CaCO₃ (Calcite)—Sol [mg] (Y2)	Fe(+2) Aq [mg] (Y2)	FeCO₃ (Siderite)—Sol [mg] (Y2)	Mn(+2) Aq [mg] (Y2)	S(+6) Aq [mg] (Y2)	Cl(−1) Aq [mg] (Y2)	C(+4) Aq [mg] (Y2)
0	6.8	584	0	18.2	18.2	21.0	729	150	24.54
200	7.0	532	150	11.0	33.2	12.6	735	150	25.76
400	7.0	459	331	9.5	36.3	10.9	735	150	26.03
600	7.1	387	512	8.1	39.3	9.3	735	150	26.34
800	7.2	315	692	6.6	42.3	7.6	735	150	26.69
1000	7.3	243	872	5.2	45.3	5.9	735	150	27.10
1199	7.4	171	1051	3.7	48.3	4.2	735	150	27.59
1399	7.6	100	1229	2.3	51.3	2.4	735	150	28.22
1599	8.1	30	1403	0.9	54.2	0.7	735	150	29.34
1799	9.4	2	1473	0.3	55.3	0.0	735	150	43.28
1998	9.6	1	1475	0.3	55.3	0.0	735	150	65.73

Table 7. Removal of remaining metals with $Ca(OH)_2$ after Fe^{3+} removal with $CaCO_3$ (OLI simulation).

$Ca(OH)_2$ [mg/L]	pH	Fe(+2) Aq [mg] (Y2)	$FeCO_3$ (Siderite)—Sol [mg] (Y2)	$Fe(OH)_2$ (Amakinite)—Sol [mg] (Y2)	Mn(+2) Aq [mg] (Y2)	$MnCO_3$ (Rhodochrosite)—Sol [mg] (Y2)	Na(+1) Aq [mg] (Y2)	Ca(+2) Aq [mg] (Y2)	C(+4) Aq [mg] (Y2)	$CaCO_3$ (Calcite)—Sol [mg] (Y2)	$CaSO_4 \cdot 2H_2O$ (Gypsum) [mg] (Y2)	Cl(−1) Aq [mg] (Y2)	$Ca(OH)_2$ (Portlandite)—Sol [mg] (Y2)
0	6.8	18.2	18.2	0.0	21.0	21.9	100	584	24.5	0	34	150	0
100	8.8	13.2	11.2	13.5	13.7	37.2	100	569	0.7	191	0	150	0
200	9.6	0.4	0.0	42.8	14.7	35.1	100	577	0.3	206	176	150	0
300	9.7	0.3	0.0	42.9	15.0	34.5	100	584	0.3	207	376	150	0
400	9.7	0.3	0.0	43.0	12.9	0.0	100	590	0.3	237	529	150	0
500	9.8	0.2	0.0	43.1	9.6	0.0	100	599	0.3	237	723	150	0
600	9.9	0.2	0.0	43.2	6.3	0.0	100	609	0.3	237	914	150	0
700	10.0	0.1	0.0	43.3	3.0	0.0	100	620	0.3	237	1099	150	0
800	10.7	0.0	0.0	43.4	0.2	0.0	100	637	0.2	237	1257	150	0
900	11.4	0.0	0.0	43.4	0.0	0.0	100	676	0.2	237	1323	150	0
999	11.6	0.0	0.0	43.4	0.0	0.0	100	717	0.2	237	1379	150	0

Tables 4 and 5 show the results when $CaCO_3$ was used for Fe^{3+} removal as $Fe(OH)_3$ (Bernalite) at pH 3.5, at 1.0 and 0.1 atm pressure, respectively. Fe^{3+} removal was achieved at a dosage of 5000 mg/L at pH 3.5, Al^{3+} at a dosage of 7265 mg/L at pH 4.0 as $Al(OH)_3$ (Gibbsite), Fe^{2+} and Mn^{2+} at a dosage of 8830 mg/L at pH 6.6. At 0.1 atm C^{4+} (CO_{2aq}) was removed to 39.3 mg/L (as C) (Table 5) compared to 592.9 mg/L (as C) (Table 4). This was confirmed by the larger mass of CO_2 vapor removed from solution at 0.1 atm (3122 mg CO_2 at a $CaCO_3$ dosage of 10,000 mg to 1 L), than in the case of 1 atm (1416 mg CO_2 from L). This finding was in line with Henry's law which stipulates that the solubility of a gas is related to the partial pressure. Fe^{2+} and Mn^{2+} were also removed faster at 0.1 atm than at 1.0 atm due to the removal of CO_2, which resulted in the shift of the equilibrium from HCO_3^- to CO_3^{2-}. Tables 8 and 9 calculate the $CaCO_3$/Metals removal eq/eq ratio needed for the removal of each metal with $CaCO_3$. For Fe^{3+} and Al^{3+} the ratio was 1, while for Fe^{2+} and Mn^{2+}, that were removed at the higher pH value of 6.6, the ratio was 2. The difference in equivalent $CaCO_3$/Metal ratios can be explained by the escape of CO_2 in the case of Fe^{3+} and Al^{3+} (Equation (4)), and by HCO_3^- that remained in solution in the case of Fe^{2+} and Mn^{2+} (Equation (5)).

When CO_2 was stripped to low levels, e.g., 0.1 atm, $FeCO_3$ (Equation (6)), $MnCO_3$ and $CaCO_3$ started to form at lower $CaCO_3$ dosages than when the atm was 1. In the case of $CaCO_3$ the remaining Ca^{2+} in solution at 0.1 atm was 590.5 mg/L compared to 756.3 mg/L at 1 atm. The 590.5 mg/L at 0.05 atm was mainly linked to SO_4^{2-}, while the difference between 756 and 590 was due to $Ca(HCO_3)_2$ in solution.

$$Fe^{3+} + 1.5CaCO_3 + 1\frac{1}{2}H_2O \rightarrow Fe(OH)_3 + 1.5CO_2 + 1.5Ca^{2+} \tag{4}$$

$$Fe^{2+} + CaCO_3 + CO_2 + H_2O \rightarrow Fe(HCO_3)_2 + Ca^{2+} \; (CO_2 \text{ present}) \tag{5}$$

$$Fe^{2+} + CaCO_3 \rightarrow FeCO_3 + Ca^{2+} \; (\text{low } CO_2) \tag{6}$$

Table 8. Mole ratio of Alkali dosage/Metals removed at 1 atm.

Parameter	Unit	Metal			
		Fe^{3+}	Al^{3+}	Fe^{2+}	Mn^{2+}
Initial conc.	mg/L	2000.0	300.0	200.0	200.0
pH		3.5	3.8	6.6	6.6
q mass	g	18.6	9.0	27.9	28.0
Acc. $CaCO_3$ dosage	mg/L	5000.0	7265.0	8047.5	8830.0
$CaCO_3$ dosage	mg/L	5000.0	2265.0	782.5	782.5
$CaCO_3$ eq mass	g	50.0	50.0	50.0	50.0
$CaCO_3$/Metal rem	mol/mol	0.9	1.4	2.2	2.2

Table 9. Mole ratio of Alkali dosage/Metals removed at 0.1 atm.

Parameter	Unit	Metal			
		Fe^{3+}	Al^{3+}	Fe^{2+}	Mn^{2+}
Initial conc.	mg/L	2000.0	300.0	200.0	200.0
pH		3.5	3.8	6.6	6.6
q mass	g	18.6	9.0	27.9	28.0
Acc. $CaCO_3$ dosage	mg/L	5000.0	7265.0	8047.5	8830.0
$CaCO_3$ dosage	mg/L	5000.0	2265.0	782.5	782.5
$CaCO_3$ eq mass	g	50.0	50.0	50.0	50.0
$CaCO_3$/Metal rem	mol/mol	0.9	1.4	2.2	2.2
CO_2 (aq)	mg/L	7.4	7.5		39.3
CO_2 (vap)	mg	2171.4	3050.7		3122.2

3.4.2. Na_2CO_3 and $Ca(OH)_2$

Section 3.4.1 showed that $CaCO_3$ can remove Fe^{3+} and Al^{3+} in quantities equivalent to the $CaCO_3$ dosage. This is due to the escape of CO_2 at the low pH where it precipitated as hydroxides. Fe^{2+} and Mn^{2+} were only removed as carbonates at excess dosages of $CaCO_3$. CO_2 stripping was needed for precipitation of $FeCO_3$ and $MnCO_3$ in the absence of excess $CaCO_3$ dosages. Ca^{2+} could only be precipitated when CO_2 stripping was applied. The aim of this section was to evaluate Na_2CO_3 and $Ca(OH)_2$ for removal of Fe^{2+}, Mn^{2+} and Ca^{2+}. $Ca(OH)_2$ or CaO is of great interest as the OH^- will convert the HCO_3^- to CO_3^{2-}, which is the ideal anion for removal of Fe^{2+}, Mn^{2+} and Ca^{2+} as carbonates. Tables 6 and 7 show the results when Na_2CO_3 and $Ca(OH)_2$ were used, respectively, for removal of 584.1 mg/L Ca^{2+}, 18.2 mg/L Fe^{2+}, 21 mg/L Mn^{2+} and 200 mg/L Mg^{2+}, the metals left in solution after Fe^{3+} and Al^{3+} were removed with $CaCO_3$ at pH 6.8. A dosage of 1600 mg/L Na_2CO_3 was needed to remove Fe^{2+}, Mn^{2+} and Ca^{2+} to low levels as $FeCO_3$ (Siderite), $MnCO_3$ (Rhodochrosite) and $CaCO_3$ (Calcite), respectively. The pH was raised to 8.1. A dosage of 800 mg/L $Ca(OH)_2$ was needed to remove Fe^{2+}, Mn^{2+} and Mg^{2+} to low levels as $Fe(OH)_2$ (Amakinite), $Mn(OH)_2$ (Pyrochroite) and $Mg(OH)_2$ (Brucite), respectively. The pH was raised to 10.5. $Ca(OH)_2$ will be the preferred alkali to use for removal of Fe^{2+}, Mn^{2+} and Mg^{2+} to low levels in the case where water is not desalinated, as no Na^+ is added to the water. Na^+ affects the suitability of treated water for further uses such as irrigation. Na_2CO_3 will be the preferred alkali to use if desalination is needed after the pre-treatment stage, as Ca^{2+} can be removed as $CaCO_3$. This way gypsum scaling of the RO membranes is avoided.

3.5. Feasibility

Table 10 compares the feasibility of pre-treatment with Na_2CO_3 for the removal of Fe^{3+}, Al^{3+}, Fe^{2+}, Mn^{2+} and Ca^{2+} with Table 11 where $CaCO_3$, $Ca(OH)_2$ and Na_2CO_3 were used, in combination with gypsum crystallization. In the latter case $CaCO_3$ was used for removal of Fe^{3+} and Al^{3+}, $Ca(OH)_2$ for the removal of Fe^{2+}, Mn^{2+}, and Na_2CO_3 for the removal of Ca^{2+} associated with SO_4^{2-}. In the case of Na_2CO_3, the TDS increased from 12,660 mg/L in the Feed to 13,684 mg/L due to the replacement of metal ions (Fe^{3+}, Al^{3+}, Fe^{2+}, Mn^{2+} and Ca^{2+}) with Na^+ in solution. In the case where $CaCO_3$ was used for the removal of Fe^{3+} and Al^{3+}, $Ca(OH)_2$ for the removal of Fe^{2+}, Mn^{2+}, and Na_2CO_3 for the removal of Ca^{2+}, the TDS dropped from 12,661 mg/L to 2288 mg/L, due to gypsum precipitation. During treatment with calcium alkalis, the Na^+ concentration increased only to 980 mg/L, compared to 4118 mg/L in the case of Na_2CO_3 treatment. Another benefit of using calcium alkalis was that of reduced chemical cost, namely $ZAR29.43/m^3$ versus $ZAR48.46/m^3$. In both cases the cost can be recovered from the value of pigment. For an Fe^{3+} concentration of 2000 mg/L, and a pigment price of ZAR20/kg, the potential income from pigment amounts to $ZAR122.71/m^3$. The capital cost in both cases was estimated at $ZAR10,000,000/(ML/d)$ or $ZAR3.65/m^3$ (term = 120 month; interest = 6%/a) and the electricity cost at $ZAR2.16/m^3$ (Electricity price = ZAR1.50/kWh).

Table 10. Chemical cost and water quality when acid mine water is treated with Na_2CO_3 for removal of Fe^{3+}, Al^{3+}, Fe^{2+}, Mn^{2+} and Ca^{2+}.

Compound	Unit	Composition			Price	Cost	Value
		Feed	Fe(OH)$_3$	Other Metals	ZAR/t	ZAR/m^3 Feed	ZAR/m^3 Feed
Flow Feed	m^3/h	40.0	40.0	40.0			
Na$_2$CO$_3$	mg/L		5958.8	3300.6	5000.0	46.30	
Na$_2$CO$_3$	mg/L				5000.0	0.00	
Product water							
pH			3.2	5.7			
TDS	mg/L	12,660.9	13,241.8	13,684.2			
H$^+$	mg/L	5.0	0.0	0.0			
Na$^+$	mg/L	100.0	2685.9	4118.3			
Mg^{2+}	mg/L	200.0	200.0	200.0			
Fe^{3+}	mg/L	2000.0	0.0	0.0			
Al^{3+}	mg/L	300.0	300.0	0.0			
Fe^{2+}	mg/L	200.0	200.0	0.0			
Mn^{2+}	mg/L	200.0	200.0	0.0			
Ca^{2+}	mg/L	300.0	300.0	10.0			
SO$_4{}^{2-}$	mg/L	9205.9	9205.9	9205.9			
Cl$^-$	mg/L	150.0	150.0	150.0			
Cations	meq/L	196.0	196.0	196.0			
Anions	meq/L	196.0	196.0	196.0			

Table 10. *Cont.*

Compound	Unit	Composition			Price	Cost	Value
		Feed	$Fe(OH)_3$	Other Metals	ZAR/t	ZAR/m^3 Feed	ZAR/m^3 Feed
Sludge							
$Fe(OH)_3$	mg/L		3826.3	0.0			
$Al(OH)_3$			0.0	866.7			
$FeCO_3$			0.0	414.9			
$MnCO_3$			0.0	418.6			
$CaCO_3$	mg/L						
Products							
Pigment	mg/L		5470.6		20,000.0		109.4
Water	mg/L				12.0		11.7
Energy usage	kWh/m^3		0.6	0.6			
Energy usage	kWh/m^3		0.6	0.6	1.2	2.16	
Total						48.46	121.1

Table 11. Chemical cost and water quality when acid mine water is treated with $CaCO_3$ for removal of Fe^{3+} and Al^{3+}, $Ca(OH)_2$ for the removal of Fe^{2+}, Mn^{2+}, and Na_2CO_3 for the removal of Ca^{2+} associated with SO_4^{2-}.

Compound	Unit	Composition						Price	Cost	Value
		Feed	$Fe(OH)_3$	$Al(OH)_3$	Other Metals	$CaSO_4$ Crystal	$CaCO_3$	ZAR/t	ZAR/m^3 Feed	ZAR/m^3 Feed
Flow Feed	m^3/h	40	40		40		40			
$CaCO_3$ OLI	mg/L		5516	1756						
$CaCO_3$ (calc)	mg/L		6031	1666				750	5.77	
$Ca(OH)_2$ (OLI)					799.0					
$Ca(OH)_2$ (calc)	mg/L				1140.4			2500	2.85	
Na_2CO_3	mg/L						2288	5000	11.44	
Inhibitor	mg/L		120.0					60,000	7.20	
Product water										
pH			3.5	4.0	10.7	10.7	10.0			
TDS	mg/L	12,660.9	13,068	13,435	13,453	2837	2853			
H^+	mg/L	5.0	0.0	0.0	0.0	0.0	0.0			
Na+	mg/L	100.0	100.0	100.0	100.0	100.0	979			
Mg^{2+}	mg/L	200.0	200.0	200.0	1.0	1.0	1.0			
Fe^{3+}	mg/L	2000.0	0.0	0.0	0.0	0.0	0.0			
Al^{3+}	mg/L	300.0	300.0	0.0	0.0	0.0	0.0			
Fe^{2+}	mg/L	200.0	200.0	200.0	0.0	0.0	0.0			
Mn^{2+}	mg/L	200.0	200.0	200.0	0.0	0.0	0.0			
Ca^{2+}	mg/L	300.0	2712.5	3379.2	3995.6	873.4	10.0			
SO_4^{2-}	mg/L	9205	9205	9205	9205	1712	1712			
Cl^-	mg/L	150.0	150.0	150.0	150.0	150.0	150.0			

Table 11. *Cont.*

Compound	Unit	Composition						Price	Cost	Value
		Feed	Fe(OH)$_3$	Al(OH)$_3$	Other Metals	CaSO$_4$ Crystal	CaCO$_3$	ZAR/t	ZAR/m^3 Feed	ZAR/m^3 Feed
HCO$_3$$^-$	mg/L			500.0	500.0	500.0	200.0			
OH$^-$	mg/L						0.0			
Cations	meq/L	196.0	204.2	204.2	204.2	48.1	43.2			
Anions	meq/L	196.0	204.2	204.2	204.2	48.1	43.2			
SO$_4$$^{2-}$ (OLI)						1503				
Sludge										
Fe(OH)$_3$	mg/L		3826.3		0.0					
Al(OH)$_3$			0.0		866.7					
FeCO$_3$			0.0		414.9					
MnCO$_3$			0.0		418.6					
CaCO$_3$	mg/L						245.9			
Products										
Pigment	mg/L		5470					20,000		109.41
Al(OH)$_3$	mg/L			866.7						
CaSO$_4$·2H$_2$O	mg/L					13,425		20.0		0.27
CaCO$_3$	mg/L						2650	500.0		1.33
Water	mg/L							12.0		11.70
Energy usage	kWh/m^3		0.6		0.6		0.6			
Energy usage	kWh/m^3		0.6		0.6		0.6	1.2	2.16	
Total									29.43	122.71

4. Conclusions

The following conclusions were made: (i) the rate of gypsum crystallization, in the absence of Fe^{3+}, is influenced by the over saturation concentration in solution, the seed crystal concentration and temperature; (ii) gypsum crystallization from an over-saturated solution, in the presence of Fe(OH)$_3$ sludge, required an inhibitor dosage of 100 mg/L to keep gypsum in solution for a period of 30 min; (iii) gypsum crystallization from an over-saturated solution, in the presence of both, Fe(OH)$_3$ sludge and CaCO$_3$ reactant, required a higher inhibitor dosage than 100 mg/L to keep gypsum in solution for a period of 30 min. A dosage of 200 mg/L kept gypsum in solution for the total reaction period; (iv) when only Fe(OH)$_3$ is present in the slurry, gypsum inhibition is more effective when Fe(OH)$_3$ sludge is allowed to settle after the initial mixing; (v) when both Fe(OH)$_3$ and CaCO$_3$ are present in the slurry, gypsum inhibition is more effective when the inhibitor is added over a period of time (10 min), rather than applying the total dosage at time zero; (vi) Fe(OH)$_3$ can be changed to yellow pigment (Goethite) by heating to 150 °C and to red pigment (Hematite) by heating to 800 °C. Pigment of nano particle size was produced; (vii) in the case of Na$_2$CO$_3$, the TDS increased from 12,660 mg/L in the Feed to 13,684 mg/L due to the replacement of metal ions (Fe^{3+}, Al^{3+}, Fe^{2+}, Mn^{2+} and Ca^{2+}) with Na$^+$ in solution. In the case where CaCO$_3$ was used for the removal of Fe^{3+} and Al^{3+}, Ca(OH)$_2$ for the removal of Fe^{2+}, Mn^{2+}, and Na$_2$CO$_3$ for the removal of Ca^{2+}, the TDS dropped from 12,661 mg/L to 2288 mg/L due to gypsum precipitation. The alkali cost in the case of calcium alkalis amounted to ZAR29.43/m^3 versus ZAR48.46/m^3 in the case of Na$_2$CO$_3$. In both cases the value of pigment recovered from mine water containing 2000 mg/L Fe^{3+} amounted to ZAR122.71/m^3 when the price of pigment was taken at ZAR20/kg.

The capital cost in both cases was estimated at ZAR10,000,000/(ML/d) or ZAR3.65/m^3 (term = 120 month; interest = 6%/a) and the electricity cost at ZAR2.16/m^3 (Electricity price = ZAR1.50/kWh.

Author Contributions: Conceptualization, T.M.M. and J.P.M.; methodology, T.M.M.; software, J.P.M.; validation, M.M., M.M.M.-M. and K.D.M.; formal analysis, T.M.M.; investigation, J.P.M.; resources, M.M.M.-M.; data curation, K.D.M.; writing—original draft preparation, T.M.M.; writing—review and editing, J.P.M.; visualization, M.M.; supervision, M.M.; project administration, J.P.M.; funding acquisition, J.P.M. All authors have read and agreed to the published version of the manuscript.

Funding: The authors express their gratitude to: NIPMO, who sponsored the grant for the innovation award of 2021 *NSTF-South32*, under the category SMME, for acknowledgment of the ROC technology, the South African *Water Research Commission* for financial support through its Wader Programme, the *Innovation Hub* for supporting business development, H Siweya for funding the OLI software in recognition for receiving the NSTF award. Tumelo M. Mogashane is grateful for the financial assistance received from Sasol Inzalo Foundation.

Data Availability Statement: The authors will make raw data available on request.

Acknowledgments: The authors express their gratitude to: Neville Nyamutswa for assisting with chemical analyses and Fritz Carlsson for proof reading and extensive editorial input.

Conflicts of Interest: The authors declare no conflict of interest.

References

1. Demers, I.; Mbonimpa, M.; Benzaazoua, M.; Bouda, M.; Awoh, S.; Lortie, S.; Gagnon, M. Use of acid mine drainage treatment sludge by combination with a natural soil as an oxygen barrier cover for mine waste reclamation: Laboratory column tests and intermediate scale field tests. *Miner. Eng.* **2017**, *107*, 43–52. [CrossRef]
2. Seo, E.; Cheong, Y.; Yim, G.; Min, K.; Geroni, J. Recovery of Fe, Al and Mn in acid coal mine drainage by sequential selective precipitation with control of pH. *Catena* **2017**, *148*, 11–16.
3. Akinwekomi, V.; Maree, J.; Masindi, V.; Zvinowanda, C.; Osman, M.; Foteinis, S.; Mpenyana-Monyatsi, L.; Chatzisymeon, E. Beneficiation of acid mine drainage (AMD): A viable option for the synthesis of goethite, hematite, magnetite, and gypsum—Gearing towards a circular economy concept. *Miner. Eng.* **2020**, *148*, 106–204. [CrossRef]
4. Demers, I.; Benzaazoua, M.; Mbonimpa, M.; Bouda, M.; Bois, D.; Gagnon, M. Valorisation of acid mine drainage treatment sludge as remediation component to control acid generation from mine wastes, part 1: Material characterization and laboratory kinetic testing. *Miner. Eng.* **2015**, *76*, 109–116.
5. Kefeni, K.K.; Msagati, T.A.; Nkambule, T.T.; Mamba, B.B. Synthesis and application of hematite nanoparticles for acid mine drainage treatment. *J. Environ. Chem. Eng.* **2018**, *6*, 1865–1874. [CrossRef]
6. Mogashane, T.M.; Maree, J.P.; Mujuru, M.; Mphahlele-Makgwane, M.M. Technologies that can be Used for the Treatment of Wastewater and Brine for the Recovery of Drinking Water and Saleable Products. In *Recovery of Byproducts from Acid Mine Drainage Treatment, Editors: Elvis Fosso-Kankeu, Jo Burgess and Christian Wolkersdorfer*; Wiley Scrivener: New York, NY, USA, 2020; pp. 97–151.
7. Munyengabe, A.; Zvinowanda, C.; Zvimba, J.N.; Ramontja, J. Characterization and reusability suggestions of the sludge generated from a synthetic acid mine drainage treatment using sodium ferrate (VI). *Heliyon* **2020**, *6*, 1–6.
8. Ang, H.M.; Muryanto, S.; Hoang, T. Gypsum scale formation control in pipe flow systems: A systematic study on the effects of process parameters and additives. In *Gypsum: Properties, Production and Applications*; Sampson, E.D.H., Ed.; Nova Science Publishers, Inc.: New York, NY, USA, 2011; pp. 1–34.
9. Amjad, Z.; Landgraf, R.T.; Penn, J.L. Calcium sulfate dihydrate (gypsum) scale inhibition by PAA, PAPEMP, and PAA/PAPEMP blend. *Int. J. Corros. Scale Inhib.* **2014**, *3*, 35–47. [CrossRef]
10. Rabizadeh, T.; Stawski, T.M.; Morgan, D.J.; Peacock, C.L.; Benning, L.G. The effects of inorganic additives on the nucleation and growth kinetics of calcium sulfate dihydrate crystals. *Cryst. Growth Des.* **2017**, *17*, 582–589.
11. Hoang, T.A.; Ang, H.M.; Rohl, A.L. Effects of temperature on the scaling of calcium sulphate in pipes. *Powder Technol.* **2007**, *179*, 31–37. [CrossRef]
12. Hamdona, S.K.; Al Hadad, O.A. Influence of additives on the precipitation of gypsum in sodium chloride solutions. *Desalination* **2008**, *228*, 277–286.
13. Fazel, M.; Chesters, S.P.; Gibson, G. Controlling Calcium Sulphate Scale Formation in Acid Mine Waters. In Proceedings of the 5th World Congress on Mechanical, Chemical, and Material Engineering (MCM'19), Lisbon, Portugal, 15–17 August 2019. Paper No. MMME 124.
14. Gregory, G.; Max, F.; Chesters, S. Membranes in Mining: Controlling CaSO4 Scale in AMD Mine waters. *Asp. Min. Miner. Sci.* **2019**, *3*, 1–3.
15. Liu, S.T.; Nancollas, G.H. The kinetics of crystal growth of calcium sulfate dihydrate. *J. Cryst. Growth* **1970**, *6*, 281–289. [CrossRef]

16. Liu, S.T.; Nancollas, G.H. A kinetics and morphological study of the seeded growth of calcium sulfate dihydrate in the presence of additives. *J. Coll. Interface Sci.* **1975**, *52*, 593–601. [CrossRef]
17. Rabizadeh, T.; Peacock, C.L.; Benning, L.G. Carboxylic acids: Effective inhibitors for calcium sulfate precipitation? *Mineral. Mag.* **2014**, *78*, 1465–1472.
18. Rabizadeh, T. The nucleation, growth kinetics and mechanism of sulfate scale minerals in the presence and absence of additives as inhibitors. Ph.D. Thesis, The University of Leeds, School of Earth and Environment, Leeds, UK, 2016; pp. 1–219.
19. Szabados, M.; Pásztor, K.; Csendes, Z.; Muráth, S.; Kónya, Z.; Kukovecz, Á.; Carlson, S.; Sipos, P.; Pálinkó, I. Synthesis of high-quality, well-characterized CaAlFe-layered triple hydroxide with the combination of dry-milling and ultrasonic irradiation in aqueous solution at elevated temperature. *Ultrason. Sonochem.* **2016**, *32*, 173–180. [PubMed]
20. Maree, J.P.; Mtombeni, T. Treatment of brine for the recovery of drinking water, calcium carbonate and sodium sulphate. In Proceedings of the 2018 WISA Biennial Conference, Cape Town, South Africa, 24–27 June 2018.
21. Akinwekomi, V.; Maree, J.P.; Strydom, A.; Zvinowanda, C. Recovery of magnetite from iron-rich mine water. In Proceedings of the WISA Biennial Conference, Durban, South Africa, 15–19 May 2016.
22. Magagane, N.; Masindi, V.; Ramakokovhu, M.M.; Shongwe, M.B.; Muedi, K.L. Facile thermal activation of non-reactive cryptocrystalline magnesite and its application on the treatment of acid mine drainage. *J. Environ. Manag.* **2019**, *236*, 499–509.
23. Cornell, R.M.; Schwertmann, U. *The Iron Oxides: Structure, Properties, Reactions, Occurrences and Uses*; Wiley-VCH: Weinheim, UK, 2000.
24. Legodi, M.; Dewaal, D. The preparation of magnetite, goethite, hematite and maghemite of pigment quality from mill scale iron waste. *Dye. Pigment.* **2007**, *74*, 161–168. [CrossRef]
25. Meng, J.; Yang, G.; Yan, L.; Wang, X. Synthesis and characterization of magnetic nanometer pigment Fe₃O₄. *Dye. Pigment.* **2004**, *66*, 109–113. [CrossRef]
26. Campos, E.A.; Pinto, D.V.B.S.; de Oliveira, J.I.S.; Mattos, E.D.C.; Dutra, R.D.C.L. Synthesis, Characterization and Applications of Iron Oxide Nanoparticles—A Short Review. *J. Aerosp. Technol. Manag.* **2015**, *7*, 267–276. [CrossRef]
27. Akinwekomi, V.; Maree, J.; Zvinowanda, C.; Masindi, V. Synthesis of magnetite from iron-rich mine water using sodium carbonate. *J. Environ. Chem. Eng.* **2017**, *5*, 2699–2707.
28. Mogashane, T.M.; Maree, J.P.; Letjiane, L.; Masindi, V.; Modibane, K.D.; Mujuru, M.; Mphahlele-Makgwane, M.M. Recovery of drinking water and nano-sized Fe2O3 pigment from iron rich acid mine water. In *Application of Nanotechnology in Mining Processes: Benefication and Sustainability*; Elvis Fosso-Kankeu, M.M.A.B.M., Ed.; Scrivener Publishing LLC: Beverly, MA, USA, 2022; pp. 237–288.
29. Xu, P.; Zeng, G.M.; Huang, D.L.; Feng, C.L.; Hu, S.; Zhao, M.H.; Lai, C.; Wei, Z.; Huang, C.; Xie, G.X.; et al. Use of iron oxide nanomaterials in wastewater treatment: A review. *Sci. Total Environ.* **2012**, *424*, 1–10.
30. Mariani, F.Q.; Borth, K.W.; Müller, M.; Dalpasquale, M.; Anaissi, F.J. Sustainable innovative method to synthesize different shades of iron oxide pigments. *Dye. Pigment.* **2017**, *137*, 403–409.
31. APHA. *Standard Methods for the Examination of Water and Wastewater (American Public Health Association)*, 22nd ed.; American Water Works Association, Water Environment Federation: Washington, DC, USA, 2012.
32. OLI. 2022. Available online: https://downloadly.net/2020/04/11634/04/oli-systems/16/?#/11634-oli-syst-122115014831.html (accessed on 15 May 2022).
33. Xu, C.; Deng, K.; Li, J.; Xu, R. Impact of environmental conditions on aggregation kinetics of hematite and goethite nanoparticles. *J. Nanoparticle Res.* **2015**, *17*, 1–13.
34. Xu, C.-Y.; Deng, K.-Y.; Li, J.-Y.; Xu, R.-K. Improved size, morphology and crystallinity of hematite (α-Fe2O3) nanoparticles synthesized via the precipitation route using ferric sulfate precursor. *Results Phys.* **2019**, *12*, 1253–1261.
35. Gupta, H.; Kumar, R.; Park, H.S.; Jeon, B.H. Geosystem Engineering. *Photocatalytic Effic. Iron Oxide Nanoparticles Degrad. Prior. Pollut. Anthracene* **2016**, *1*, 21–27.
36. Mashao, G.; Modibane, K.D.; Mdluli, S.B.; Hato, M.J.; Makgopa, K.; Molapo, K.M. Polyaniline-Cobalt Benzimidazolate Zeolitic Metal-Organic Framework Composite Material for Electrochemical Hydrogen Gas Sensing. *Electrocatalysis* **2019**, *10*, 406–419.

Article

Industrial Ceramics: From Waste to New Resources for Eco-Sustainable Building Materials

Maura Fugazzotto [1], Paolo Mazzoleni [1,*], Isabella Lancellotti [2], Rachel Camerini [3], Pamela Ferrari [3], Maria Rosaria Tiné [4], Irene Centauro [5], Teresa Salvatici [5] and Germana Barone [1]

[1] Department of Biological, Geological and Environmental Sciences, University of Catania, Corso Italia, 57, 95129 Catania, Italy; maura.fugazzotto@unict.it (M.F.); gbarone@unict.it (G.B.)

[2] Department of Engineering "Enzo Ferrari", University of Modena and Reggio Emilia, Via Pietro Vivarelli, 10, 41125 Modena, Italy; isabella.lancellotti@unimore.it

[3] Center for Colloid and Surface Science (CSGI) and Department of Chemistry, University of Florence, Via Della Lastruccia, 3, 50019 Sesto Fiorentino, Italy; camerini@csgi.unifi.it (R.C.); pamelaferrari.84@gmail.com (P.F.)

[4] Department of Chemistry and Industrial Chemistry, University of Pisa, Via Giuseppe Moruzzi, 13, 56124 Pisa, Italy; mariarosaria.tine@unipi.it

[5] Department of Earth Science, University of Florence, Via La Pira, 4, 50121 Firenze, Italy; irene.centauro@unifi.it (I.C.); teresa.salvatici@unifi.it (T.S.)

* Correspondence: pmazzol@unict.it

Abstract: Today, the need to dispose of a huge amount of ceramic industrial waste represents an important problem for production plants. Contextually, it is increasingly difficult to retrieve new mineral resources for the realization of building materials. Reusing ceramic industrial waste as precursors for building blocks/binders, exploiting their aluminosilicate composition for an alkaline activation process, could solve the problem. This chemical process facilitates the consolidation of new binders/blocks without thermal treatments and with less CO_2 emissions if compared with traditional cements/ceramics. The alkali-activated materials (AAMs) are today thought as the materials of the future, eco-sustainable and technically advanced. In this study, six different kind of industrial ceramic waste are compared in their chemical and mineralogical composition, together with their thermal behaviour, reactivity in an alkaline environment and surface area characteristics, with the aim of converting them from waste into new resources. Preliminary tests of AAM synthesis by using 80%–100% of ceramic waste as a precursor show promising results. Workability, porosity and mechanical strengths in particular are measured, showing as, notwithstanding the presence of carbonate components, consolidated materials are obtained, with similar results. The main factors which affect the characteristics of the synthetized AAMs are the precursors' granulometry, curing temperature and the proportions of the activating solutions.

Keywords: recycling; alkali-activated materials; ceramic; construction waste; geopolymers; sustainability

Citation: Fugazzotto, M.; Mazzoleni, P.; Lancellotti, I.; Camerini, R.; Ferrari, P.; Tiné, M.R.; Centauro, I.; Salvatici, T.; Barone, G. Industrial Ceramics: From Waste to New Resources for Eco-Sustainable Building Materials. *Minerals* **2023**, *13*, 815. https://doi.org/10.3390/min13060815

Academic Editors: Hendrik Gideon Brink, Pierfranco Lattanzi, Elisabetta Dore and Fabio Perlatti

Received: 4 May 2023
Revised: 3 June 2023
Accepted: 13 June 2023
Published: 15 June 2023

1. Introduction

Nowadays, the abundance of ceramic industrial waste constitutes an issue of global concern, and it requires a sustainable solution [1–3]. According to the current literature [4–7], about 45%–50% of the total construction and demolition (C&D) waste in the world is due to ceramic materials, while C&D waste in turn represents one third of the total waste generated by economic activities and households [8]. In particular, Italy produced 68 million tons of waste from construction and demolition before the pandemic event of the year 2020 [9]. Additionally, it should be considered that these official data are underestimated, since undesirable yet possible illegal dumping is not accounted for [8]. Besides the waste management problem, a further question arises, which must be faced urgently. Indeed, while the demand for construction materials grows [6,10,11], the availability of clay supply in Italy and in other European countries is dramatically reduced [2,3,10]. As the most common of construction materials (cement, concrete, bricks, tiles, etc.) are obtained from clays, a consequent deep crisis in the entire field is obvious. In this scenario, the reuse of ceramic industrial waste as new feedstock for the production of building materials is attracting increasing interest, with the considerable advantage of minimizing the need to dispose of waste and exploit virgin clay deposits, as well as minimizing the negative environmental impact [1,3,12–15]. Due to their aluminosilicate composition and their partial amorphous phase, ceramics are cited among those waste materials which offer potential for valorization by alkali-activation [4,12,16,17]. Alkali-activated materials (AAMs) are hydraulic binders with amorphous structure and ceramic-like properties, chemically synthetized by the reaction between an alkaline solution and an aluminosilicate powder [18–21]. Consolidating at room temperature, or, at least, at much lower temperature compared to those required for making traditional cements or ceramics, the AAMs allow the abatement of CO_2 emissions. Furthermore, the use of ceramic waste supplied by local industries also promotes a process of circular economy [16]. As eco-sustainable materials, technically advanced and with performances similar to those of the traditional cements [18,19]—sometimes superior [22,23]—the AAMs could be considered the building materials of the future. Various studies have investigated the possibility to obtain AAMs from C&D waste [24,25], but the mix design, or formulation step, still remains challenging because of the heterogeneity of these products [14,26]. Undoubtedly, the use of construction debris is much easier than that of the demolition ones, because it is also easier to control their chemical and mineralogical composition [12]. However, the structure and the performance of AAMs strictly depend on the characteristics of aluminosilicate powders used as precursors [27–29]. Studying how different kinds of ceramics are alkali-activated is therefore important. An extensive experimental investigation was undertaken to compare six different kind of industrial ceramic waste (three kind of tiles, solid brick, hollow brick and stoneware) in terms of chemical and mineralogical composition, together with thermal behaviour, reactivity in an alkaline environment and surface area, with the aim of converting them from waste into new resources for the synthesis of alkali-activated products. Preliminary tests of synthesis, laboratory observations, compressive and porosimetric analysis are also performed. All the ceramic products selected are obtained from local clays and they are available in large quantities as waste, due to damage during the manufacturing process, decoration or during transport. They have therefore become unusable materials and, therefore, materials that need to be disposed of in landfills.

2. Materials and Methods

2.1. Materials

Different types of industrial ceramic wastes were selected as potential precursors for alkaline activation (Figure 1). Industrial and handmade tiles (labelled LBCa, LBCb, LBCc and LBCf) were provided by the local industry La Bottega Calatina, located in Caltagirone (CT), Italy; two types of red clay bricks (CWF and CWM) were recovered from the local industry Laquattro, located in Rometta (ME), Italy. All the ceramic products selected are pre-consumer waste.

LBCa is a tile waste characterized by a fine, homogeneous grain and pinkish body, made industrially by mixing commercial clay with 3% of water and fired at around 1100 °C, after natural drying. LBCb is a tile waste characterized by a coarse grain, dark red body, handmade by mixing local clay with a higher water percentage; volcanic aggregates and sand are present. The firing temperature to obtain it was between 900 and 1100 °C. LBCc is tile waste of medium grain size, semi-industrially produced, of reddish colour and homogeneous texture. LBCf is a stoneware tile waste, greyish in colour, perfectly homogeneous and smooth. LBCf was produced by an industrial process starting from kaolinitic clay fired at a temperature over 1300 °C. CWF is waste hollow brick, composed of ceramic materials fired at low temperature, around 870 °C, originating from local clay of the Plio-Pleistocene Rometta Formation retrieved from the areas of Fondachelli, San

Pier Niceto and Barcellona Pozzo di Gotto (Italy), and local sand aggregates. The texture is quite heterogeneous, light red in colour with fine grain size with white spots, probably of carbonate nature, and dark grains of millimetric size. Lastly, CWM is waste from solid brick, light red in colour, with a fine grain size and a homogenous texture. Sporadic silver/gold-like millimetric lamellae are visible together with some white spots. According to the literature [26,30–35], sodium hydroxide (NaOH) and sodium silicate (Na_2SiO_3) were selected as alkaline activating solutions. A concentration of 8 M NaOH was chosen in order to have an enough alkaline medium, preventing possible excessive efflorescence, without being excessively aggressive towards the environment. The desired molarity was obtained by diluting a commercial 10M NaOH solution, supplied by Carlo Erba, Milan (Italy). The sodium silicate used was provided by Ingessil srl, Verona (Italy); it is characterized by a module $SiO_2/Na_2O = 3.3$ and pH = 11.5. Metakaolin (MK), one of the main raw materials commonly used for the production of geopolymers because of its high reactivity [36–38], is used here in small amounts (maximum 20% of the solid precursor) in order to increase the reactivity and the performance of the ceramic-based geopolymers. The MK used is ARGICAL[TM] M-1000, supplied by Imerys (France). Prompt (P), a natural cement prepared and supplied by Vicat Group (France), was also used as additive (in substitution of MK) in order to promote the reactivity of the ceramic precursors in the form of binary mixtures. It is commonly known as Roman cement and it is characterized by rapid setting (2–3 min) and hardening. The chemical composition of MK, according to the literature [36], and of P, according to the data sheet, are shown in Table 1.

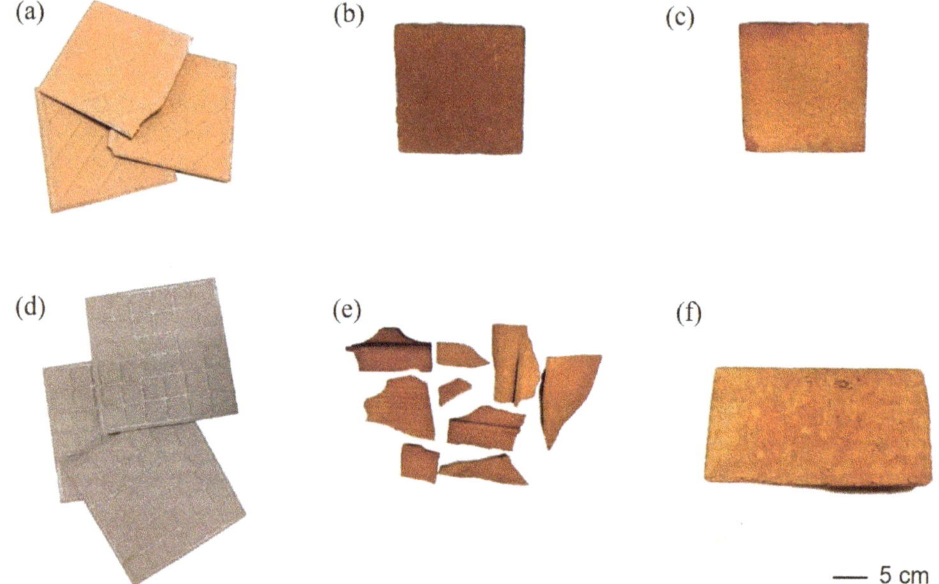

Figure 1. Ceramic waste retrieved by local industries: (**a**) LBCa; (**b**) LBCb; (**c**) LBCc; (**d**) LBCf; (**e**) CWF; and (**f**) CWM.

Table 1. XRF results on additives MK and P. nd = not determined.

Major Oxides (wt%)	SiO_2	Al_2O_3	Fe_2O_3	MgO	CaO	Na_2O	K_2O	TiO_2	P_2O_5	SO_3	LOI	Tot.
MK	58.56	34.03	2.57	-	2.08	-	0.69	1.88	0.2	-	nd	100
P	18.09	7.24	3.2	3.84	53.07	0.28	1.16	1.13	0.35	3.24	9.28	100

2.2. Synthesis Parameters

According to the suggestions of the producers, LBCa and CWF were selected for experimenting with alkali activation. Among all the ceramic waste supplied, indeed, these are the most abundant, and thus, those mainly responsible for the largest disposal problem. They are furthermore the most representative as industrial products and they are available in greater amounts for the tests. The synthesis process was carried out after preparing the solid and liquid raw materials. The LBCa and CWF raw materials were ground in porcelain jars with Al_2O_3 balls in order to reach the desired granulometry. The first tests were performed with a LBCa ground to around 75 µm. A granulometry similar to that of commercial MK was also chosen (~10 µm). For experimenting with binary mixture, MK or P were dried in an oven and sieved before adding them, separately, to the ceramic powders, in order to avoid the formation of agglomerates. The powdered components were carefully homogenized in a porcelain jar before the addition of the activating solution. Preliminary tests using only sodium hydroxide yielded poor results in setting and curing of the pastes. Thus, alkali-activated mixtures of sodium hydroxide and sodium silicate were tested, as well as activation by using only sodium silicate. When a mixture of sodium hydroxide and sodium silicate was used, it was allowed to rest for some minutes before the synthesis, in order to avoid the effect of the exothermic reaction between the two liquids on the geopolymerization [39,40]. The liquid was then poured on the powders, and immediately subjected to mechanical mixing for 5 min. The so-obtained slurries were then poured into molds and manually agitated for 1 min in order to facilitate the escape of air bubbles from the mixture [8,26,33]. After 24 h at room temperature, the samples were demolded and wrapped with polyethylene film for the remaining curing time (fixed at 28 days). Maturation was carried out at room temperature, around 25 °C, avoiding firing procedures in order to reduce environmental impact. However, some formulation replicates were also subjected to a low-temperature treatment (65 °C) for the first 24 h after synthesis [41], in order to assess the effect of different curing temperatures. The best liquid/solid (L/S) ratio for each formulation was defined according to the workability of the slurry during the synthesis, while also considering the setting and curing times.

2.3. Analytical Methods

The as-received industrial ceramic wastes were characterized by X-ray Fluorescence (XRF), X-ray Diffractometry (XRD), Reactivity Test + Inductively Coupled Plasma—Optical Emission Spectroscopy (ICP-OES) and Thermogravimetric Analysis (TGA). The ground samples were characterized via laser granulometry and Brunauer–Emmett–Teller analysis (BET), while the obtained geopolymeric samples were studied using integrity tests, visual critical evaluations and tested by Mercury Intrusion Porosimetry (MIP) and uniaxial compressive test. Where present, the efflorescence was investigated by XRD. Chemical analyses by XRF were performed on pearls by a PANalytical instrument, model Zetium, with ceramic tube with Rh anode; ultra-fine high transmission Beryllium front window, at least 75 µm and tube geometry below the sample; High Stability Power 4 kW X-ray Generator; decoupled goniometer $\theta/2\theta$ with optical positioning system and high angular reproducibility (0.0001°) (CIC—University of Granada, Spain). Mineralogical investigations were performed by a PANalytical X'Pert PRO X-ray Diffractometer, with Cu Kα radiation and operating at 45 kV and 40 mA; the following operative conditions were used: time 20 s, step 0.04 in a range of 3–70 2θ (Department of Mineralogy and Petrology—University of Granada, Spain). The qualitative analysis was then conducted by using High Score Plus software v.4.8. The reactivity test was performed following the experiments according to the literature [42,43]; however, while they usually use HF, in this research the tests were carried out reproducing the solubility conditions chosen for the alkaline activating process, namely, ambient T and pH of a NaOH 8 M solution. An amount of 1 ± 0.05 g powder of each material was subjected to a reactivity attack by using NaOH 8 M (NaOH, 99%, J. T. Baker; water Millipore purified by Milli-Q UV, resistivity > 18 MΩ·cm) in 100 mL of solution. The system was mechanically agitated (300 rpm) for 12 h at room temperature. The analysis of the eluate allowed the determination of the solid residue using the gravimetric method, and obtained the amount of soluble Si and Al by ICP-OES. In particular, after the basic attack, the eluates were filtered with Whatman paper and diluted with Milli-Q water until reaching the volume of 1 L. The washing of the filters was carried out until reaching a neutral pH. The solid residues were then obtained after drying in an oven at 110 °C for 1 h. A Varian 720-ES ICP—OES was used for the chemical analysis of the eluates by introducing the sample via a spray chamber. The quantification was obtained by using the internal standard method of Ge (1 mg/L); the calibration curve was realized considering 6 points in the 2–200 µg/L interval (CSGI and Department of Chemistry—University of Florence, Italy). TGA was performed by using a thermobalance TA Instrument, model Q5000IR, according to the procedure reported in [44,45]. The analyses were performed on about 10–15 mg of sample in a nitrogen flux, with gradual increases of 20 °C/minute, considering a temperature range from the environmental temperature to a maximum of 900 °C (Department of Chemistry and Industrial Chemistry—University of Pisa, Italy). Granulometric measurements were carried out by using the Laser Granulometer Master-sizer 2000, Hydro 2000S model (Malvern Instrument). The measurements were conducted in humid conditions; after the acquisition of a "white" reference, 10 successive acquisitions for each sample were elaborated by the software, which returned an average curve. The results were then expressed with a cumulative curve and a curve representing the granulometric distribution. BET analyses were performed by using the Micromeritics Instrument, Gemini Model 2380, with the following technical specifications: N_2 analysis adsorptive, 10 s equilibration time and pressure of 1.03979×10^5 Pascal (Department of Engineering "Enzo Ferrari"—University of Modena and Reggio Emilia, Italy). Preliminary evaluations of the pastes were conducted by simple observation of a few criteria concerning the curing time required to demold the geopolymers, their complete hardening (when the surfaces are completely dry), the tendency to shrink, the appearance in terms of homogeneity and the efflorescence crystallization. The integrity test, on the other hand, is a preliminary test generally adopted by the scientific community working with AAMs in order to check the stability of the material in water, when soaked for 24 h [39]. The integrity test could be, thus, considered a first step in the evaluation of the advancement of the geopolymerization process, i.e., the formation of the alkaline gel [39]. The visible parameters considered to check the sample's integrity were the appearance of water (clear, with residues, turbid) and the tendency of the geopolymeric fragment to be broken by a metallic pincer with hand pressure. A formulation passes the test if the water remains clear and the geopolymer does not break with the pincer (Figure 2).

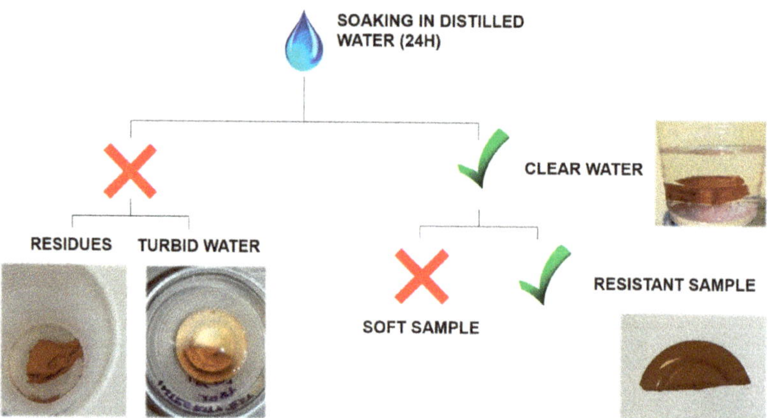

Figure 2. Integrity test scheme.

Three samples of $2 \times 2 \times 2$ cm^3 for preliminary compressive tests were analysed by the Instron Model 5592 instrumentation, maximum charge of 600 kN and a velocity of 0.5 MPa/s, according to CEN EN 1926:2006 standard [46] (Department of Earth Science—University of Florence, Italy). For MIP, a Thermoquest Pascal 240 porosimeter was used in order to explore the pore size distribution with radii between 0.0074 mm and 15 mm, and a Thermoquest Pascal 140 porosimeter to investigate the pore radii range 3.8 mm–116 mm (Department of Biological, Geological and Environmental Sciences—University of Catania, Italy). Before the analysis, the fragments were dried in an oven for 24 h at 100 °C.

3. Results and Discussion

3.1. Chemical Composition of the Ceramic Industrial Wastes

The XRF results are shown in Table 2. Data related to LBCa major oxides are reported from the literature [16].

Table 2. XRF results on the ceramic waste materials selected as precursors.

Major Oxides (wt%)	SiO$_2$	Al$_2$O$_3$	Fe$_2$O$_3$	MnO	MgO	CaO	Na$_2$O	K$_2$O	TiO$_2$	P$_2$O$_5$	Others	LOI	Tot.
LBCa	60.67	16.30	5.69	0.08	2.31	8.73	1.35	3.27	0.69	0.14	0.30	0.47	100
LBCb	57.91	17.33	7.53	0.11	2.58	7.41	2.35	2.30	1.13	0.35	0.53	0.47	100
LBCc	61.17	17.03	5.63	0.04	2.08	7.81	0.49	4.02	0.76	0.17	0.29	0.51	100
LBCf	72.37	18.37	0.99	0.01	0.54	0.80	3.22	2.37	0.68	0.12	0.18	0.35	100
CWF	57.33	14.32	5.51	0.08	2.43	11.99	1.16	2.49	0.73	0.17	0.33	3.46	100
CWM	56.83	15.20	5.63	0.07	2.52	10.15	1.39	2.80	0.70	0.18	0.21	4.32	100

The materials show similar chemistry, with the exception of sample LBCf. In general, all the materials show a SiO$_2$ content between 57 and 61 wt%. The second important chemical component is Al$_2$O$_3$, with values between 14 and 18 wt%. CaO is the third component of the chemistry, with values between 7 and 12 wt%. Another important element is Fe$_2$O$_3$ (more than 5 wt%). The LBCf sample is the exception, showing the highest values of silica reaching 72 wt%, while it could be considered free of calcium and iron components. All the materials show a different LOI%, with very low values for the LBC samples and higher values for CWF and CWM, which suggests the presence of carbonate phases in these latter samples due to the lower firing temperature of bricks. This is confirmed below by XRD analysis. In this context, the relatively high CaO content of the LBC samples is attributable to non-carbonate phases, as calcium-rich silicates. The amounts of the three major oxides, SiO$_2$%, Al$_2$O$_3$% and CaO% are plotted in the triangular diagram (Figure 3), showing the usual range of the composition of the aluminosilicate materials most commonly used as precursors in the field of AAMs, and OPC as comparison—sketched approximately according to [20]. It is evident that the ceramic waste has an intermediate composition, with a high percentages of silica in comparison to the other materials, somewhat less alumina and an intermediate content of CaO. Relative to this component, the studied ceramic wastes are placed between the materials known as "low-calcium" (as metakaolin and fly ash) and those of "high calcium" (as blast furnace slag) [20]. Particularly, they are located near the limit area of fly ash, with a higher silica content. LBCf, on the other hand, falls much closer to the SiO$_2$-Al$_2$O$_3$ axis, because of the absence of the CaO component. The data are consistent with further studies on the composition of aluminosilicate materials used for making AAMs [47].

3.2. Mineralogical Composition of the Ceramic Industrial Wastes

The individuated phases for each sample are shown in Table 3, where the data related to LBCa come from the literature [16]. Mineral abbreviations are reported as in [48]. As expected from the XRF data, LBC samples are distinguished from the CWF-CWM samples because of the absence of calcite and muscovite/illite, which instead characterize the latter. Quartz, feldspars and haematite are present everywhere, while diopside, gehlenite, wollastonite and montmorillonite differentiate the mineralogical compositions of the LBC samples among themselves. Furthermore, the results for LBCf, being a different kind of ceramic product, are very different. It is composed almost completely of quartz, feldspars and mullite. It is the only sample where haematite is not present.

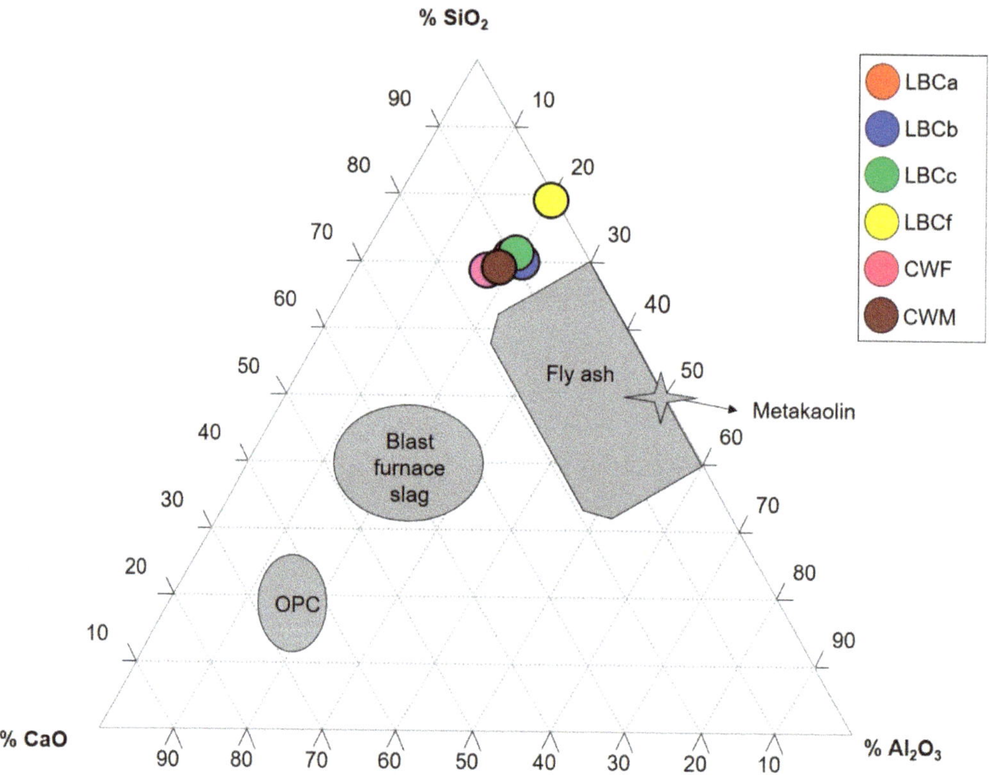

Figure 3. Triangular diagram CaO-SiO$_2$-Al$_2$O$_3$ for the ceramic waste composition. The colored circles represent the ceramic waste analyzed, while the grey areas indicate the compositional range of typical raw materials used for geopolymerization. The compositional range of typical OPC is also marked on the diagram.

Table 3. Mineralogical phases detected on ceramic waste samples; Qz = quartz, Fsp = feldspar, Hem = hematite, Di = diopside, Gh = gehlenite, Wo = wollastonite, Mnt = montmorillonite, Cal = calcite, Ms = muscovite, and Mul = mullite; X = present, / = absent.

Sample	Qz	Fsp	Hem	Di	Gh	Wo	Mnt	Cal	Ms	Mul
LBCa	X	X	X	X	X	X	/	/	/	/
LBCb	X	X	X	X	/	/	X	/	/	/
LBCc	X	X	X	X	X	/	/	/	/	/
LBCf	X	X	/	/	/	/	/	/	/	X
CWF	X	X	X	X	X	/	/	X	X	/
CWM	X	X	X	/	/	/	/	X	X	/

3.3. Thermogravimetric Behavior of the Ceramic Industrial Waste

From the TG curves reported in Figure 4 and the values of Table 4, it is possible to distinguish two groups corresponding to the observed groups of the LOI%. A high mass loss residue is registered, about 99% at 850–900 °C for the LBC samples, while relatively lower values are observed for the CWF and CWM samples, respectively, of 97% and 95.8%. It is possible to note how the two groups are characterized by different patterns of mass loss. Apart from the mass loss occurring in the temperature interval 80–200 °C, attributed to surface or hygroscopic water loss, several steps attributable to the dehydroxylation of clay minerals are visible in the LBC samples. The CWF and CWM TG curves are dominated by the step at 600–900 °C, typical of a decarbonation process, attributable to the calcite already detected by XRD and LOI%.

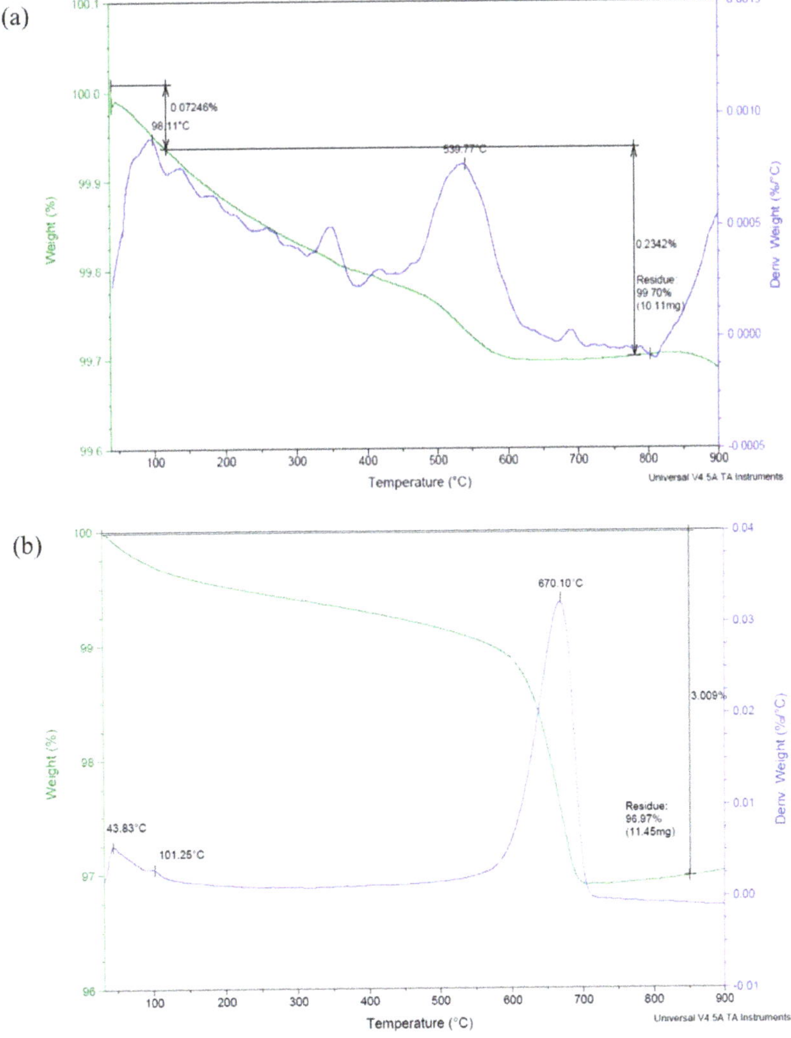

Figure 4. TG curves of two representative ceramic waste materials: (**a**) LBCa, representative of the tiles group LBC; and (**b**) CWF, representative of the bricks group CWF-CWM.

Table 4. Decomposition temperatures and mass loss of the ceramic industrial wastes under study. Residual mass at 850 °C.

Sample	0–80 °C	80–130 °C	130–200 °C	200–500 °C	500–600 °C	600–900 °C	Residue (%)
LBCa		0.15		0.05	0.09		99.7
LBCb		0.14			0.19		99.7
LBCc			0.4				99.6
LBCf	0.12			0.4			99.5
CWF *	0.36					2.7	97
CWM *	0.78	0.24				3.2	95.8

* Data partially reported from the literature [45].

3.4. Reactivity Test in Alkali of Ceramic Industrial Waste

As it is well known, the total amount of silica and alumina present in a ceramic material could not be considered reactive [20,49]. In order to investigate how much of these components are potentially reactive in alkaline media [20], a reactivity test was performed on all the ceramic industrial waste, evaluating their solubility in 8 M NaOH solution at room temperature, and the obtained solution was analysed by ICP-OES. LBCb was analysed as representative of the group of handmade tiles LBCb-LBCc. By the gravimetric measurement of the solid residue, the soluble phase (that means potentially reactive) was obtained (Table 5). The sample with the highest value of soluble phase was CWM, with 30 wt% of soluble phase, while LBCa and CWF showed the lowest value of soluble phases, respectively, with a solid residue of 90 and 94 wt%.

Table 5. Data results of ICP-OES on the eluates after reactivity test on ceramic waste: solid residue (wt%), soluble fraction (wt%), analytical results of the Si and Al concentration (mg/L) and comparison between the Si and Al amount solubilized with the reactivity test and the soluble fraction (% as solid completion).

Sample	Solid Residue (wt%)	Soluble Fraction (wt%)	Si (mg/L)	Al (mg/L)	Si + Al (mg/L)	(Si + Al)/Soluble Fraction (wt%)
LBCa	90	10	9.5	3.1	12.63	12
LBCb	84	16	8.1	3	11.07	7
LBCf	83	17	15.7	3.8	19.48	11
CWF	94	6	10	3.6	13.61	22
CWM	70	30	22.1	14.6	36.7	12

Table 5 also shows the Si and Al concentrations of the eluates determined by the ICP-OES. As expected, the solubility results were higher for Si than for Al. The contribution sum of Si and Al and their percentage in the soluble phase brings to light the highest value calculated for CWF. The high percentage of solid residue after NaOH treatment was observed in the literature. Different studies indeed demonstrated that the alkali attack, even performed under different conditions of sodium hydroxide concentration, temperature or treatment time, give only a partial dissolution of the Si and Al of the aluminosilicate material [50–52]. Their soluble fraction in alkaline solution is also linked to the nature of the aluminosilicate materials; for example, it seems that tectosilicates are more sensitive to this kind of treatment than phyllosilicates, inosilicates or cyclosilicates [50]. Furthermore, the ability of the aluminosilicate material to exchange cations, the Al^{3+} coordination and the surface area influence the Si and Al dissolution [51,52]. In order to understand the percentage of potentially reactive silica and alumina, the obtained results were recalculated on the total silica and alumina amount measured from the bulk chemical composition of the ceramic precursors. The final data are shown in Table 6.

Table 6. Amount of silica and alumina dissolved after basic attack, calculated on the total silica and alumina amount measured from the bulk chemical composition of ceramic precursors tested. $SiO_2\%$ s Tot = soluble $SiO_2\%$ with respect to the total chemical composition of the sample; $Al_2O_3\%$ s Tot = soluble $Al_2O_3\%$ with respect to the total chemical composition of the sample; $SiO_2\%$ s Silica Tot = soluble $SiO_2\%$ with respect to the total $SiO_2\%$ present in the sample; $Al_2O_3\%$ s Alumina Tot = soluble $Al_2O_3\%$ with respect to the total $Al_2O_3\%$ present in the sample.

Sample	$SiO_2\%$ s Tot	$Al_2O_3\%$ s Tot	$SiO_2\%$ s Silica Tot	$Al_2O_3\%$ s Alumina Tot	$[SiO_2]/[Al_2O_3]_{reactive}$
LBCa	0.002	0.001	0.328	0.359	3.397
LBCb	0.002	0.001	0.299	0.327	3.057
LBCf	0.003	0.001	0.464	0.391	4.678
CWF	0.002	0.001	0.373	0.475	3.145
CWM	0.005	0.003	0.832	1.815	1.714

The amount of the soluble silica and alumina of the ceramic precursors under these conditions (NaOH 8M solution at room temperature) is very low, less than 1% of the total silica and alumina detected by XRF. Only the CWM sample is an exception, with a value of around 2%. The low measured value could be highly affected by the reprecipitation of SiO_2 and Al_2O_3 as, for instance, sodium silicates and aluminates. Nevertheless, the $([SiO_2]/[Al_2O_3])_{reactive}$ ratio of the studied ceramic industrial wastes is on average around 3. This is considered a critical parameter, which in order to obtain a good mechanical performance in geopolymers, should be between 2 and 4 [20]. Only the CWM sample is quite below the threshold.

3.5. Granulometric Analysis on Ceramic Waste (Laser Granulometry and BET)

As preliminary synthesis tests were performed using LBCa and CWF as raw materials, a granulometric investigation was performed on these two ceramic wastes. While preliminary tests were performed with a granulometry of 75 µm, a lower granulometry, at least 15 µm, was preferred for the larger synthesis set. Starting from centimetre-sized fragments of ceramics (1–2 cm), different grinding procedures were tested in order to reach the desired granulometry. The preferred individuated procedure consisted of 40 min grinding in a porcelain jar with alumina spheres, in accordance with the literature [41]. About 50% by volume of the particles had a size of about 16 µm and 12 µm, measured by laser diffraction, obtained for LBCa and CWF ceramics, respectively (Figure 5).

As is immediately visible, furthermore, the LBCa particle size distribution shows a unimodal pattern, with a mean particle size of around 10 µm, while CWF is characterized by a bimodal particle size distribution, with mean particle sizes around 1 µm and 10 µm. The obtained powder fineness is similar to that of Portland cement, having particles in the size range 1–100 µm [35], and to that of MK, which was also used in this research project. During material processing, its specific surface could change; thus, BET analyses were performed in order to evaluate the correlation between the granulometric analysis and the superficial area [53]. It was found that when the granulometry decreases, the specific area of the LBCa ceramic increases, while that of CWF decreases (Table 7), even if there is a significant difference between the sintered ceramic product fired at 1100 °C compared to porous bricks fired under 1000 °C.

3.6. Visual Observations and Integrity Tests

Around fifty formulations were synthesized and, after 28 days of curing, submitted to preliminary considerations and to the integrity test. In the Supplementary Materials (Table S1), all the formulations and their synthesis parameters are shown. Table 8, on the other hand, summarizes the preliminary evaluations and the results of the integrity tests.

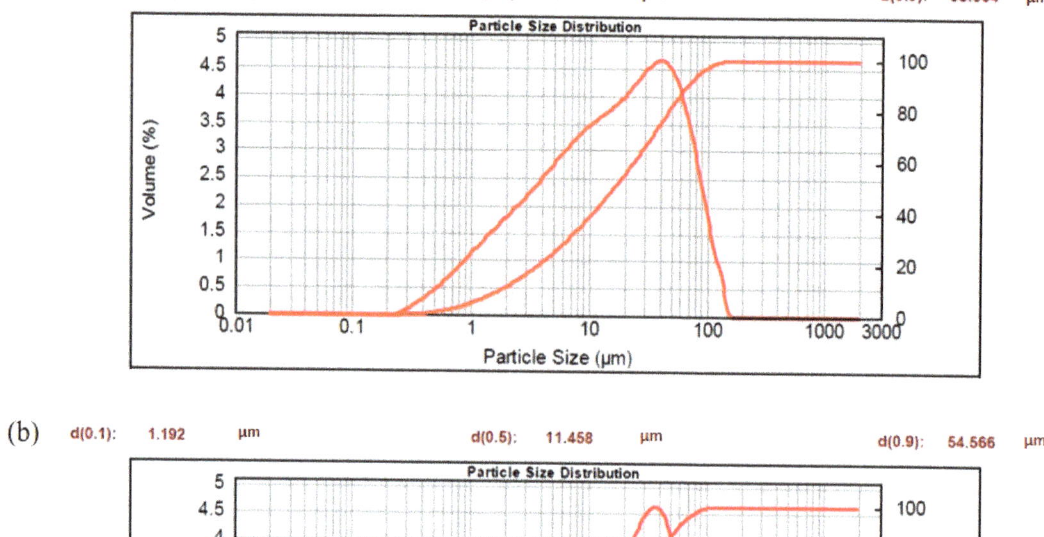

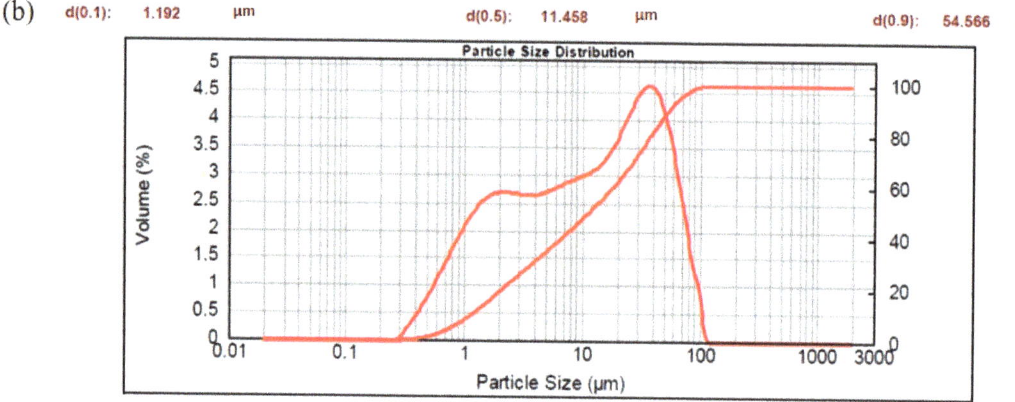

Figure 5. (**a**) Particle size distributions of LBCa; and (**b**) particle size distributions of CWF; measured by laser granulometry. d(0.5) = 50% of volume with the particle size less than this dimension; and d(0.9) = 90% of volume with the particle size less than this dimension.

Table 7. BET analysis results of the selected precursors.

Sample	D_{50}	Surface Area (m^2/g)
LBCa	1–2 mm	0.5272
	~15 μm	1.0155
CWF	>125 μm	47.8880
	~12 μm	43.1090

Table 8. Preliminary observations and integrity test results of the experimental AAMs: St = setting time and Ct = curing time.

Sample	St (min)	Ct (Days)	Homogeneity	Shrinkage	Cracks	Salts	Integrity Tests Stability in Water	Resistance to Pincer
LBCa$_{75}$ 1	/	28	no	yes	no	no	no	no
LBCa$_{75}$ 2	/	28	no	yes	no	no	no	no
LBCa$_{75}$ 3+50P	/	28	yes	yes	no	no	yes	no
LBCa$_{75}$ 4	/	28	no	yes	no	no	no	no

Table 8. *Cont.*

Sample	St (min)	Ct (Days)	Homogeneity	Shrinkage	Cracks	Salts	Integrity Tests	
							Stability in Water	Resistance to Pincer
LBCa$_{10}$ 1	5	28	yes	yes	no	no	no	no
LBCa$_{10}$ 2	5	28	yes	yes	no	no	no	no
LBCa$_{10}$ 3+50P	5	1	yes	yes	no	no	yes	no
LBCa$_{10}$ 4	5	28	no	yes	no	no	no	no
LBCa 5	5	1	no	yes	no	no	no	no
LBCa 5A	5	1	no	yes	no	yes	yes	no
LBCa 5+10MK	5	1	yes	yes	no	no	yes	no
LBCa 5+10MK_A	5	1	no	yes	no	yes	yes	no
LBCa 5+20MK	5	1	yes	yes	no	no	yes	no
LBCa 6	5	1	yes	yes	no	no	no	no
LBCa 6+10MK	5	1	no	yes	no	no	no	no
LBCa 6+20MK	5	1	yes	yes	no	no	no	no
LBCa 7	5	1	yes	yes	no	no	no	no
LBCa 8	5	1	yes	yes	no	no	no	no
LBCa 9	5	7	yes	yes	no	no	no	no
LBCa 10	5	7	yes	yes	no	no	no	no
LBCa 11	5	1	no	yes	no	no	yes	no
LBCa 12	5	1	yes	yes	no	yes	yes	no
LBCa 12A	5	1	yes	yes	no	yes	yes	no
LBCa 13	5	7	yes	yes	no	no	yes	yes
LBCa 13+10MK	5	1	yes	yes	no	yes	yes	yes
LBCa 13+20MK	5	1	yes	no	no	no	yes	yes
LBCa 14	5	7	yes	yes	no	no	yes	yes
LBCa 14+10MK	5	1	yes	yes	no	no	yes	yes
LBCa 14+20MK	5	1	yes	no	no	no	yes	yes
LBCa 14+5P	5	<1	yes	yes	no	no	yes	yes
LBCa 14+10P	2	<1	yes	yes	no	no	yes	yes
LBCa 14+20P	2	<1	yes	yes	no	no	yes	yes
LBCa 15	5	7	yes	yes	no	yes	yes	no
LBCa 15+10MK	5	1	yes	yes	no	yes	yes	yes
LBCa 15+20MK	5	1	yes	no	no	yes	yes	yes
CWF 13	5	7	yes	yes	no	no	yes	yes
CWF 13+10MK	5	1	yes	yes	no	no	yes	yes
CWF 13+20MK	5	1	no	no	no	no	yes	yes
CWF 14	5	7	yes	yes	no	no	yes	yes
CWF 14+10MK	5	1	yes	yes	no	no	yes	yes
CWF 14+20 MK	5	1	yes	no	no	no	yes	yes
CWF 15	5	7	yes	yes	no	yes	no	yes
CWF 15+10MK	5	1	yes	yes	no	yes	no	yes
CWF 15+20MK	5	1	yes	no	no	yes	no	yes

The first observation is the failure of the formulations made with 75 μm powders, which do not set, but harden just because of drying after very long times. Indeed, they totally failed the integrity tests, with the exception of the sample with 50% of P. The setting time was then accelerated to 5 min by the decrease in the granulometry of the ceramic precursor. As reported in the literature [54], indeed, the powders' reactivity is also determined by its fineness, increasing with the decrease in the granulometry. The curing time instead ranges from 1 to 7 days, likely depending on the liquid content of the fresh paste and on the degree of reactivity of the ceramic industrial waste [34,55]. The majority of the samples show a fairly homogeneous visual appearance, with some exceptions probably due to insufficient mixing of the different components in binary mixtures, or due to a separation of the sodium silicate, which tends to stratify at the top of the poured samples, when in excess. In some cases, the lack of workability observed during the synthesis makes the surface of the dried samples rough because of an increased difficulty in pouring the slurry into the mold. The tendency to shrink characterizes all the samples, with the exception of those with the highest amount of MK. Nevertheless, the extent of the shrinkage upon drying could be considered low, as it did not produce any cracks nor micro-cracks, and it should be considered as an intrinsic characteristic of this kind of material [56]. In order to reduce drying shrinkage in the final products, inert aggregates should be added in the mix design [57]. Regarding the efflorescence, it is evident that this appears when the sodium hydroxide proportion in the liquid component is higher (e.g., sodium hydroxide/sodium silicate = 2.33), or when moderate amounts of sodium hydroxide (e.g., sodium hydroxide/sodium silicate ~ 0.5) is combined with high amounts of water added during the synthesis. Furthermore, efflorescence appears in samples which were subjected to a low-temperature firing step, probably because of the faster evaporation of water compared to the replicates cured entirely at room temperature [26,27,33]. Regarding the stability in water and the resistance to the pincer cut after soaking, it is possible to see in Table 8 how the first synthesis tests totally failed both criteria of the integrity test, with the exception of the fired samples and of the binary mixtures. The apparent quick setting and apparent consolidation of all the samples which tend to disintegrate in water could probably be ascribed to the hardening action of the drying of sodium silicate [26,40,58], generally present in high amounts, without the occurrence of an efficient geopolymerization process. In the second part of the table, the results of the integrity tests show samples leaving the water clear and resisting the pincer cut. Among these, the formulations obtained by adding a percentage of P are characterized by a very quick setting time, which is a very promising result for the scale-up of building materials. After considering the preliminary results of the empirically adjusted slurries in the lab, a batch of LBCa-based formulations (highlighted in grey in Table 8) was selected to be replicated using CWF as a precursor. The last lines of Table 8 show that by changing the ceramic precursor, similar results are obtained in terms of preliminary visual observation and the results of the integrity test.

3.7. Mechanical Resistance and Porosity of Alkali-Activated Samples

The formulations tested both with curing at room temperature and with a 65 °C treatment were analyzed. Uniaxial compressive tests and mercury porosimetry were performed in order to investigate the effect of the curing temperature step on the performance of the final hardened product. The results are summarized in the scheme of Figure 6, and shown in detail in the Supplementary Materials (Tables S2 and S3).

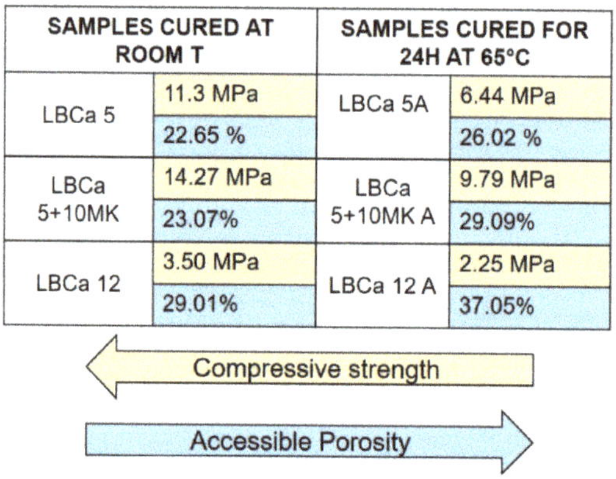

SAMPLES CURED AT ROOM T		SAMPLES CURED FOR 24H AT 65°C	
LBCa 5	11.3 MPa	LBCa 5A	6.44 MPa
	22.65 %		26.02 %
LBCa 5+10MK	14.27 MPa	LBCa 5+10MK A	9.79 MPa
	23.07%		29.09%
LBCa 12	3.50 MPa	LBCa 12 A	2.25 MPa
	29.01%		37.05%

Compressive strength

Accessible Porosity

Figure 6. Porosimetric and mechanical results obtained from samples LBCa 5, LBCa 5+10MK and LBCa 12, cured at ambient temperature, and on the corresponding samples cured for 24 h at 65 °C and then allowed to mature at room temperature: LBCa 5A, LBCa 5+10MK_A and LBCa 12A.

The samples LBCa 5 and LBCa 5+10MK do not show visible efflorescence, while the same formulations after the firing step (LBCa 5A and LBCa 5+10MK_A) are characterized by visible efflorescence as well as a decrease in compressive strength of about 5 MPa. The simultaneous occurrence of these features could be correlated to the same event, i.e., the crystallization of crypto-efflorescence would interrupt the structural continuity of the alkaline gel in the samples, compromising their compactness and resistance [59]. Decreases in compressive strength are also individuated in LBCa 12A (same formulation as LBCa 12, but cured at 65 °C), but in this case efflorescence is already visible in the non-fired sample. In general, the resistance of both samples is very poor, while the strength value for the other samples could be considered within the range of other AAMs [60]. Concerning porosity, the samples show values of accessible porosity between those of traditional cement and typical geopolymers [61]. It is furthermore possible to notice an increase in the accessible porosity after firing, which is an inverse pattern with respect to that observed for the compressive strength values. Again, the fired samples show worse performances than those cured at room temperature. The higher accessible porosity volume would indeed determine a higher vulnerability of the products to environmental decay. From the results shown in Figure 6, it is furthermore possible to appreciate how the addition of only 10% of MK to the solid precursor determines an increase in the compressive strength of the final product.

3.8. Mineralogical Investigation of the Efflorescence

In order to investigate the efflorescence formed on the samples, a few samplings were carried out on the powdery deposit of salts when present; otherwise, the salt samples were scratched from the surface with a scalpel. In this latter case, we must consider the possibility that the geopolymer was sampled as well. XRD was carried out on samples LBCa 12A, LBCa 13+10MK, LBCa 15, LBCa 15+10MK and LBCa 15+20MK. The diffraction pattern of salts collected from sample LBCa 15+20MK is shown in Figure 7 as representative. From this analysis, mainly sodium carbonates with different hydration grades were detected, as expected [62–64]. Some further peaks cannot be attributed to sodium carbonates, and are mainly compatible with the presence of gypsum; however, its presence cannot be confirmed by XRF analysis, since sulfur was not revealed. In detail, trona $(Na_3(CO_3)(HCO_3)*2(H_2O))$ and thermonatrite $(Na_2CO_3(H_2O))$ were individuated in sample LBCa 15, in samples LBCa 13+10MK and LBCa 15+20MK (with the possible presence of gypsum $(CaSO_4*2H_2O)$) and in sample LBCa 12A (with the possible presence of ettringite $(Ca_6Al_2(SO_4)_3(OH)_{12}*26(H_2O))$; only thermonatrite was detected in sample LBCa 15+10MK. In some samples, peaks attributable to quartz, most likely due to the geopolymeric substrate scratched unintentionally during sampling, are also present.

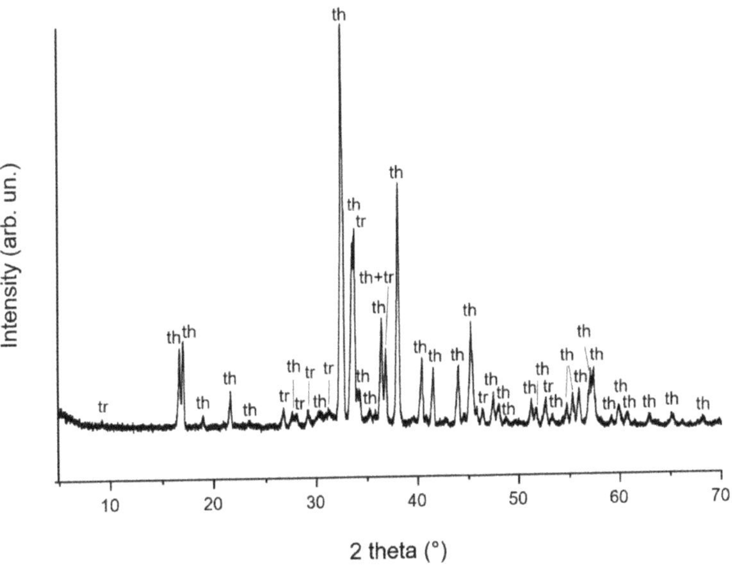

Figure 7. Representative diffraction pattern of the salts collected from the samples LBCa 12A, LBCa 13+10MK, LBCa 15, LBCa 15+10MK and LBCa 15+20MK. The diffraction pattern of the sample LBCa 15+20MK in particular is shown in the image. th = thermonatrite; tr = trona.

4. Conclusions

Six different industrial waste samples of ceramic materials were analyzed to determine their chemistry and mineralogy, thermal behavior, and reactivity in the alkaline environment required for the synthesis of AAMs. On two selected ceramic wastes, selected according to their representativeness and higher abundance in the industries, the effect of particle size on the properties of the final products was investigated. A plethora of formulations were designed and prepared from the same kind of ceramic by varying the type of alkaline solutions, L/S ratio, amount and kind of additives, and curing conditions. A set of formulations were replicated by changing the waste used as an aluminosilicate precursor. Visual observations of fresh samples were recorded during the synthesis, and preliminary tests were carried out after 28 days. The main important conclusions can be summarized as follows:

- The comparison of performances among the ceramic wastes distinguished two groups related to the presence or absence of calcite. This demonstrates, as we could expect, a different behavior at high temperatures and in alkaline conditions. Nevertheless, by using two ceramics representative of the two individuated groups (LBCa and CWF) for the synthesis, similar results were obtained in terms of workability, curing time, possible efflorescence formations, cracks or shrinkage and water resistance (integrity tests);
- The results obtained reveal that construction industrial wastes of ceramic nature can be activated by using proportioned mixtures of sodium hydroxide (8 M) and sodium silicate (R = 3.3), in accordance with the literature [61], or with only sodium silicate;
- AAMs tend to consolidate between 1 day and 7 days, depending on the formulation and particularly on the amount of additives;
- P and MK, even if added in small amounts (5%–10%), are able to promote the alkaline gel building, and thus the consolidation. Such additives also improve the strength and counteract the formation of efflorescence;
- Where present, the efflorescence seems to be strictly linked to the higher sodium hydroxide or water amount; therefore, as well as with the addition of additives, the formation of efflorescence could be avoided by balancing formulations in a stoichiometric way;
- In contrast to the current literature [4,26,65], this study highlighted that curing at room temperature (around 25 °C) is preferable, whereas thermal curing at 65 °C for 24 h resulted in lower performance in terms of compressive strength and an increase in porosity, and facilitated efflorescence crystallization. Similar results were also found in [39];
- LBCa and CWF proved to be the least reactive ceramics among all the samples studied. Despite that, good results for the synthesis of AAMs were obtained, allowing for better results by using the other, more reactive, ceramic wastes characterized here.

Supplementary Materials: The following supporting information can be downloaded at https://www.mdpi.com/article/10.3390/min13060815/s1. Table S1: Synthesis parameters of AAMs experimented: *the % is calculated on total solid; **the % is calculated on total liquid; Tc = curing temperature; Table S2: Results of uniaxial compressive strength tests performed on examples of geopolymers cured at room temperature and at 65 °C for 24 h; Table S3: Porosimetric data of geopolymeric examples cured at room temperature and at 65 °C for 24 h.

Author Contributions: Visualization, M.F.; Conceptualization, M.F., P.M. and G.B.; Data curation, M.F., I.L., R.C., P.F., M.R.T., I.C. and T.S.; Formal analysis, M.F., P.M. and G.B.; Funding acquisition, P.M. and G.B.; Investigation, M.F., I.L., R.C., P.F., M.R.T., I.C. and T.S.; Methodology, M.F., P.M. and G.B.; Project administration, P.M. and G.B.; Supervision, P.M. and G.B.; Validation, M.F., P.M., I.L., R.C., P.F., M.R.T., I.C., T.S. and G.B.; Visualization, M.F.; Writing—original draft, M.F., I.L., R.C., P.F., M.R.T., I.C. and T.S.; Writing—review and editing, M.F., G.B. and P.M. All authors have read and agreed to the published version of the manuscript.

Funding: This research was funded by the Advanced Green Materials for Cultural Heritage (AGM for CuHe) project (MIUR PNR fund with code: ARS01_00697; CUP E66C18000380005).

Data Availability Statement: Data are contained within the article or Supplementary Materials.

Acknowledgments: The authors wish to thank La Bottega Calatina and Laquattro for having supplied ceramic wastes used as geopolymeric precursors. The authors also thank, for the support in measurements and investigation, Giuseppe Cultrone (Department of Mineralogy and Petrology, University of Granada—Spain), and CIC (Centro de Instrumentación Científica—University of Granada); Piero Baglioni and Emiliano Carretti (Department of Chemistry, University of Florence), and CSGI (Center for Colloid and Surface Science—Department of Chemistry, University of Florence); Carlo Alberto Garzonio (Department of Earth Science, University of Firenze) and Cristina Leonelli (Department of Engineering "Enzo Ferrari", University of Modena and Reggio Emilia).

Conflicts of Interest: The authors declare no conflict of interest.

References

1. Pathak, A.; Kumar, S.; Jha, V.K. Development of Building Material from Geopolymerization of Construction and Demolition Waste (CDW). *Trans. Indian Ceram. Soc.* **2014**, *73*, 133–137. [CrossRef]
2. Allaoui, D.; Nadi, M.; Hattani, F.; Majdoubi, H.; Haddaji, Y.; Mansouri, S.; Oumam, M.; Hannache, H.; Manoun, B. Eco-Friendly Geopolymer Concrete Based on Metakaolin and Ceramics Sanitaryware Wastes. *Ceram. Int.* **2022**, *48*, 34793–34802. [CrossRef]
3. Tan, J.; Cai, J.; Li, J. Recycling of Unseparated Construction and Demolition Waste (UCDW) through Geopolymer Technology. *Constr. Build. Mater.* **2022**, *341*, 127771. [CrossRef]
4. Reig, L.; Tashima, M.M.; Soriano, L.; Borrachero, M.V.; Monzó, J.; Payá, J. Alkaline Activation of Ceramic Waste Materials. *Waste Biomass Valoriz.* **2013**, *4*, 729–736. [CrossRef]
5. Robayo-Salazar, R.; Valencia-Saavedra, W.; de Gutiérrez, R.M. Reuse of Powders and Recycled Aggregates from Mixed Construction and Demolition Waste in Alkali-Activated Materials and Precast Concrete Units. *Sustainability* **2022**, *14*, 9685. [CrossRef]
6. Hwang, C.L.; Yehualaw, M.D.; Vo, D.H.; Huynh, T.P.; Largo, A. Performance Evaluation of Alkali Activated Mortar Containing High Volume of Waste Brick Powder Blended with Ground Granulated Blast Furnace Slag Cured at Ambient Temperature. *Constr. Build. Mater.* **2019**, *223*, 657–667. [CrossRef]
7. Wong, C.L.; Mo, K.H.; Yap, S.P.; Alengaram, U.J.; Ling, T.C. Potential Use of Brick Waste as Alternate Concrete-Making Materials: A Review. *J. Clean. Prod.* **2018**, *195*, 226–239. [CrossRef]
8. Panizza, M.; Natali, M.; Garbin, E.; Tamburini, S.; Secco, M. Assessment of Geopolymers with Construction and Demolition Waste (CDW) Aggregates as a Building Material. *Constr. Build. Mater.* **2018**, *181*, 119–133. [CrossRef]
9. Frittelloni, V. *Rapporto Rifiuti Speciali Edizione 2022*; Istituto Superiore per la Protezione e la Ricerca Ambientale: Rome, Italy, 2022; ISBN 9788844811167.
10. Sarkar, M.; Dana, K. Partial Replacement of Metakaolin with Red Ceramic Waste in Geopolymer. *Ceram. Int.* **2021**, *47*, 3473–3483. [CrossRef]
11. Mir, N.; Khan, S.A.; Kul, A.; Sahin, O.; Lachemi, M.; Sahmaran, M.; Koç, M. Life Cycle Assessment of Binary Recycled Ceramic Tile and Recycled Brick Waste-Based Geopolymers. *Clean. Mater.* **2022**, *5*, 100116. [CrossRef]
12. Bernal, S.A.; Rodríguez, E.D.; Kirchheim, A.P.; Provis, J.L. Management and Valorisation of Wastes through Use in Producing Alkali-Activated Cement Materials. *J. Chem. Technol. Biotechnol.* **2016**, *91*, 2365–2388. [CrossRef]
13. Almutairi, A.L.; Tayeh, B.A.; Adesina, A.; Isleem, H.F.; Zeyad, A.M. Potential Applications of Geopolymer Concrete in Construction: A Review. *Case Stud. Constr. Mater.* **2021**, *15*, e00733. [CrossRef]
14. Bassani, M.; Tefa, L.; Russo, A.; Palmero, P. Alkali-Activation of Recycled Construction and Demolition Waste Aggregate with No Added Binder. *Constr. Build. Mater.* **2019**, *205*, 398–413. [CrossRef]
15. Coletti, C.; Maritan, L.; Cultrone, G.; Mazzoli, C. Use of Industrial Ceramic Sludge in Brick Production: Effect on Aesthetic Quality and Physical Properties. *Constr. Build. Mater.* **2016**, *124*, 219–227. [CrossRef]
16. Fugazzotto, M.; Cultrone, G.; Mazzoleni, P.; Barone, G. Suitability of Ceramic Industrial Waste Recycling by Alkaline Activation for Use as Construction and Restoration Materials. *Ceram. Int.* **2023**, *49*, 9465–9478. [CrossRef]
17. Azevedo, A.R.G.; Vieira, C.M.F.; Ferreira, W.M.; Faria, K.C.P.; Pedroti, L.G.; Mendes, B.C. Potential Use of Ceramic Waste as Precursor in the Geopolymerization Reaction for the Production of Ceramic Roof Tiles. *J. Build. Eng.* **2020**, *29*, 101156. [CrossRef]
18. Van Deventer, J.S.J.; Provis, J.L.; Duxson, P.; Brice, D.G. Chemical Research and Climate Change as Drivers in the Commercial Adoption of Alkali Activated Materials. *Waste Biomass Valoriz.* **2010**, *1*, 145–155. [CrossRef]
19. Davidovits, J. Geopolymers—Inorganic Polymeric New Materials. *J. Therm. Anal.* **1991**, *37*, 1633–1656. [CrossRef]
20. Pachego-Torgal, F.; Labrincha, J.A.; Leonelli, C.; Palomo, A.; Chindaprasirt, P. *Handbook of Alkali-Activated Cements, Mortars and Concretes*; Elsevier: Cambridge, UK, 2015; ISBN 9781782422884.
21. Provis, J.L.; Bernal, S.A. Geopolymers and Related Alkali-Activated Materials. *Annu. Rev. Mater. Res.* **2014**, *44*, 299–327. [CrossRef]
22. Allahverdi, A.; Kani, E.N. Construction Wastes as Raw Materials for Geopolymer Binders. *Int. J. Civ. Eng.* **2009**, *7*, 154–160.

23. Albitar, M.; Mohamed Ali, M.S.; Visintin, P.; Drechsler, M. Durability Evaluation of Geopolymer and Conventional Concretes. *Constr. Build. Mater.* **2017**, *136*, 374–385. [CrossRef]

24. Alhawat, M.; Ashour, A.; Yildirim, G.; Aldemir, A.; Sahmaran, M. Properties of Geopolymers Sourced from Construction and Demolition Waste: A Review. *J. Build. Eng.* **2022**, *50*, 104104. [CrossRef]

25. Lancellotti, I.; Vezzali, V.; Barbieri, L.; Leonelli, C.; Grillenzoni, A. Construction and Demolition Waste (Cdw) Valorization in Alkali Activated Bricks. In *Wastes: Solutions, Treatments and Opportunities III*; Taylor and Francis Group: London, UK, 2020; pp. 495–499, ISBN 9780367257774.

26. Komnitsas, K.; Zaharaki, D.; Vlachou, A.; Bartzas, G.; Galetakis, M. Effect of Synthesis Parameters on the Quality of Construction and Demolition Wastes (CDW) Geopolymers. *Adv. Powder Technol.* **2015**, *26*, 368–376. [CrossRef]

27. Amin, S.K.; El-Sherbiny, S.A.; El-Magd, A.A.M.A.; Belal, A.; Abadir, M.F. Fabrication of Geopolymer Bricks Using Ceramic Dust Waste. *Constr. Build. Mater.* **2017**, *157*, 610–620. [CrossRef]

28. Leonelli, C.; Romagnoli, M. *Geopolimeri: Polimeri Inorganici Attivati Chimicamente*, 2nd ed.; ICerS: Bologna, Italy, 2013.

29. Duxson, P.; Fernández-Jiménez, A.; Provis, J.L.; Lukey, G.C.; Palomo, A.; Van Deventer, J.S.J. Geopolymer Technology: The Current State of the Art. *J. Mater. Sci.* **2007**, *42*, 2917–2933. [CrossRef]

30. Geraldes, C.F.M.; Lima, A.M.; Delgado-Rodrigues, J.; Mimoso, J.M.; Pereira, S.R.M. Geopolymers as Potential Repair Material in Tiles Conservation. *Appl. Phys. A Mater. Sci. Process.* **2016**, *122*, 197. [CrossRef]

31. Moutinho, S.; Costa, C.; Cerqueira, Â.; Rocha, F.; Velosa, A. Geopolymers and Polymers in the Conservation of Tile Facades. *Constr. Build. Mater.* **2019**, *197*, 175–184. [CrossRef]

32. Ricciotti, L.; Molino, A.J.; Roviello, V.; Chianese, E.; Cennamo, P.; Roviello, G. Geopolymer Composites for Potential Applications in Cultural Heritage. *Environments* **2017**, *4*, 91. [CrossRef]

33. Robayo, R.A.; Mulford, A.; Munera, J.; de Gutiérrez, R.M. Alternative Cements Based on Alkali-Activated Red Clay Brick Waste. *Constr. Build. Mater.* **2016**, *128*, 163–169. [CrossRef]

34. Rovnaník, P.; Rovnaníková, P.; Vyšvařil, M.; Grzeszczyk, S.; Janowska-Renkas, E. Rheological Properties and Microstructure of Binary Waste Red Brick Powder/Metakaolin Geopolymer. *Constr. Build. Mater.* **2018**, *188*, 924–933. [CrossRef]

35. Tuyan, M.; Andiç-Çakir, Ö.; Ramyar, K. Effect of Alkali Activator Concentration and Curing Condition on Strength and Microstructure of Waste Clay Brick Powder-Based Geopolymer. *Compos. Part B Eng.* **2018**, *135*, 242–252. [CrossRef]

36. Barone, G.; Caggiani, M.C.; Coccato, A.; Finocchiaro, C.; Fugazzotto, M.; Lanzafame, G.; Occhipinti, R.; Stroscio, A.; Mazzoleni, P. Geopolymer Production for Conservation-Restoration Using Sicilian Raw Materials: Feasibility Studies. *IOP Conf. Ser. Mater. Sci. Eng.* **2020**, *777*, 012001. [CrossRef]

37. Djobo, J.N.Y.; Tchadjié, L.N.; Tchakoute, H.K.; Kenne, B.B.D.; Elimbi, A.; Njopwouo, D. Synthesis of Geopolymer Composites from a Mixture of Volcanic Scoria and Metakaolin. *J. Asian Ceram. Soc.* **2014**, *2*, 387–398. [CrossRef]

38. Yip, C.; Lukey, G.C.; van Deventer, J.S.J. The Coexistence of Geopolymeric Gel and Calcium Silicate Hydrate at the Early Stage of Alkaline Activation. *Cem. Concr. Res.* **2005**, *35*, 1688–1697. [CrossRef]

39. Lancellotti, I.; Catauro, M.; Ponzoni, C.; Bollino, F.; Leonelli, C. Inorganic Polymers from Alkali Activation of Metakaolin: Effect of Setting and Curing on Structure. *J. Solid State Chem.* **2013**, *200*, 341–348. [CrossRef]

40. Ahmari, S.; Ren, X.; Toufigh, V.; Zhang, L. Production of Geopolymeric Binder from Blended Waste Concrete Powder and Fly Ash. *Constr. Build. Mater.* **2012**, *35*, 718–729. [CrossRef]

41. Reig, L.; Tashima, M.M.; Borrachero, M.V.; Monzó, J.; Cheeseman, C.R.; Payá, J. Properties and Microstructure of Alkali-Activated Red Clay Brick Waste. *Constr. Build. Mater.* **2013**, *43*, 98–106. [CrossRef]

42. Ruiz-Santaquiteria, C.; Fernández-Jiménez, A.; Skibsted, J.; Palomo, A. Clay Reactivity: Production of Alkali Activated Cements. *Appl. Clay Sci.* **2013**, *73*, 11–16. [CrossRef]

43. Fernández-Jiménez, A.; Palomo, A. Characterisation of Fly Ashes. Potential Reactivity as Alkaline Cements. *Fuel* **2003**, *82*, 2259–2265. [CrossRef]

44. Pulidori, E.; Lluveras-Tenorio, A.; Carosi, R.; Bernazzani, L.; Duce, C.; Pagnotta, S.; Lezzerini, M.; Barone, G.; Mazzoleni, P.; Tiné, M.R. Building Geopolymers for CuHe Part I: Thermal Properties of Raw Materials as Precursors for Geopolymers. *J. Therm. Anal. Calorim.* **2022**, *147*, 5323–5335. [CrossRef]

45. Pelosi, C.; Occhipinti, R.; Finocchiaro, C.; Lanzafame, G.; Pulidori, E.; Lezzerini, M.; Barone, G.; Mazzoleni, P.; Tiné, M.R. Thermal and Morphological Investigations of Alkali Activated Materials Based on Sicilian Volcanic Precursors (Italy). *Mater. Lett.* **2023**, *335*, 133773. [CrossRef]

46. EN 1926 2006; Natural Stone Test Methods—Determination of Uniaxial Compressive Strength. European Commission: Brussels, Belgium, 2006.

47. Buchwald, A.; Kaps, C.; Hohmann, M. Alkali-Activated Binders and Pozzolan Cement Binders—Compete Binder Reaction or Two Sides of the Same Story ? In Proceedings of the 11th International Congress on the Chemistry of Cement, Durban, South Africa, 11–16 May 2003; pp. 1238–1247.

48. Warr, L.N. IMA–CNMNC Approved Mineral Symbols. *Miner. Mag.* **2021**, *85*, 291–320. [CrossRef]

49. Lancellotti, I.; Ponzoni, C.; Barbieri, L.; Leonelli, C. Alkali Activation Processes for Incinerator Residues Management. *Waste Manag.* **2013**, *33*, 1740–1749. [CrossRef]

50. Xu, H.; Van Deventer, J.S.J. The Geopolymerisation of Alumino-Silicate Minerals. *Int. J. Miner. Process.* **2000**, *59*, 247–266. [CrossRef]

51. Mostafa, N.Y.; El-Hemaly, S.A.S.; Al-Wakeel, E.I.; El-Korashy, S.A.; Brown, P.W. Characterization and Evaluation of the Pozzolanic Activity of Egyptian Industrial By-Products: I: Silica Fume and Dealuminated Kaolin. *Cem. Concr. Res.* **2001**, *31*, 467–474. [CrossRef]
52. Panagiotopoulou, C.; Kontori, E.; Perraki, T.; Kakali, G. Dissolution of Aluminosilicate Minerals and By-Products in Alkaline Media. *J. Mater. Sci.* **2007**, *42*, 2967–2973. [CrossRef]
53. Fagerlund, G. Determination of Specific Surface by the BET Method. *Matériaux Constr.* **1973**, *6*, 239–245. [CrossRef]
54. Nazari, A.; Bagheri, A.; Riahi, S. Properties of Geopolymer with Seeded Fly Ash and Rice Husk Bark Ash. *Mater. Sci. Eng. A* **2011**, *528*, 7395–7401. [CrossRef]
55. Azevedo, A.G.D.S.; Strecker, K.; Lombardi, C.T. Produção de Geopolímeros à Base de Metacaulim e Cerâmica Vermelha. *Cerâmica* **2018**, *64*, 388–396. [CrossRef]
56. Komnitsas, K.; Zaharaki, D. Geopolymerisation: A Review and Prospects for the Minerals Industry. *Miner. Eng.* **2007**, *20*, 1261–1277. [CrossRef]
57. Pecchioni, E.; Fratini, F.; Cantisani, E. *Le Malte Antiche e Moderne Tra Tradizione Ed Innovazione*; Pàtron Editore: Bologna, Italy, 2008.
58. Pacheco-Torgal, F.; Tam, V.W.Y.; Labrincha, J.A.; Ding, Y.; de Brito, J. *Handbook of Recycled Concrete and Demolition Waste*; Woodhead Publishing: Cambridge, UK, 2013.
59. Rowles, M.R.; O'Connor, B.H. Chemical and Structural Microanalysis of Aluminosilicate Geopolymers Synthesized by Sodium Silicate Activation of Metakaolinite. *J. Am. Ceram. Soc.* **2009**, *92*, 2354–2361. [CrossRef]
60. Hassan, A.; Arif, M.; Shariq, M. Use of Geopolymer Concrete for a Cleaner and Sustainable Environment e A Review of Mechanical Properties and Microstructure. *J. Clean. Prod.* **2019**, *223*, 704–728. [CrossRef]
61. Fořt, J.; Vejmelková, E.; Koňáková, D.; Alblová, N.; Čáchová, M.; Keppert, M.; Rovnaníková, P.; Černý, R. Application of Waste Brick Powder in Alkali Activated Aluminosilicates: Functional and Environmental Aspects. *J. Clean. Prod.* **2018**, *194*, 714–725. [CrossRef]
62. Najafi Kani, E.; Allahverdi, A.; Provis, J.L. Efflorescence Control in Geopolymer Binders Based on Natural Pozzolan. *Cem. Concr. Compos.* **2012**, *34*, 25–33. [CrossRef]
63. Zhang, Z.; Provis, J.L.; Ma, X.; Reid, A.; Wang, H. Efflorescence and Subflorescence Induced Microstructural and Mechanical Evolution in Fly Ash-Based Geopolymers. *Cem. Concr. Compos.* **2018**, *92*, 165–177. [CrossRef]
64. Leonelli, C. Definizione, Preparazione, Proprietà Ed Applicazioni. In *Geopolimeri: Polimeri Inorganici Attivati Chimicamente*; Leonelli, C., Romagnoli, M., Eds.; ICerS: Bologna, Italy, 2013; pp. 1–22.
65. Sun, Z.; Cui, H.; An, H.; Tao, D.; Xu, Y.; Zhai, J.; Li, Q. Synthesis and Thermal Behavior of Geopolymer-Type Material from Waste Ceramic. *Constr. Build. Mater.* **2013**, *49*, 281–287. [CrossRef]

Article

Increase in Recovery Efficiency of Iron-Containing Components from Ash and Slag Material (Coal Combustion Waste) by Magnetic Separation

Tatiana Aleksandrova [1], Nadezhda Nikolaeva [1], Anastasia Afanasova [1,*], Duan Chenlong [2], Artyem Romashev [1], Valeriya Aburova [1] and Evgeniya Prokhorova [1]

[1] Department of Mineral Processing, Saint Petersburg Mining University, 199106 St. Petersburg, Russia; aleksandrova_tn@pers.spmi.ru (T.A.); nadegdaspb@mail.ru (N.N.); romashev_ao@pers.spmi.ru (A.R.); lera.aburova@mail.ru (V.A.); proh98jane@gmail.com (E.P.)

[2] Key Laboratory of Coal Processing and Efficient Utilization of Ministry of Education, School of Chemical Engineering and Technology, China University of Mining and Technology, Xuzhou 221116, China; clduan@cumt.edu.cn

* Correspondence: afanasova_av@pers.spmi.ru

Abstract: This article presents the results of research aimed at optimizing the process of recovery of valuable components from ash and slag waste from thermal power plants. In this work, both experimental and theoretical studies were carried out to substantiate the use of magnetic separation methods for ash and slag waste processing. Ash and slag wastes were chosen as an object of research due to the presence of valuable components such as iron, aluminum, etc., in them. The research results showed that the method of magnetic separation, including high-gradient magnetic separation, can be effectively used in ash and slag waste processing. As a result, the topology of a magnetic beneficiation technological scheme has been proposed to obtain high-value-added products such as high-magnetic iron minerals, low-magnetic iron minerals, and aluminosilicate microspheres. By using magnetic separation in a weak magnetic field, magnetic microspheres containing high-magnetic iron minerals associated with intermetallics, ranging in size from 20 to 80 μm, were recovered. In the second stage of magnetic separation (high-gradient magnetic separation), an iron ore product with an iron content of 50% with a recovery of 92.07% could be obtained. By using scanning electron microscopy, it was found that the main part of microspheres, which contain low-magnetic iron minerals and aluminosilicates, with sizes from 2 to 15 microns, was recovered in the magnetic fraction. This paper proposes a new approach to the enrichment of ash and slag materials using magnetic separation, which will increase the efficiency of their processing and make the process environmentally sustainable.

Keywords: high-gradient magnetic separation; ash and slag waste from thermal power plants; microspheres; intermetallics; iron; aluminum

Citation: Aleksandrova, T.; Nikolaeva, N.; Afanasova, A.; Chenlong, D.; Romashev, A.; Aburova, V.; Prokhorova, E. Increase in Recovery Efficiency of Iron-Containing Components from Ash and Slag Material (Coal Combustion Waste) by Magnetic Separation. *Minerals* **2024**, *14*, 136. https://doi.org/10.3390/min14020136

Academic Editors: Elvis Fosso-Kankeu, Pierfranco Lattanzi, Elisabetta Dore, Fabio Perlatti and Hendrik Gideon Brink

Received: 17 October 2023
Revised: 11 January 2024
Accepted: 23 January 2024
Published: 26 January 2024

1. Introduction

Due to the growth of the world population and its needs, the demand for electricity is increasing [1,2]. Access to both energy resources and electricity itself is a necessary condition for the growth of the world economy. The main sources of electricity can be divided into three main groups: fossil fuels (oil, gas, coal, and oil shale), nuclear and thermonuclear fusion energy, and renewable energy sources (water, wind, solar energy, etc.) [3–7]. At present, a large portion of electricity generation is provided by fossil fuels, while the portion of renewable energy sources is still insignificant [8,9]. In 2021, at the 26th UN Climate Change Conference (COP26), commitments were accepted to reduce the carbon footprint and gradually reduce the use of coal energy [10].

Currently, coal power generation is one of the largest sources of electricity in the world, accounting for 36% of global electricity generation in 2022 [11]. In the present conditions of the world economy, even considering the tendency to reduce the use of fossil fuels, coal power will retain its leading position [12]. Consequently, there is a huge number of power plants in the world that are operated by coal. At coal power plants, electricity is generated by coal combustion. Which generates a huge amount of waste (ash and slag waste) [13–15].

Under these conditions, the involvement of ash and slag wastes in economic turnover becomes a priority, which will allow us to obtain additional economic effects and reduce the environmental impact in the regions of placement [16,17]. Depending on the deposit and quality of coal, as well as the conditions of its combustion, the generated ash and slag wastes have different compositions and different physical and chemical properties [18]. The main directions of ash and slag waste processing can be highlighted as follows: production of building materials, various fill materials, and soil stabilizers; road building; recultivation of mined-out pits; as well as use as secondary raw materials for the recovery of valuable components [19,20].

Critical reviews of previous studies have shown that in ash and slag wastes, elements (Fe, Si, Ti, Al, Ni, Mo, V, and many others) are present in significant amounts, some of them being strategic metals for a number of industries [21–25]. Various beneficiation processes are used to recover these valuable components: flotation and gravity concentration [26,27], magnetic separation [28–30], and leaching processes [31–33].

Magnetic and aluminosilicate microspheres are one of the components included in ash and slag waste, which have unique technological properties, and they are of industrial interest for recovery [34]. Magnetic and flotation methods are most often used for their separation, but their efficiency is not high enough, which is due to the size of the material (most of it is in the class of 40 μm) [35]. A perspective direction for the recovery of magnetic and aluminosilicate microspheres from fine classes is the use of high-gradient magnetic separation [36–40].

Thus, the aim of this work was to determine and investigate the dependence of magnetic separation process efficiency from the technological parameters of a high-gradient magnetic separator for the recovery of valuable components in the processing of ash and slag wastes as an additional commercial product.

2. Materials and Methods

2.1. Characteristics of Research Objects

Ash and slag waste (ASW) of coal heat power plants (CHPP) weighing 100 kg was chosen as the object of research. For all investigations, representative samples of the required mass were taken in the amount of three samples for each test. Representative samples of 500 g were taken for sieve analysis. Sieve analysis was performed to determine the particle size distribution of the sample. The sample of each size class was analyzed using a Shimadzu EDX 700 X-ray fluorescence analyzer (Shimadzu Corporation, Kyoto, Japan). The results of the analysis are presented in Table 1.

Table 1. Results of X-ray fluorescence analysis of size classes after sieve analysis of the sample.

Size Class, mm	Yield, %	Content, %												
		Fe	Si	S	As	Al	Ca	Sb	K	Mn	Zn	Cu	Ni	Sr
+0.800	0.13	31.55	16.23	3.53	2.75	1.17	2.18	1.03	0.81	0.14	0.076	0.069	0.034	0.011
−0.800 + 0.425	0.14	30.82	15.84	3.24	3.05	1.44	3.19	1.41	1.02	0.18	0.073	0.121	0.039	0.014
−0.425 + 0.212	1.16	29.35	16.43	2.58	3.38	1.65	3.48	2.22	0.77	0.18	0.076	0.068	0.038	0.013
−0.212 + 0.106	0.80	28.52	15.85	2.89	2.77	2.32	3.94	1.88	1.10	0.18	0.073	0.069	0.033	0.014
−0.106 + 0.045	4.40	23.79	16.85	6.33	3.07	2.28	1.84	0.82	1.78	0.13	0.049	0.047	0.021	0.010
−0.045 + 0	93.37	29.97	15.80	2.45	3.44	2.31	3.37	2.02	0.83	0.19	0.074	0.074	0.037	0.012
Total	100	29.68	15.85	2.63	3.42	2.30	3.31	1.97	0.87	0.19	0.07	0.073	0.036	0.012

As shown in Table 1, ASW has a significant yield of fine grades, with 93.37% represented in the −45 μm fraction. When beneficiating material in this size range, it is very difficult to produce a concentrate of suitable quality. Analysis of the various size grades was carried out using a Vega 3 LMH electron microscope (TESCAN, Brno, Czech Republic) equipped with an Oxford Instruments (Oxfordshire, UK) INCA Energy 250/X-max 20 energy-dispersive microanalyzer. Micrographs of the particles are shown in Figure 1.

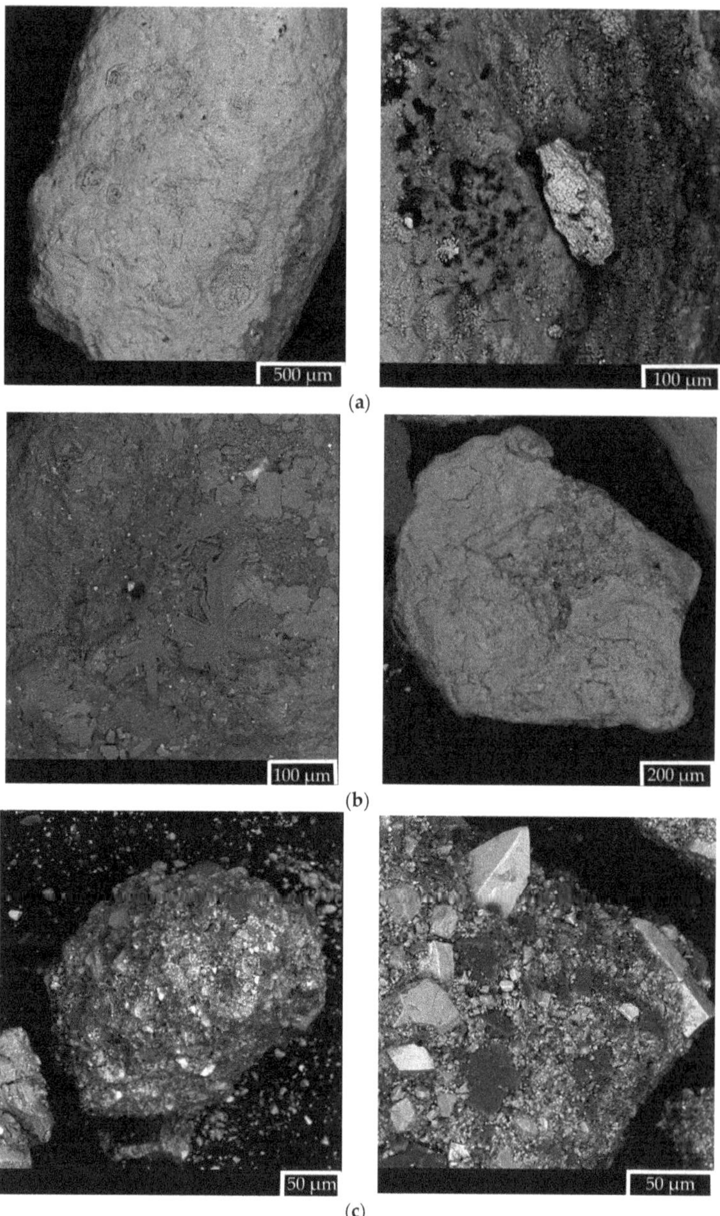

Figure 1. *Cont.*

Figure 1. Microphotographs of ASW by size grade (BSE): (**a**) + 0.8 mm; (**b**) -0.8 + 0.425 mm; (**c**) −0.425 + 0.212 mm; (**d**) −0.212 + 0.106 mm; (**e**) −0.106 + 0.045 mm; (**f**) −0.045 +0 mm.

Analysis of the data presented in Figure 1 by size classes allows us to evaluate the forms of aggregates in the studied material, as well as show the presence of microspheres in fine classes. The main component of ash and slag composition is slag of black, gray, and less often whitish-gray colors; porous, pumice, spongy, and dense textures; as well as slag in the form of fragments.

The particles in the ash composition are heterogeneous both in shape and surface condition; the heterogeneity is preserved in different groups of size fractions. All particles can be divided into two types:

- Spheroids of various diameters formed as a result of the solidification of molten particles suspended in the flue gas stream (several types depending on the composition are fixed). It is established that in the fine class, there is a significant amount of particles smaller than 5 microns;
- unmelted and partially melted.

In the composition of ash and slag, up to 20% is magnetic particles, and aluminosilicate particles are also present. Additionally, such elements as arsenic, sulfur, potassium, manganese, titanium, copper, nickel, etc., are also present.

2.2. Magnetic Separation

Magnetic fractionation was carried out on a Davis Tube Tester 298 SE (NPK "Mekhanobr-Tekhnika" (JSC), St. Petersburg, Russia) to evaluate the distribution of Fe into fractions depending on the magnetic induction and to determine the possibility of using high-gradient magnetic separation. The initial sample of size −0.5 + 0 mm weighing 50 g in pulp form was separated at different current intensities (1, 2, 3, 4, 5, and 6 A, which correspond to 0.14, 0.25, 0.345, 0.422, 0.483, and 0.54 Tesla). In the first experiment (I = 1 A), magnetic and non-magnetic fractions are separated. The feed of subsequent experiments (with increasing current intensity) is the nonmagnetic fraction of the previous experiment. As a result, 7 products were obtained (6 magnetic fractions and 1 non-magnetic fraction), which were dried and analyzed.

Experiments on high-gradient magnetic separation were carried out on a vertical pulsation high-gradient magnetic separator SLon 100 (Outotec, Espoo, Finland) (Figure 2). Samples of size -0.8 + 0 mm weighing 100 g were prepared.

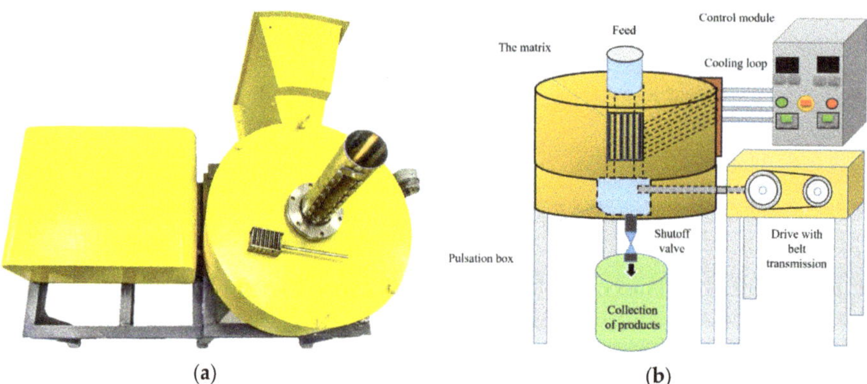

Figure 2. Vertical pulsation high-gradient magnetic separator SLon 100: (**a**) general view, (**b**) separator scheme.

The separation of the magnetic fraction is carried out by passing the pulp through a rod matrix (magnetic particles are retained) under a strong magnetic field (Figure 3). The magnetic system of the SLon 100 separator uses a rod matrix made of rods with a diameter of 1–6 mm for different feed material sizes. The rods are arranged perpendicular to the magnetic field at an equal distance in order to ensure optimal field intensity and reduce the possibility of particle entrapment.

Figure 3. Matrices with rods arranged in staggered order.

The magnetic force (F_m) acting on the particle from the rod side (one of the main factors of magnetic separation) in the cylindrical coordinate system can be calculated by the following formula [41–43]:

$$F_m = \sqrt{F_r^2 + F_\tau^2} \tag{1}$$

where F_r and F_τ are the radial and tangential components of the magnetic force F_m.

$$F_r = -\frac{1}{2}\mu_0 K_p V_p M H \frac{a^2}{r^3}\left(cos2\alpha + \frac{M}{H}\frac{a^2}{r^2}\right) \tag{2}$$

$$F_\tau = -\frac{1}{2}\mu_0 K_p V_p M H \frac{a^2}{r^3} sin2\alpha \tag{3}$$

where M is the induced magnetization of the particle; r is the distance from the axis of the rod to the center of the particle; μ_0 is the permittivity of the media; a is the radius of the rod; H is the value of magnetic field intensity; K_p is the magnetic susceptibility of the particle; V_p is the volume of the particle; α is the angle between the direction of the radial component of the magnetic force and the x-axis; and H is the magnetic field intensity.

For the simplest condition, the sample grain size is homogeneous, and the particles differ only in composition; it is possible to denote the parameters characterizing the particle by

$$W_p = K_p M V_p, \text{ then } W_p = f(H) \tag{4}$$

To understand the character of the dependence ($W^p = f(H)$), studies have been carried out based on the accumulated database of investigations of the action of magnetic forces on a mineral particle in the working zone of a high gradient magnetic separator (rods) and using the given Equations (1)–(4) (Figure 4).

Investigations on magnetic separation were carried out in one stage by varying the following parameters: magnetic induction (0.2, 0.5, 1.1 Tesla), diameter of matrix rods (1, 3, 6 mm), and pulp pulsation (200, 250, 300 min^{-1}). Also, to increase the separation efficiency, Polypam and Flotfloc flocculants from Flotent Chemicals were used with varying consumptions of 1, 10, and 100 g/t.

All experimental investigations were carried out at least three times to increase reliability. The tables summarize the average results of the measurements of the elemental composition of the samples.

All the obtained products were dried at 105 °C, weighed, and analyzed using a Shimadzu EDX 700 X-ray fluorescence analyzer and a Vega 3 LMH scanning electron microscope. To analyze the samples using X-ray fluorescence analysis, representative samples for analysis were taken from the initial samples and from all beneficiation products with a mass of 5 g. The samples were put into special cuvettes and covered with mylar film. At least three samples were taken for each product. The obtained samples were analyzed at least three times. For the investigation of samples using scanning electron microscopy, representative samples of 2 g were taken from the initial ash and beneficiation products. The obtained samples were put on carbon tape and carburized to avoid sample lightening. The obtained samples were placed in a microscope and analyzed by local XRF.

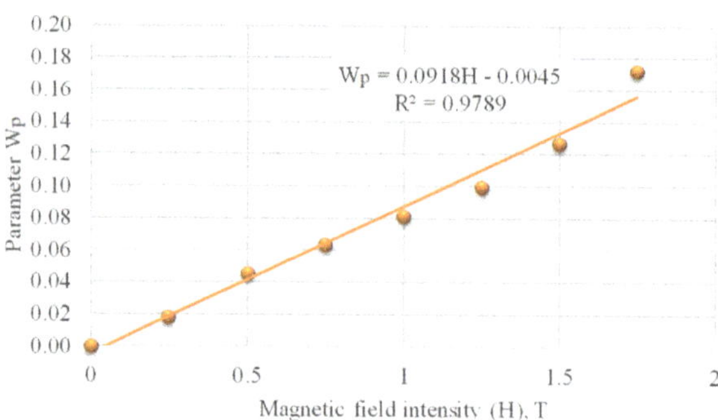

Figure 4. Dependence of the parameter characterizing the particle property on the magnetic field intensity ($W_p = f(H)$).

3. Results

Magnetic fractionation was carried out on a Davis Tube Tester at different current intensities. These investigations are necessary to assess the possibility and feasibility of the recovery of valuable iron-containing components from ash and slag by magnetic methods. The results are presented in Table 2 and Figure 5.

Table 2. Results of X-ray fluorescence analysis of magnetic fractionation products.

Product	Yield, %	Fe	Si	S	As	Al	Ca	Sb	K	Mn	Zn	Cu	Ni	Sr
								Content, %						
Magnetic fraction (1 A)	1.04	29.55	16.23	2.53	2.75	1.17	2.18	1.03	0.81	0.14	0.076	0.069	0.034	0.011
Magnetic fraction (2 A)	4.32	32.82	15.84	3.29	3.05	1.44	3.19	1.41	1.02	0.21	0.073	0.121	0.039	0.014
Magnetic fraction (3 A)	2.02	33.65	16.43	2.58	3.38	1.61	3.48	2.22	0.77	0.25	0.071	0.068	0.038	0.013
Magnetic fraction (4 A)	3.68	30.52	15.85	2.89	2.77	2.32	3.94	1.88	1.05	0.19	0.073	0.069	0.033	0.014
Magnetic fraction (5 A)	4.96	23.79	16.87	4.17	3.07	2.28	1.84	0.82	1.75	0.13	0.049	0.047	0.021	0.01
Magnetic fraction (6 A)	12.6	28.41	14.59	3.13	3.21	1.94	2.06	1.61	0.84	0.17	0.084	0.08	0.032	0
Non-magnetic fraction	71.38	29.97	15.98	2.38	3.55	2.45	3.62	2.16	0.8	0.19	0.073	0.071	0.037	0.014
Total:	100	29.68	15.85	2.63	3.42	2.30	3.31	1.97	0.87	0.19	0.07	0.073	0.036	0.012
Product	**Yield, %**							Recovery, %						
Magnetic fraction (1 A)	1.04	1.035	1.065	1.002	0.836	0.529	0.605	0.543	0.968	0.783	1.080	0.984	0.996	0.954
Magnetic fraction (2 A)	4.32	4.777	4.317	5.410	3.851	2.706	4.163	3.090	5.063	4.876	4.309	7.166	4.745	5.046
Magnetic fraction (3 A)	2.02	2.290	2.094	1.984	1.996	1.415	2.123	2.275	1.787	2.714	1.960	1.883	2.162	2.191
Magnetic fraction (4 A)	3.68	3.784	3.680	4.048	2.979	3.714	4.380	3.510	4.440	3.758	3.671	3.481	3.420	4.298
Magnetic fraction (5 A)	4.96	3.976	5.279	7.873	4.451	4.920	2.757	2.064	9.973	3.466	3.321	3.196	2.934	4.138
Magnetic fraction (6 A)	12.6	12.061	11.598	15.013	11.822	10.634	7.840	10.292	12.160	11.512	14.462	13.818	11.356	0.000
Non-magnetic fraction	71.38	72.077	71.966	64.669	74.065	76.081	78.052	78.225	65.609	72.892	71.198	69.473	74.386	83.373
Total:	100	100	100	100	100	100	100	100	100	100	100	100	100	100

The magnetic fraction is represented by small crystals of iron-bearing minerals and their fragments, often melted from the edges or pelletized, as well as small fragments of flattened veins. Interpretation of the results presented in Figure 5 shows that the microspheres observed in the concentrates consist mainly of Fe. Si, S, Ca (5a), and Si, S (5b) are noted as impurities. The detected microspheres have a sufficiently large size of 20–80 microns. It is worth noting that at a current strength of 2 A, the recovery of larger particles containing more iron occurs than at a current strength of 3 A.

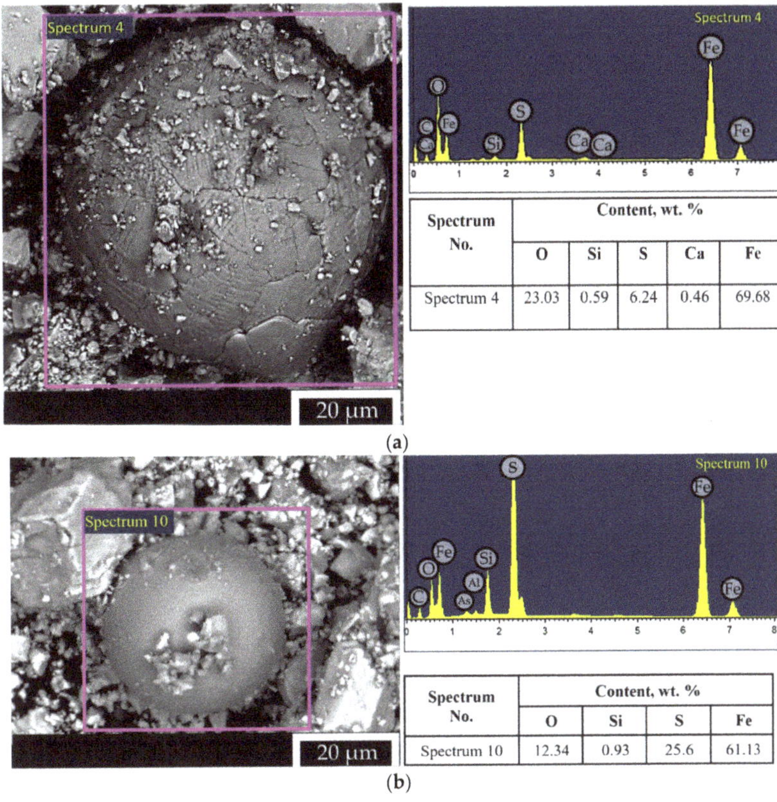

Figure 5. Electron image of magnetic microspheres in the separation products: (**a**) iron-containing sphere associated with intermetallics in the magnetic fraction obtained at a current strength of 2 A, (**b**) iron-containing sphere associated with intermetallics in the magnetic fraction obtained at a current of 3 A.

The obtained results showed that the distribution of iron is quite equal in all fractions, but microspheres, which include high-magnetic iron minerals associated with intermetallics, are mainly concentrated in the magnetic fractions obtained at current values of 2, 3, and 4 A (0.25, 0.345, and 0.422 Tesla, respectively). It can be supposed that iron in the compounds has different valence forms and different magnetic properties [44]. At the same time, particles containing low magnetic iron and aluminum were not found in the separated magnetic fractions. The iron content in the non-magnetic fraction is explained by the presence of complex particles including Ca, Fe, K, Al, Si, S, and rare metals. These particles have a reduced magnetic susceptibility. High-gradient magnetic separation has been proposed to separate such particles (microspheres) from fine materials.

Investigations on the influence of different parameters and settings of the magnetic separator (including matrix size, field strength, and pulsation frequency) on the characteristics of recovery and concentration of the target component were carried out on a high-gradient magnetic separator with variation of technological parameters and modes.

The I optimal experiment plan with additional central points was made (Table 3). Plans of this type are best suited for determining the optimal parameters and initializing the response surface. As a response was used, the recovery parameter was calculated as the arithmetic mean of three measurements of the obtained concentrate (the obtained values lie within three sigma).

Table 3. Experimental design.

		Factor 1	Factor 2	Factor 3	Response 1
Group	Run	a: Field Density	b: Pulsation Mode	C: Matrix	Recovery
		T	1/min		%
1	1	0.2	200	Matrix 6 mm	71.82
1	2	0.2	200	Matrix 3 mm	66.81
1	3	0.2	200	Matrix 1.5 mm	61.38
2	4	1.1	200	Matrix 3 mm	76.81
2	5	1.1	200	Matrix 1.5 mm	74.52
3	6	1.2	250	Matrix 6 mm	77.04
3	7	1.2	250	Matrix 1.5 mm	75.62
4	8	0.2	300	Matrix 3 mm	72.47
4	9	0.2	300	Matrix 6 mm	74.69
4	10	0.2	300	Matrix 1.5 mm	65.03
5	11	1.1	300	Matrix 6 mm	81.92
5	12	1.1	300	Matrix 3 mm	78.79
5	13	1.1	300	Matrix 1.5 mm	76.99
6	14	0.5	250	Matrix 1.5 mm	70.23
6	15	0.5	250	Matrix 3 mm	73.51
7	16	0.5	200	Matrix 6 mm	75.11
7	17	0.5	200	Matrix 1.5 mm	64.68
8	18	0.5	250	Matrix 3 mm	73.29
8	19	0.5	250	Matrix 6 mm	75.23

The Restricted Maximum Likelihood (REML) analysis and calculated Kenward–Roger p-values are summarized in Table 4.

Table 4. Analyzing the significance of the model coefficients.

Source	Term df	Error df	F-Value	p-Value	
Whole-plot	3	2.95	24.14	0.0140	significant
a-Field density	1	3.24	53.50	0.0040	
b-Pulsation mode	1	3.28	10.25	0.0436	
ab	1	3.15	0.0015	0.9713	
Subplot	8	3.65	11.51	0.0208	significant
C-Matrix	2	4.29	26.06	0.0040	
aC	2	4.10	8.00	0.0385	
bC	2	4.02	0.0934	0.9127	
abC	2	3.86	1.23	0.3855	

p-values less than 0.0500 indicate that the model conditions are significant. In this case, a, b, C, and aC are significant model terms. Values greater than 0.1000 indicate that the model terms are not significant.

As a result, adequate process models are obtained:

- matrix rod diameter = 6 mm

$$Recovery = 71.906 - 11.396a + 0.0047b + 0.062a \cdot b, \tag{5}$$

где a—field density, b—pulsation mode;

- matrix rod diameter = 3 mm

$$Recovery = 52.127 + 19.041a + 0.065b - 0.0401a \cdot b, \tag{6}$$

- matrix rod diameter = 1.5 mm

$$Recovery = 49.460 + 17.179a + 0.047b - 0.017a{\cdot}b. \tag{7}$$

Figure 6 shows the results of high-gradient magnetic separation with variations of technological parameters.

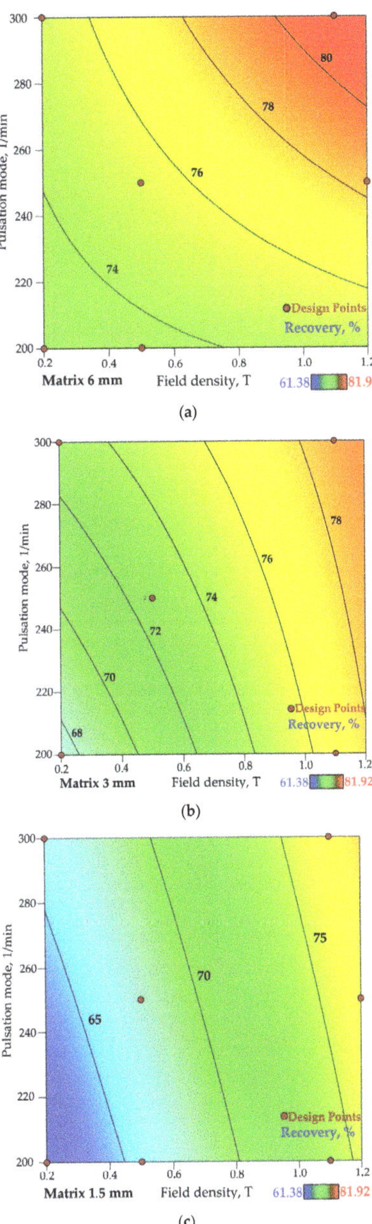

Figure 6. Graphical interpretation of the obtained results of high-gradient magnetic separation: (**a**) diameter of matrix rods = 6 mm; (**b**) diameter of matrix rods = 3 mm; (**c**) diameter of matrix rods = 1.5 mm.

The influence of technological parameters on iron recovery is presented in Figure 7.

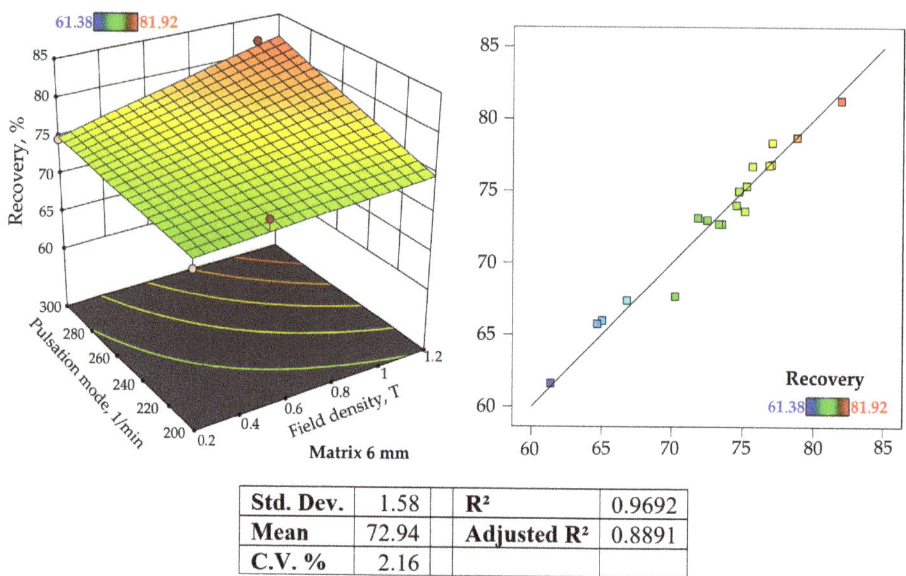

Std. Dev.	1.58	R²	0.9692
Mean	72.94	Adjusted R²	0.8891
C.V. %	2.16		

Figure 7. Three-dimensional response surface and statistical characteristics of the model.

The search for the largest response value was conducted using the Levenberg–Marquardt method (Figure 8).

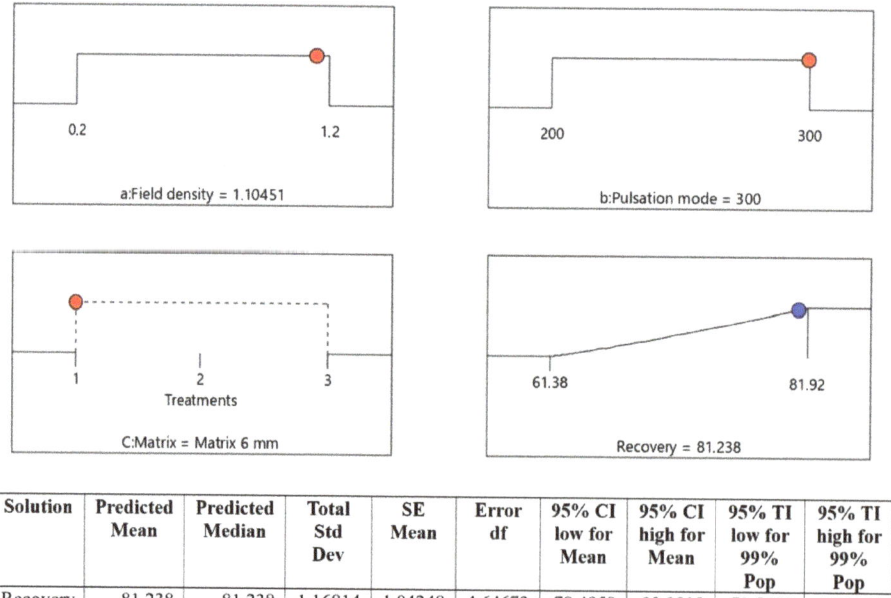

Solution	Predicted Mean	Predicted Median	Total Std Dev	SE Mean	Error df	95% CI low for Mean	95% CI high for Mean	95% TI low for 99% Pop	95% TI high for 99% Pop
Recovery	81.238	81.238	1.16814	1.04248	4.64673	78.4958	83.9802	72.5653	89.9107

Figure 8. Results of optimization of iron recovery process in high-gradient magnetic separation.

The highest magnetic force can be obtained by using a 1.5 mm matrix, while the highest gradient can be obtained by using a 6 mm matrix. In [45], it is shown that the use of a larger matrix results in a purer concentrate in terms of the iron content of the concentrate, which is particularly evident for particles smaller than 37 μm. This agrees well with the obtained results, as the recovered microspheres are 2–15 μm in size.

According to the results of our investigation, iron recovery in the concentrate increases with the increase in size due to the increase in content. The above beneficiation parameters confirm that the unique design of the Slon separator, as well as the higher magnetic field strength, allows us to significantly increase the recovery of iron in the concentrate. Thus, because of this research, it was established that the best results are achieved with the following parameters of operation of the high-gradient magnetic separator: magnetic induction 1.1 Tesla, diameter of matrix rods 6 mm, pulsation frequency 300 min^{-1}.

In order to increase the recovery of iron in the concentrate, another series of experiments with the use of flocculants was performed. The analysis of previously published works showed that during the magnetic separation of fine-grained ash and slag materials, the addition of flocculants allows an increase in the recovery of valuable components in the magnetic fraction [46]. When using flocculants and magnetic particles together, the mechanism of heteroflocculation is intensified. This occurs because of the binding of fine sludge by flocculant molecules, as well as the formation of "soft" flocculates around magnetic centers due to the selective action of the reagent, which increases the rate of coagulation [47]. Due to the increase in tension on additional particles of magnetite, there is an increase in the degree of the magnetic susceptibility of magnetic particles in the initial material. The flocculants Polypam and Flotfloc were chosen for this research because they are universal and can be used in rather complex slurries (acidic, neutral, and alkaline media). The consumption rate was varied to 1, 10, and 100 g/t. The results are presented in Figure 9.

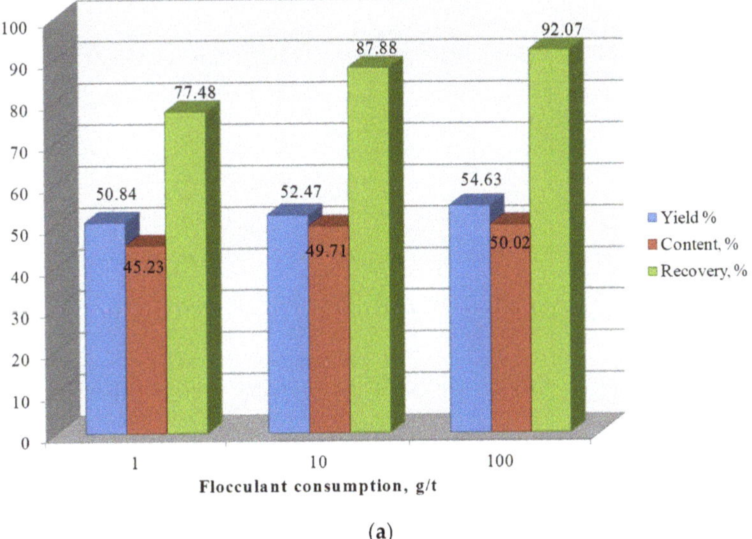

(a)

Figure 9. *Cont.*

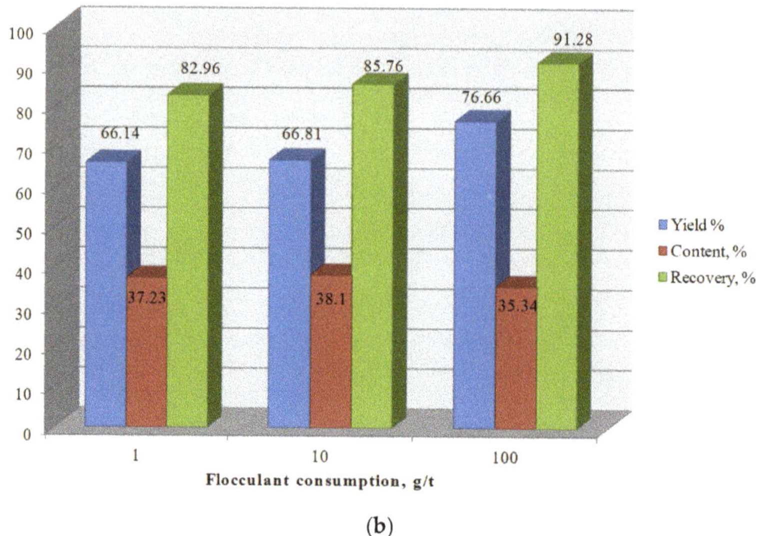

(**b**)

Figure 9. Investigation of the influence of flocculant addition: (**a**) Flotifloc; (**b**) Polypam.

The obtained results showed that the addition of the Flotifloc flocculant in an amount of 100 g/t allows us to obtain an iron content in the concentrate of 50% with a recovery of 92%. The addition of the Polypam flocculant increases the yield of the magnetic fraction, but the content decreases.

The conducted investigations made it possible to establish the optimal mode of magnetic concentrate production (Figure 10 and Table 5) on the high-gradient magnetic separator SLon.

Table 5. Results of elemental composition of ash after magnetic beneficiation (Figure 10).

	Content, wt. %				
	O	Al	Si	Ca	Fe
Spectrum 1	32.34	9.2	21.31	3.13	34.02
Spectrum 2	30.83	6.62	16.91	2.2	43.44
Spectrum 3	33.05	9.84	16.17	0.85	40.09
Spectrum 4	30.12				69.88

Analysis of the data shown in Figure 10 shows that low-magnetic and aluminosilicate iron-containing microspheres were detected in the concentrates obtained on the established regime for high-gradient separation with the application of scanning electron microscopy. The composition of the microspheres is summarized in Table 3. Interpretation of the results shows that the detected microspheres are mainly composed of Al, Si, Ca, and Fe. At the same time, microspheres representing iron oxides were detected (Figure 10, spectrum 4). Thus, the presence of low-magnetic and aluminosilicate iron-containing microspheres is confirmed, which agrees well with the results of earlier studies [22,48]. It is also worth noting that the sizes of extracted particles of valuable components with low values of magnetic susceptibility (less than 15 microns) are in good agreement with the work [49], in which the application of a high-gradient separator for sludge beneficiation is justified. The possibility of their extraction into the magnetic product using high-gradient magnetic separation is also proven.

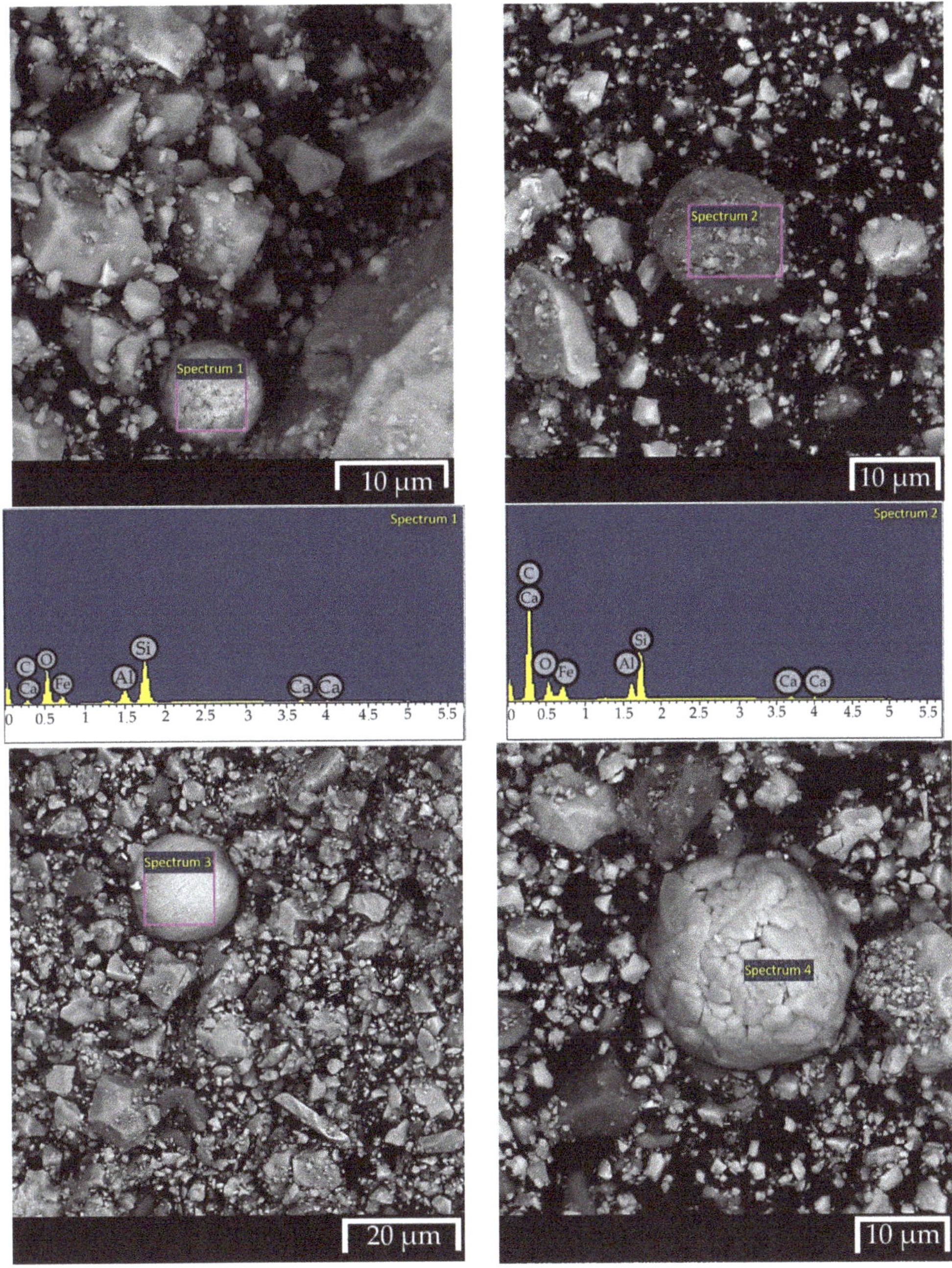

Figure 10. *Cont.*

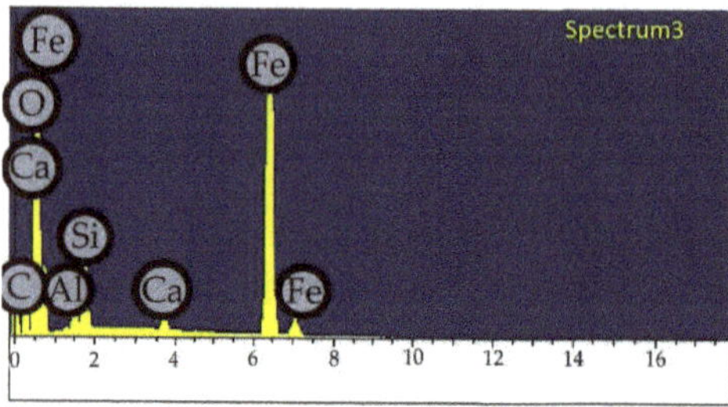
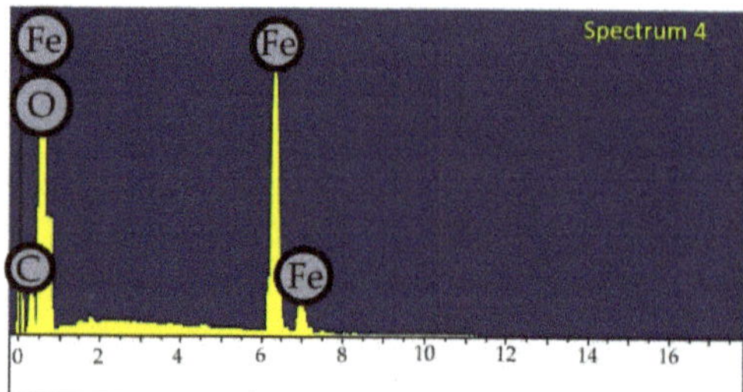

Figure 10. Results of research of ash and slag after magnetic beneficiation under optimal separation conditions using scanning electron microscopy.

The concentrate contains magnetic (low-magnetic iron) and aluminosilicate microspheres with sizes from 2 to 15 microns. It is practically impossible to recover such materials by standard beneficiation methods [35]. As a result of this research, it was established that for the beneficiation of ash and slag material by the method of high-gradient magnetic separation, the optimal parameters are magnetic induction 1.1 Tesla, diameter of matrix rods 6 mm, pulsation frequency 300 min^{-1}, and consumption of flocculant Flotifloc 100 g/t.

Thus, to recover microspheres of different compositions and sizes, a sequential scheme of magnetic concentration is recommended: magnetic separation in a weak magnetic field (concentrate contains high-magnetic iron-containing microspheres associated with intermetallics) and high-gradient separation of tailings (concentrate contains low-magnetic iron-containing and aluminosilicate microspheres ranging in size from 2 to 15 microns) (Figure 11).

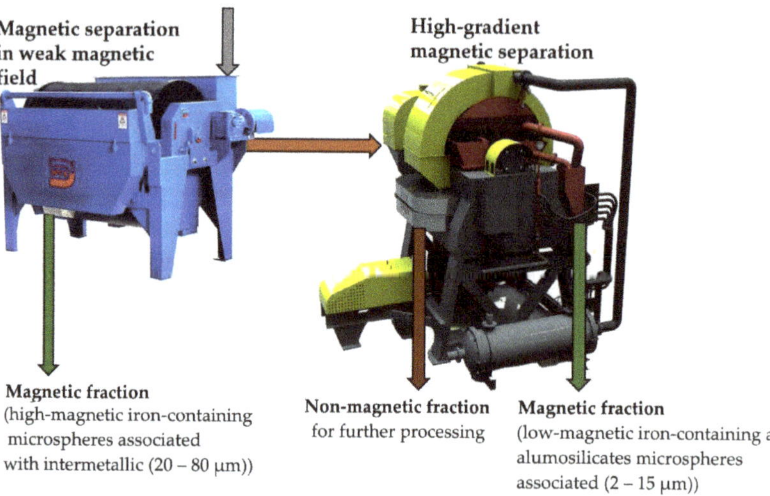

Figure 11. Topology of the scheme of magnetic beneficiation of ash and slag waste from CHPP.

The materials obtained by this technology have unique properties and can serve as raw materials to produce sorbents, magnetic carriers, etc.

4. Conclusions

This paper presents the research results aimed at substantiating the possibility of using magnetic beneficiation methods (including high-gradient magnetic separation) in the processing of ash and slag wastes from CHPPs, as well as the optimization of the parameters of the magnetic beneficiation process. Ash and slag wastes were chosen as objects of research on the principle of the presence of valuable components such as iron, aluminum, etc.

Based on magnetic fractioning studies on a Davis tube tester, it was found that iron is present in all fractions in approximately the same amount, but only three fractions obtained at current values of 2, 3, and 4 A (0.25, 0.345, and 0.422 Tesla, respectively), concentrated magnetic microspheres, contained high-magnetic iron minerals associated with intermetallics ranging in size from 20 to 80 microns.

This is due to the fact that iron is included in the compounds in different valence forms and has different magnetic properties. The results of the conducted research proved the principal possibility of using the pulsation high-gradient magnetic separator Slon. Its unique design (high magnetic field intensity, pulsation of the pulp) allows us to separate finely accumulated particles quite effectively. New features of magnetic and aluminosilicate microsphere recovery were established during the complex study of the parameters of operation of a high-gradient magnetic separator (joint influence of the size of matrix rods and magnetic induction).

The proposed regime (magnetic induction 1.1 Tesla, diameter of matrix rods 6 mm, pulsation frequency 300 min^{-1}, Flotifloc 100 g/t flocculant consumption) allows for one stage to obtain an iron ore product with an iron content of 50% and a recovery of 92%. With the use of scanning electron microscopy, it was established that at these parameters, the main portion of microspheres, which contain low-magnetic iron minerals and aluminosilicates with sizes ranging from 2 to 15 microns, is recovered in the magnetic fraction.

As a result of the performed research, the topology of the technological scheme of magnetic beneficiation for obtaining products with high added value (high- and low-magnetic iron-containing and aluminosilicate microspheres) was proposed. The proposed solution will not only allow us to obtain materials unique in their technological properties but also to reduce the environmental load in areas of ash dumps.

Author Contributions: T.A. conceived and designed the experiments and analyzed the data; N.N. and A.R. implemented and processed the analysis results; A.A., V.A. and E.P. performed the experiments; D.C. analyzed the data. All authors have read and agreed to the published version of the manuscript.

Funding: This work was carried out with a grant from the Russian Science Foundation (Project N 23-47-00109).

Data Availability Statement: Data are contained within the article.

Conflicts of Interest: The authors declare no conflicts of interest.

References

1. Meng, J.; Liao, W.; Zhang, G. Emerging CO_2-Mineralization Technologies for Co-Utilization of Industrial Solid Waste and Carbon Resources in China. *Minerals* **2021**, *11*, 274. [CrossRef]
2. Grigoryev, L.M.; Kheifets, E.A. Oil market: Conflict between recovery and energy transition. *Vopr. Ekon.* **2022**, *9*, 5–33. [CrossRef]
3. Aleksandrova, T.; Nikolaeva, N.; Kuznetsov, V. Thermodynamic and Experimental Substantiation of the Possibility of Formation and Extraction of Organometallic Compounds as Indicators of Deep Naphthogenesis. *Energies* **2023**, *16*, 3862. [CrossRef]
4. Nepsha, F.S.; Varnavskiy, K.A.; Voronin, V.A.; Zaslavskiy, L.S.; Liven, A.S. Integration of renewable energy at coal mining enterprises: Problems and prospects. *J. Min. Inst.* **2023**, *261*, 455–469.
5. Dvoynikov, M.V.; Leusheva, E.L. Modern trends in hydrocarbon resources development. *J. Min. Inst.* **2022**, *258*, 879–880.
6. Bazhin, V.Y.; Kuskov, V.B.; Kuskova, Y.V. Processing of low-demand coal and other carbon-containing materials for energy production purposes. *Inz. Miner.* **2019**, *21*, 195–198. [CrossRef]
7. Shklyarskiy, Y.E.; Skamyin, A.N.; Jiménez Carrizosa, M. Energy efficiency in the mineral resources and raw materials complex. *J. Min. Inst.* **2023**, *261*, 323–324.
8. Litvinenko, V.S.; Petrov, E.I.; Vasilevskaya, D.V.; Yakovenko, A.V.; Naumov, I.A.; Ratnikov, M.A. Assessment of the role of the state in the management of mineral resources. *J. Min. Inst.* **2022**, *259*, 95–111. [CrossRef]
9. Litvinenko, V.S. Digital Economy as a Factor in the Technological Development of the Mineral Sector. *Nat. Resour. Res.* **2020**, *29*, 1521–1541. [CrossRef]

10. Albalate, D.; Bel, G.; Teixidó, J.J. The Influence of Population Aging on Global Climate Policy. *Popul. Environ.* **2023**, *45*, 13. [CrossRef]
11. Our World in Data. Available online: https://ourworldindata.org/electricity-mix (accessed on 19 November 2023).
12. Plakitkin, Y.A.; Plakitkina, L.S.; Dyachenko, K.I. Coal as the basis of a great civilization leap and new opportunities for world development. *Ugol'* **2022**, *8*, 77–83. [CrossRef]
13. Chukaeva, M.A.; Matveeva, V.A.; Sverchkov, I.P. Complex processing of high-carbon ash and slag waste. *J. Min. Inst.* **2022**, *253*, 97–104. [CrossRef]
14. Yatsenko, E.A.; Goltsman, B.M.; Trofimov, S.V.; Lazorenko, G.I. Processing of Ash and Slag Waste from Coal Fuel Combustion at CHPPs in the Arctic Zone of Russia with Obtaining Porous Geopolymer Materials. *Therm. Eng.* **2022**, *69*, 615–623. [CrossRef]
15. Yatsenko, E.A.; Smolii, V.A.; Klimova, L.V.; Gol'tsman, B.M.; Ryabova, A.V.; Golovko, D.A.; Chumakov, A.A. Solid Fuel Combustion Wastes at CHPP in the Arctic Zone of the Russian Federation: Utility in Eco-Geopolymer Technology. *Glass Ceram.* **2022**, *78*, 374–377. [CrossRef]
16. Marinina, O.; Nevskaya, M.; Jonek-Kowalska, I.; Wolniak, R.; Marinin, M. Recycling of Coal Fly Ash as an Example of an Efficient Circular Economy: A Stakeholder Approach. *Energies* **2021**, *14*, 3597. [CrossRef]
17. Pukhov, S.A.; Kiseleva, S.P. Involvement in the economic turnover of ash and slag waste of thermal powerplants in the interests of eco-oriented economic development. *Russ. J. Resour. Conserv. Recycl.* **2020**, *7*, 10. (In Russian) [CrossRef]
18. Dyk, J.C.V.; Benson, S.A.; Laumb, M.L.; Waanders, B. Coal and Coal Ash Characteristics to Understand Mineral Transformations and Slag Formation. *Fuel* **2009**, *88*, 1057–1063. [CrossRef]
19. Uliasz-Bochenczyk, A.; Mokrzycki, E. Fly Ashes from Polish Power Plants and Combined Heat and Power Plants and Conditions of Their Application for Carbon Dioxide Utilization. *Chem. Eng. Res. Des.* **2006**, *84*, 837–842. [CrossRef]
20. Basu, M.; Pande, M.; Bhadoria, P.B.S.; Mahapatra, S.C. Potential Fly-Ash Utilization in Agriculture: A Global Review. *Prog. Nat. Sci.* **2009**, *19*, 1173–1186. [CrossRef]
21. Bhatt, A.; Priyadarshini, S.; Mohanakrishnan, A.A.; Abri, A.; Sattler, M.; Techapaphawit, S. Physical, Chemical, and Geotechnical Properties of Coal Fly Ash: A Global Review. *Case Stud. Constr. Mater.* **2019**, *11*, e00263. [CrossRef]
22. Vereshchak, M.; Manakova, I.; Shokanov, A.; Sakhiyev, S. Mössbauer Studies of Narrow Fractions of Fly Ash Formed after Combustion of Ekibastuz Coal. *Materials* **2021**, *14*, 7473. [CrossRef] [PubMed]
23. Ma, Z.; Shan, X.; Cheng, F. Distribution Characteristics of Valuable Elements, Al, Li, and Ga, and Rare Earth Elements in Feed Coal, Fly Ash, and Bottom Ash from a 300 MW Circulating Fluidized Bed Boiler. *ACS Omega* **2019**, *4*, 6854–6863. [CrossRef] [PubMed]
24. Hu, X.; Huang, X.; Zhao, H.; Liu, F.; Wang, L.; Zhao, X.; Gao, P.; Li, X.; Ji, P. Possibility of Using Modified Fly Ash and Organic Fertilizers for Remediation of Heavy-Metal-Contaminated Soils. *J. Clean. Prod.* **2021**, *284*, 124713. [CrossRef]
25. Loya, M.I.M.; Rawani, A.M. A review: Promising applications for utilization of fly ash. *Int. J. Adv. Technol. Eng. Sci.* **2014**, *2*, 143–149.
26. Bazhin, V.Y.; Kuskov, V.B.; Kuskova, Y.V. Problems of using unclaimed coal and other carbon-containing materials as energy briquettes. *Ugol* **2019**, *4*, 50–54. [CrossRef]
27. Huang, J.; Li, Z.; Chen, B.; Cui, S.; Lu, Z.; Dai, W.; Zhao, Y.; Duan, C.; Dong, L. Rapid detection of coal ash based on machine learning and X-ray fluorescence. *J. Min. Inst.* **2022**, *256*, 663–676. [CrossRef]
28. Korchevenkov, S.A.; Aleksandrova, T.N. Preparation of Standard Iron Concentrates from Non-Traditional Forms of Raw Material Using a Pulsed Magnetic Field. *Metallurgist* **2017**, *61*, 375–381. [CrossRef]
29. Svoboda, J.; Fujita, T. Recent Developments in Magnetic Methods of Material Separation. *Miner. Eng.* **2003**, *16*, 785–792. [CrossRef]
30. Andronov, G.P.; Zakharova, I.B.; Filimonova, N.M.; L'vov, V.V.; Aleksandrova, T.N. Magnetic Separation of Eudialyte Ore under Pulp Pulsation. *J. Min. Sci.* **2016**, *52*, 1190–1194. [CrossRef]
31. Fokina, S.B.; Petrov, G.V.; Sizyakova, E.V.; Andreev, Y.V.; Kozlovskaya, A.E. Process solutions of zinc-containing waste disposal in steel industry. *Int. J. Civ. Eng. Technol.* **2019**, *10*, 2083–2089.
32. Kairakbaev, A.K.; Abdrakhimov, V.Z.; Abdrakhimova, E.S. The use of ash material of east Kazakhstan in the production of porous aggregate on the basis of liquid-glass compositions. *Ugol'* **2019**, *1*, 70–73. [CrossRef]
33. Lan, M.; He, Z.; Hu, X. Optimization of Iron Recovery from BOF Slag by Oxidation and Magnetic Separation. *Metals* **2022**, *12*, 742. [CrossRef]
34. Petropavlovskaya, V.; Zavadko, M.; Novichenkova, T.; Petropavlovskii, K.; Sulman, M. The Use of Aluminosilicate Ash Microspheres from Waste Ash and Slag Mixtures in Gypsum-Lime Compositions. *Materials* **2023**, *16*, 4213. [CrossRef] [PubMed]
35. Valeev, D.; Kunilova, I.; Alpatov, A.; Varnavskaya, A.; Ju, D. Magnetite and Carbon Extraction from Coal Fly Ash Using Magnetic Separation and Flotation Methods. *Minerals* **2019**, *9*, 320. [CrossRef]
36. Li, W.; Han, Y.; Xu, R.; Gong, E. A Preliminary Investigation into Separating Performance and Magnetic Field Characteristic Analysis Based on a Novel Matrix. *Minerals* **2018**, *8*, 94. [CrossRef]
37. Chen, L.; Xiong, D.; Huang, H. Pulsating High-Gradient Magnetic Separation of Fine Hematite from Tailings. *Min. Metall. Explor.* **2009**, *26*, 163–168. [CrossRef]
38. Luo, L.; Zhang, J.; Yu, Y. Recovering Limonite from Australia Iron Ores by Flocculation-High Intensity Magnetic Separation. *J. Cent. South Univ. Technol.* **2005**, *12*, 682–687. [CrossRef]

39. Wills, B.A.; Napier-Munn, T.J. *Mineral Processing Technology*, 7th ed.; Elsevier Science & Technology Books: Amsterdam, The Netherlands, 2006.
40. Chanturia, V.A.; Minenko, V.G.; Samusev, A.L.; Koporulina, E.V.; Kozhevnikov, G.A. Stimulation of Leaching of Rare Earth Elements from Ash and Slag by Energy Impacts. *J. Min. Sci.* **2022**, *58*, 278–288. [CrossRef]
41. Chen, L.; Qian, Z.; Wen, S.; Huang, S. High-gradient magnetic separation of ultrafine particles with rod matrix. *Miner. Process. Extr. Metall. Rev.* **2013**, *34*, 340–347. [CrossRef]
42. Jin, J.X.; Liu, H.K.; Zeng, R.; Dou, S.X. Developing a HTS Magnet for High Gradient Magnetic Separation Techniques. *Phys. C Supercond.* **2000**, *341–348*, 2611–2612. [CrossRef]
43. Xiong, D.; Liu, S.; Chen, J. New Technology of Pulsating High Gradient Magnetic Separation. *Int. J. Miner. Process.* **1998**, *54*, 111–127. [CrossRef]
44. Bessais, L. Structure and Magnetic Properties of Intermetallic Rare-Earth-Transition-Metal Compounds: A Review. *Materials* **2022**, *15*, 201. [CrossRef] [PubMed]
45. Ding, L.; Chen, L.; Zeng, J. Investigation of Combination of Variable Diameter Rod Elements in Rod Matrix on High Gradient Magnetic Separation Performance. *Adv. Mater. Res.* **2014**, *1030*, 1193–1196. [CrossRef]
46. Song, S.; Lopez-Valdivieso, A.; Ding, Y. Effects of Nonpolar Oil on Hydrophobic Flocculation of Hematite and Rhodochrosite Fines. *Powder Technol.* **1999**, *101*, 73–80. [CrossRef]
47. Su, T.; Chen, T.; Zhang, Y.; Hu, P. Selective Flocculation Enhanced Magnetic Separation of Ultrafine Disseminated Magnetite Ores. *Minerals* **2016**, *6*, 86. [CrossRef]
48. Drozhzhin, V.S.; Shpirt, M.Y.; Danilin, L.D.; Kuvaev, M.D.; Pikulin, I.V.; Potemkin, G.A.; Redyushev, S.A. Formation processes and main properties of hollow aluminosilicate microspheres in fly ash from thermal power stations. *Solid Fuel Chem.* **2008**, *42*, 107–119. [CrossRef]
49. Xiong, T.; Ren, X.; Xie, M.; Rao, Y.; Peng, Y.; Chen, L. Recovery of Ultra-Fine Tungsten and Tin from Slimes Using Large-Scale SLon-2400 Centrifugal Separator. *Minerals* **2020**, *10*, 694. [CrossRef]

Article

Efficient Aqueous Copper Removal by Burnt Tire-Derived Carbon-Based Nanostructures and Their Utilization as Catalysts

Iviwe Cwaita Arunachellan [1,2], Madhumita Bhaumik [3], Hendrik Gideon Brink [3,*], Kriveshini Pillay [1] and Arjun Maity [2,3,*]

1 Department of Chemical Science, University of Johannesburg, Doornfontein, Johannesburg 2028, South Africa; ivienokalipa@gmail.com (I.C.A.); kriveshinip@uj.ac.za (K.P.)
2 DST/CSIR, Centre for Nanostructure and Advanced Materials (CeNAM), Council for Scientific and Industrial Research (CSIR), 1-Meiring Naude Road, Pretoria 0081, South Africa
3 Department of Chemical Engineering, Faculty of Engineering, Built Environment and Information Technology, University of Pretoria, Private Bag X20, Pretoria 0028, South Africa; bhaumikmadhu@gmail.com
* Correspondence: deon.brink@up.ac.za (H.G.B.); amaity@csir.co.za (A.M.); Tel.: +27-12-841-2658 (A.M.); Fax: +27-12-841-3553 (A.M.)

Abstract: This research focuses on valorising waste burnt tires (BTs) through a two-phase oxidation process, leading to the production of onion-like carbon-based nanostructures. The initial carbonization of BTs yielded activated carbon (AC), denoted as "BTSA", followed by further oxidation using the modified Hummer's method to produce onion-like carbon designated as "BTHM". Brunauer–Emmett–Teller (BET) surface area measurements showed 5.49 m^2/g, 19.88 m^2/g, and 71.08 m^2/g for raw BT, BTSA, and BTHM, respectively. Additional surface functionalization oxidations were observed through Fourier-Transform Infrared (FTIR), X-ray diffraction (XRD), Scanning Electron Microscopy (SEM), and Transmission Electron Microscopy (TEM) analyses. Raman spectroscopy indicated an increased graphitic nature during each oxidation stage. BTHM was assessed in batch adsorption studies for cupric wastewater remediation, revealing a two-phase pseudo-first-order behaviour dominated by mass transfer to BTHM. The maximum adsorption capacity for Cu^{2+} on BTHM was determined as 136.1 mg/g at 25 °C. Langmuir adsorption isotherm best described BTHM at a solution pH of 6, while kinetics studies suggested pseudo-second-order kinetics. Furthermore, BTHM, laden with Cu^{2+}, served as a catalyst in a model coupling reaction of para-idoanisole and phenol, successfully yielding the desired product. This study highlights the promising potential of BTHM for both environmental remediation and catalytic reuse applications to avoid the generation of secondary environmental waste by the spent adsorbent.

Keywords: carbon-based nanostructures; copper adsorption; used car tyres; coupling reactions; wastewater; spent adsorbent reuse

Citation: Arunachellan, I.C.; Bhaumik, M.; Brink, H.G.; Pillay, K.; Maity, A. Efficient Aqueous Copper Removal by Burnt Tire-Derived Carbon-Based Nanostructures and Their Utilization as Catalysts. *Minerals* **2024**, *14*, 302. https://doi.org/10.3390/min14030302

Academic Editor: Jean-François Blais

Received: 4 December 2023
Revised: 6 March 2024
Accepted: 8 March 2024
Published: 13 March 2024

1. Introduction

Copper, as the third most widely utilized metal across diverse industries, plays a pivotal role in economic activities. However, the established correlation between copper mining and adverse effects on surrounding water sources, including groundwater aquifers, wildlife, and farmland, raises significant environmental and health concerns [1]. Epidemiological studies have linked copper mining activities to various health issues, ranging from headaches to serious conditions such as cirrhosis, kidney failure, and cancer [2–4]. The substantial waste production, with approximately 99% of mined rock ending up as waste, contributes significantly to copper contamination [1]. Additionally, copper refining demands substantial water input, leading to massive production of copper-contaminated wastewater that requires effective treatment [5]. Current treatment technologies necessitate further research to develop sustainable and efficient methods for copper recovery from aqueous streams [2].

Activated carbon (AC), a versatile material, can be derived from various sources, including high-purity carbon feedstocks or low-cost agricultural waste such as fruit seeds, nut shells, vegetable waste, and wheat straw [6,7]. However, traditional methods for AC synthesis are often energy-intensive and involve high temperatures, catalysts, and expensive chemicals [8]. Carbon nanostructured materials (CNs), encompassing materials like carbon nanotubes (CNTs), carbon fibres, graphene, fullerenes, and onion-like carbon (OLC), offer unique properties but are typically produced using energy-intensive and expensive methods [9]. Some of these methods include hydrothermal carbonisation (HTC) [10], chemical vapour deposition (CVD), and thermal activation [11]. OLCs are usually produced through high-temperature annealing under a vacuum [12]. Graphene is produced using mechanical cleavage of graphite crystals, exfoliation of graphite through its intercalation compounds, CVD on various substrates, or utilising an organic synthesis process [13,14]. Synthetic design is crucial for CNs, enabling the fine-tuning of properties, but the associated methods are often costly and environmentally hazardous [15]. Natural materials such as agricultural waste, tyre waste, plastics, eggshells, leaves, newspaper, vegetable and fruit waste, agricultural waste, and sewage waste have been explored for CN [11], CNT [15], and graphene [16] production.

The wastewater treatment industry, a key source of recycled water globally, employs various techniques to ensure water safety [17]. Current wastewater treatment technologies have limitations [18], prompting the exploration of CN adsorbents for their economic and environmental benefits. CN adsorbents, with metal ion-attracting functional groups on their surfaces, exhibit enhanced sensitivity and specificity in the removal of toxic heavy metals from wastewater [19]. These adsorbents are expected to efficiently remove metal ions and allow for multiple reuse cycles.

However, the saturated adsorbent often presents an environmental problem when it loses its adsorption efficiency and is discarded back into the environment. Hence, it has become imperative to explore reuse applications of the spent adsorbent. Among such reuse applications, catalysis has become increasingly attractive since heavy metals have been effectively used as catalysts in industrial reactions.

Coupling reactions in organic chemistry, facilitated by metal-containing catalysts, produce valuable compounds by joining two fragments. Carbon-supported copper catalysts have proven effective in various coupling reactions, including Grignard, Sonagashira, Suzuki, and Ullmann coupling reactions [20,21].

This study therefore aims to synthesise CNs via a sophisticated two-step oxidation process, utilizing an economically viable carbonaceous precursor—burnt tyre residue. The procedure commences with a low-temperature carbonization of the residue employing a strong acid, resulting in the production of activated carbon (AC) [22]. Subsequently, a refined Hummer's method is applied for further oxidation [23,24], followed by treatment with hydrogen peroxide to eradicate any lingering contaminating ions, culminating in the creation of CNs. Notably, the innovation presented transcends conventional AC synthesis methods, addressing the energy-intensive and environmentally detrimental characteristics associated with prevalent approaches to CN production. This investigation underscores the potential of harnessing natural waste materials, including agricultural and tyre waste, for the synthesis of CNs. By reimagining abundant and economical precursors, the study introduces an ecologically sound strategy, harmonizing with the principles of sustainable material synthesis.

The innovative thrust further extends to the practical application of the synthesized CNs in batch adsorption studies for the treatment of synthetic cupric wastewater, thereby highlighting their promise in overcoming the limitations inherent in prevailing wastewater treatment technologies.

Finally, the assessment of the copper-ion laden material as a catalyst in model coupling organic reactions accentuates the multifaceted nature of the synthesized CNs. This inventive exploration of their catalytic potential introduces a fresh dimension to the myriad applications of CNs, transcending the conventional realm of wastewater treatment.

2. Materials and Methods

2.1. Materials and Reagents

Burnt tyre residue (collected from a recycling site where old tyres are recycled or burnt), 98% H_2SO_4 and $NaHCO_3$, $NaNO_3$, 98% H_2SO_4, $KMnO_4$, and HCl were used in this study. All materials (except for the burnt tyre residue) were of analytical grade and were obtained in <90% purity from Sigma Aldrich (Burlington, MA, USA).

2.2. Preparation of Sulfuric Acid Modified Burnt Tyre Material

The burnt tyre residue underwent multiple washes with deionized water and was subsequently dried in a vacuum oven at 60 °C. The resulting residue, now in fine powder form (5 g), was subjected to treatment with 98% H_2SO_4 at a 1:1 ratio (tyre waste to H_2SO_4, w/w) and placed in a vacuum oven at 100 °C for 24 h for carbonization. Following this, the material was mixed with deionized water and agitated for 2 h to eliminate any excess acid, followed by filtration through a vacuum. The material underwent repeated washings with deionized water until a clear filtrate was obtained and was then immersed in a 1% $NaHCO_3$ solution to eliminate any residual acid. The carbonized adsorbent was subsequently dried in a vacuum oven at 100 °C, ground, and stored in an airtight container [25].

2.3. Oxidation via Modified Hummer's Method

A total of 2 g of carbonized burnt tyre and 2 g of $NaNO_3$ were combined in a 1000 mL Erlenmeyer flask containing 100 mL of 98% H_2SO_4, where continuous stirring was maintained overnight. Subsequently, the mixture underwent stirring for 4 h at temperatures ranging from 0 to 5 °C, facilitated by an ice bath. During this process, 12 g of $KMnO_4$ was slowly introduced into the suspension, ensuring that the temperature remained below 15 °C to prevent overheating. The mixture was then gradually diluted by adding 100 mL of water, with stirring continuing for an additional 2 h. After removing the ice, the stirring persisted at 35 °C for another 2 h. The temperature was then raised to 98 °C for 10–15 min, gradually lowered, and then maintained at 35 °C for a final 2 h.

To complete the process, 6 mL of H_2O_2 (30%), diluted in 200 mL of water, was divided into beakers, with equal portions poured and stirred for 1 h. Subsequently, the solution was left unstirred for the remaining 3–4 h, allowing particles to settle, and was then subjected to filtration. The particles underwent repeated washing via centrifugation with 10% HCl and deionized water. Following centrifugation, the particles were vacuum-dried at 60 °C for a duration of 12 h [26].

2.4. Characterisation

The visualization of surface morphology and structure of the materials was conducted using a JEOL JSM 7500F microscope (JEOL, Tokyo, Japan) and a Zeiss field emission scanning electron microscope (FE-SEM) (FE-SEM, LEO Zeiss SEM, ZEISS, Oberkochen, Germany). For more in-depth size and morphology studies, a high-resolution transmission electron microscope (HR-TEM), specifically the JEOL JEM-2100 instrument (JEOL, Tokyo, Japan) with a LAB6 filament, operated at 200 kV.

The determination of surface area, pore volume, and pore diameter for each synthesized material was carried out using the Brunauer, Emmett, and Teller (BET) method. This analysis utilized a Micromeritics ASAP 2020 gas adsorption apparatus (Micromeritics, Norcross, GA, USA). Crystalline behaviour of all the powdered samples was explored by employing a PANalytical X'Pert PRO diffractometer (Malvern Panalytical, Malvern, UK). X-ray diffraction (XRD) patterns were obtained using CuKα radiation with a generator voltage of 45 kV along with a turbo current of 40 mA. The diffractograms were recorded in the range of $2\theta = 5–90°$ with a scanning step size of $0.026°$.

Functional groups present on the surfaces of the various materials were identified through Fourier-Transform Infrared Spectroscopy (FTIR), which was performed using a PerkinElmer Spectrum 100 spectrometer in the range of 600–4000 cm^{-1} with spectral resolution.

For the evaluation of vibrational, rotational, and other low-frequency modes resulting from the materials' structure, Raman spectroscopy was employed. The elemental composition of the materials before and after adsorption was determined using X-ray photoelectron spectroscopy (XPS) (PHOIBOS 150, SPECS, Berlin, Germany). Inductively coupled plasma mass spectrometry (ICP-MS) (iCAP RQplus, ThermoScientific, Waltham, MA, USA) was utilized to assess the remaining concentration of the solution after the adsorption process. Lastly, Nuclear Magnetic Resonance (NMR) was performed on a Bruker Fourier 300 MHz spectrometer (Bruker, Billerica, MA, USA) for the comprehensive analysis of the synthesized compounds.

2.5. Batch Adsorption Studies

A solution stock was prepared by dissolving 3.8019 g of copper(II) nitrate trihydrate in deionized water, resulting in a 1000 mg/L solution in a flask. Experimental solutions containing Cu^{2+} ions for adsorption studies were then derived by diluting the stock solution (1000 mg/L) to various concentrations, such as 20, 50, 100, and 200 mg/L. Adsorption equilibrium studies were conducted in a thermostatic shaker at 200 rpm for 24 h. This involved introducing a known quantity of BTHM into 50 mL glass bottles containing a pre-determined concentration of Cu^{2+} solution. Subsequently, the solutions were analysed using ICP-MS to determine the remaining concentration post adsorption.

To investigate the impact of pH on Cu^{2+} adsorption, the pH of the initial Cu^{2+} solution was varied between pH 1 and 6. pH adjustments were made using prepared solutions of HCl or NaOH at concentrations ranging from 1 M to 0.1 M. The quantity of BTHM adsorbent was systematically varied from 10 mg to 100 mg to assess the optimal dose for maximum adsorbate removal. The Cu^{2+} percentage removal and adsorption capacity were calculated using Equations (1) and (2):

$$\%removal = \left(\frac{C_0 - C_e}{C_o} \right) \times 100 \tag{1}$$

$$q_e = \left(\frac{C_0 - C_e}{m} \right) v \tag{2}$$

where C_0 and C_e are the initial and equilibrium Cu^{2+} concentrations in mg/L, respectively. The adsorption kinetics studies were carried out by varying the initial Cu^{2+} concentration (20, 50, 100 mg/L). The adsorbents and Cu^{2+} solution were stirred in a 1 L beaker for 24 h at 200 rpm, while pH and dose were kept at a constant optimum amount. At allocated time intervals, aliquots of 2 mL were collected and filtered using a 0.45 µm syringe filter. The filtrate was analysed for residual Cu^{2+} concentration. Furthermore, to analyse the adsorption kinetics for the adsorption of Cu^{2+} ion by the adsorbents, the experimental data were then fitted to the models: the pseudo-first-order, pseudo-second-order [27], and two-phase pseudo-first-order [28] models represented in Equation (3) to Equation (5), respectively.

$$q_t = q_e[1 - exp(-k_1 t)] \tag{3}$$

$$q(t) = \frac{q_e^2 k_2 t}{1 + q_e k_2 t} \tag{4}$$

$$q_t = q_e \left\{ \phi_{fast}[1 - exp(-k_{1,\text{fast}}t)] + \left[1 - \phi_{fast}\right][1 - exp(-k_{1,\text{slow}}t)] \right\} \tag{5}$$

where q_e (mg/g) and q_t (mg/g) are the adsorption capacity at equilibrium and at a determined time t (min), respectively; k_1 (1/min), k_2 (g/mg/min) are the pseudo-first-order and pseudo-second-order rate constants, respectively; ϕ_{fast} is the fraction of q_e that adsorbs rapidly; $k_{1,fast}$ and $k_{1,slow}$ are the rapid and slow first order reaction rate constants in Equation (5).

The rate determining step for adsorption can be determined Using the Crank and Weber and Morris intra-particle models [27]. The Crank model is represented by Equation (6) and the simplified solution of this model can be calculated using Equation (7).

$$\frac{\partial q_t}{\partial t} = \frac{D_e}{R^2}\frac{\partial}{\partial R}\left(R^2\frac{\partial q_t}{\partial R}\right)$$ (6)

$$\frac{q_t}{q_e} = \begin{cases} 6\left(\frac{D_e t}{R^2}\right)^{\frac{1}{2}}\left[\pi^{-\frac{1}{2}} - \left(\frac{1}{2}\right)\left(\frac{-D_e t}{R^2}\right)^{\frac{1}{2}}\right], & \frac{q_t}{q_e} < 0.8 \\ 1 - \frac{6}{\pi^2}\exp\left(\frac{-D_e \pi^2 t}{R^2}\right), & \frac{q_t}{q_e} \geq 0.8 \end{cases}$$ (7)

The intra-particle model makes use of Equation (8):

$$q_t = k_i t^{0.5} + C,$$
$$k_i = 6q_e\left(\frac{D_e}{\pi R^2}\right)^{0.5}$$ (8)

where D_e is the effective diffusivity of the adsorbate within the adsorbent intraparticle space, R is the radius of the agglomerated particle, k_i is the intra-particle diffusion rate constant $(mg/g/min^{0.5})$, and C is the intercept related to boundary layer thickness (mg/g). The Webber and Morris model can be interpreted using a plot of q_t vs. $t^{0.5}$ which is represented as 3 linear regions separated from the curve [29].

The isotherm experiments were conducted where the pH, concentration, and dose were kept constant while the temperature was varied at 25, 35, and 45 °C. The mixtures were shaken in a temperature-controlled thermostat at 200 rpm in 50 mL solutions for 24 h. The effect of temperature on the Cu^{2+} adsorption capacity of the adsorbents was studied by fitting the data from the adsorption equilibrium studies to two models: Langmuir and Freundlich, represented by Equations (9) and (10), respectively [30].

$$\frac{q_e}{q_m} = \frac{bC_e}{1 + bC_e}$$ (9)

$$q_e = K_F C_e^{1/n}$$ (10)

where q_m is the maximum adsorption capacity, and b is the Langmuir constant. The Freundlich constants k_F (mg/g) and $1/n$ signify the adsorption capacity and adsorption intensity.

To determine the thermodynamic parameters in the adsorption process, the isotherm studies were used. The Gibbs free energy change ($\Delta G°$), enthalpy change ($\Delta H°$), and entropy change ($\Delta S°$) parameters can be used to determine the nature of the thermodynamics observed in the adsorption of Cu^{2+} by BTHM. These make use of Equations (11) and (12):

$$\Delta G° = -RT In K_c$$ (11)

$$In K_c = \frac{\Delta S°}{R} - \frac{\Delta H°}{RT} \quad K_c = \frac{m}{V}\frac{q_e}{C_e}$$ (12)

where K_c is the equilibrium constant (L/mol), R is the ideal gas constant (J/mol/K), m/V is the adsorbent concentration (g/L), and T is temperature (K).

2.6. Catalysis

The copper catalyst was used for C-O cross coupling reaction as presented in Scheme 1.

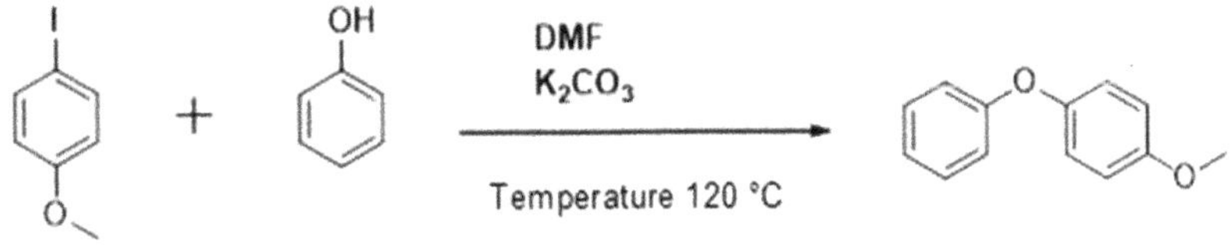

Scheme 1. Cross coupling reaction of para-idoanisole and phenol.

In a 25 mL two-neck round-bottom flask, a mixture comprising 1 mmol of para-idoanisole, 1.2 mmol of phenol, and 10 mg of copper adsorbed BTHM (Cu@BTHM) was stirred. Subsequently, K_2CO_3 (3 mmol) and 5 mL of DMF were added once stirring commenced. The entire reaction mixture was stirred at 120 °C for 24 h, with continuous monitoring using thin-layer chromatography. Upon completion of the reaction, the mixture was cooled to room temperature, and the catalyst was filtered using Whatman A filter paper. The filtrate was quenched with water (15 mL), and the product was subjected to extraction with ethyl acetate (3×, 10 mL).

The combined organic layers underwent washing with saturated sodium bicarbonate, a brine solution, and, finally, the organic layer was dried using anhydrous sodium sulphate. The solution was then concentrated under vacuum at 45 °C to obtain the crude product. Purification of the crude product was achieved through column chromatography (60–120 silica gel) using hexane and ethyl acetate as the mobile phase in a ratio of 9:1, resulting in the recovery of the pure compound [31].

3. Results and Discussion

3.1. Characterisation of Carbonaceous Adsorbents before and after Adsorption

The SEM images of the raw burnt tyre material "raw BT" are shown in Figure 1a; it was carbonised at 100 °C using a strong acid to obtain clumps of AC, shown in Figure 1b, denoted as "BTSA". The AC was then further oxidised using the modified Hummer's method to produce amorphous, smooth and homogenous, nanostructured spheres, "BTHM", shown in Figure 1c. The images shown in Figure 1d–f of all the materials depict the change from micro-size spheres of the BT to nanostructured spheres observed after the oxidation steps. The spheres were analysed to determine the average particle size and are graphically represented in a histogram in Figure 1g; an average particle size of 25.3 nm was calculated from the particle size distribution. The diffraction pattern in Figure 1h confirms the amorphous structure of the nanostructured spheres.

The characterisation using Raman spectroscopy shown in Figure 2a depicts the D band's increasing intensity from raw BT to BTHM with each oxidation stage until it was equal to the G band, indicating an increase in graphitic nature. The active D- and G-bands from BTHM at ~1340 cm^{-1} and ~1590 cm^{-1}, respectively, were due to the planar graphite. This is further indicative of the presence of sp2 hybridised carbons [32]. Illustrated in Figure 2b are the diffraction patterns using XRD for the materials. The Raw BT, BTSA, and BTHM included a prominent broad amorphous hump at $2\theta \approx 20°$ and a broad peak around $2\theta \approx 43°$, which are indicative of the amorphous carbonaceous structure of the burnt tyre material [33,34]. In addition, the Raw BT XRD pattern included the characteristic XRD patterns for CuCl and ZnS, as identified using the built in database (obtained from the "Crystallography Open Database" (COD)) within the Match! Phase Analysis software published by Crystal Impact (Bonn, Germany). This is likely a result of the inherent heavy metal content of tyres [35]. However, these metals were likely leached out during the sulphuric acid treatment step in which BTSA was obtained.

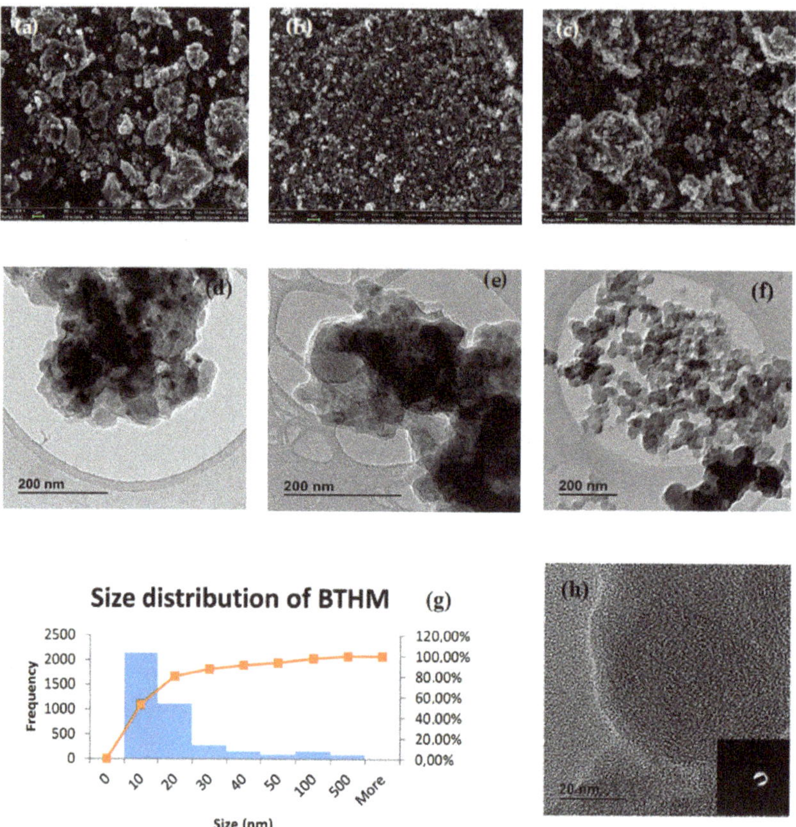

Figure 1. SEM images of (**a**) raw BT, (**b**) BTSA, and (**c**) BTHM. (**d**–**f**) Show TEM of the adsorbents. (**g**) Size distribution of BTHM and (**h**) HR-TEM nanostructure with diffraction pattern inserted.

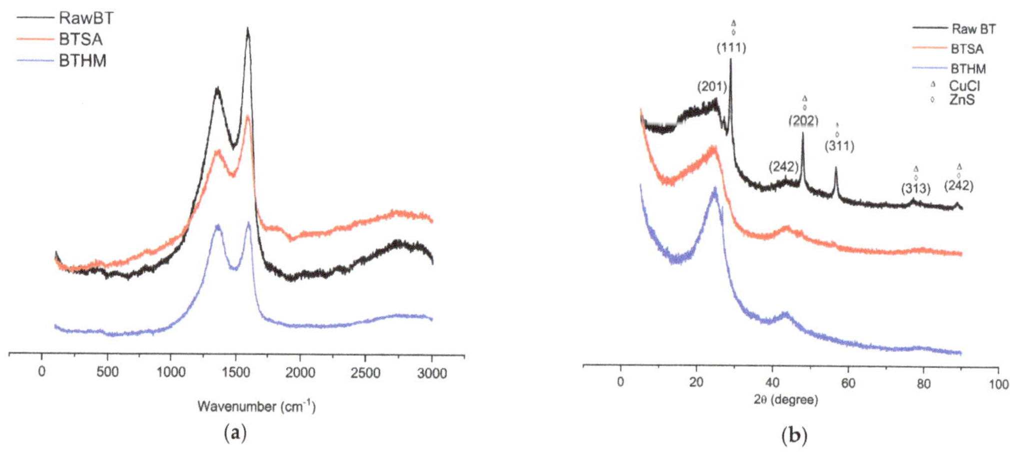

Figure 2. (**a**) Raman shifts and (**b**) XRD patterns of all the materials.

The BET analysis is shown in Table 1. The materials all followed a trend of increasing surface area and pore volume after each oxidation synthetic step, while the pore diameters varied after oxidation. Shown in Table 1 is the analysis of the raw BT, with 5.49 m^2/g, 0.035 cm^3/g, and 25.55 nm, BTSA, with 19.88 m^2/g, 0.088 cm^3/g, and 17.77 nm, and BTHM with 71.08 m^2/g, 0.380 cm^3/g, and 21.42 nm surface area, pore volume, and pore diameter, respectively. This indicates the mesoporous nature of the adsorbent. Analyses of the FTIR before adsorption of the raw BT, BTSA, and BTHM are shown in Figure S1 (Supporting Information). BTSA and BTHM before adsorption show the C–O stretching vibration (1633–1705 cm^{-1}), C–C from sp^2 C–C bonds (1596–1633 cm^{-1}), and C–O vibrations (1100–1144 cm^{-1}) [36].

Table 1. BET analysis of Raw BT, BTSA, and BTHM.

	Surface Area (m^2/g)	Pore Volume (cm^3/g)	Pore Diameter (nm)
Raw BT	5.49	0.035	25.55
BTSA	19.88	0.088	17.77
BTHM	71.08	0.380	21.42

For characterisation before the adsorption of Cu, the XPS analysis, with the data illustrated in Figure 3, was conducted. The BTHM survey that scanned the spectrum before adsorption, shown in Figure 3a,b, displayed clear signals of C 1s and O 1s. Before adsorption, the high-resolution XPS spectrum, shown in Figure 3a, of C 1s, was deconvoluted to four peaks due to the different chemical states of the C atom. The characteristic peaks located at ~284.8, ~286.1, ~287.8, and ~289.9 eV are contributions by carbons in C−C, C=O, C−O, O−C=O, respectively. Figure 3b represents the O 1s spectrum prior to adsorption. This O 1s spectrum can be divided into four major peaks situated at ~530.7, ~531.6, ~532.4, ~533.4, and ~534.1 eV, which correspond to oxygen water, organic O−C=O, C=O, C−OH, C–O–C, and organic O, respectively.

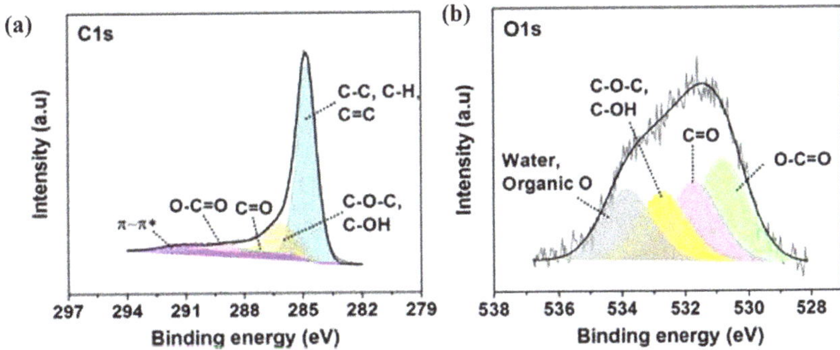

Figure 3. XPS of BTHM before adsorption showing (**a**) the C 1s peaks, (**b**) the O 1s peaks.

To evaluate the thermal stability, the materials BTSA and BTHM were tested under an air atmosphere from room temperature to 800 °C, and the TGA curve is presented in Figure S2 (Supporting Information). A small fraction of weight was lost between 23 °C and 100 °C, which can be attributed to the removal of surface-adsorbed water from the materials. From 100 to 200 °C, there was no observable significant mass loss. The majority of the decomposition took place in the range of 200 to 800 °C, where there was primary carbonization which presented a considerably greater weight loss for both BTSA and BTHM, due to the elimination of large amounts of volatile matter. Both materials have significant thermal stability, which is a great property for adsorption applications.

After adsorption, the Cu^{2+} laden BTHM was characterised using XPS, SEM-EDS, and HR-TEM to visualise or quantify the presence of Cu in the material. The BTHM survey which scanned the spectrum after adsorption is shown in Figure 4. It displayed clear signals of C 1s, O 1s, and S 2p at 164, 284, and, 530 eV, respectively.

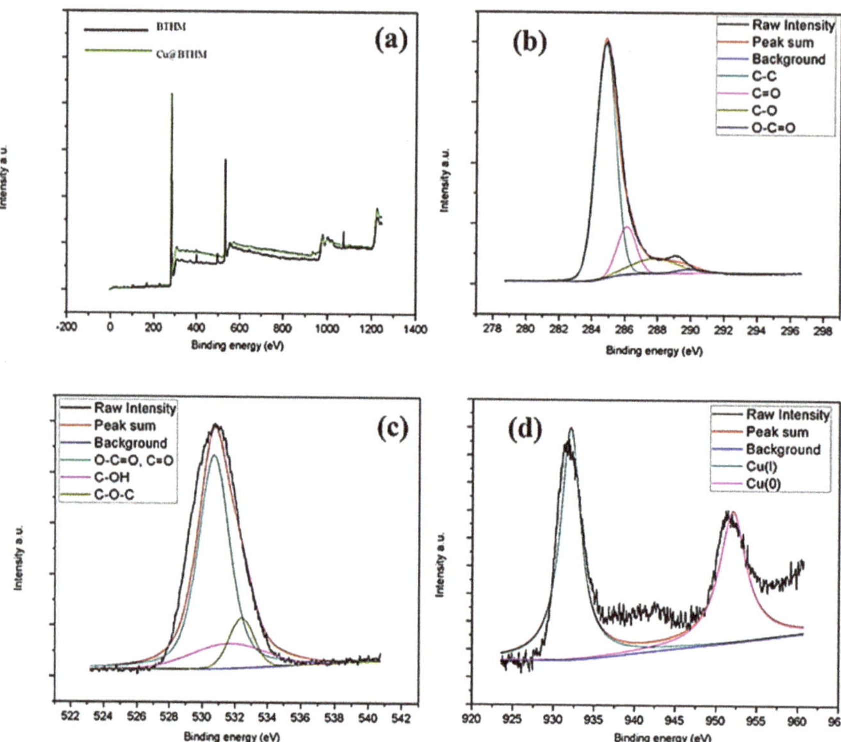

Figure 4. XPS spectra after adsorption for Cu@BTHM and BTHM for (**a**) the survey scan, (**b**) the C 1s peaks, (**c**) the O 1s peaks and (**d**) the Cu 2p3/2 peaks.

The high-resolution XPS spectrum of C 1s after adsorption is shown in Figure 4b. The characteristic peaks located at 284.8, 285.7, 287.4, and 288.8 eV are contributions by carbons in C−C, C=O, C−O, and O−C=O, respectively. After adsorption, the O 1s spectrum, shown in Figure 4c, can be divided into three major peaks situated at 530.7, 531.8, and 532.4 corresponding to oxygen O−C=O/C=O, O−H, and C−C−O, respectively. Finally, in Figure 4d, the high-resolution XPS spectrum with a binding energy of 368.2 eV, which corresponds to the Cu^{I} and Cu^{0} valence state of deposited Cu, respectively, is shown. The spin energy separation of 6.0 eV is representative of the ionic copper (Cu^{2+}). The after adsorption characterisation of materials using HR-TEM and SEM-EDS is shown in Figure 5. The HR-TEM in Figure 5a shows the Cu particles in the carbonaceous structure of BTHM. In Figure 5b, the SEM-EDS of Cu@BTHM was determined to contain majority carbon and oxygen groups with some sulphur, aluminium, and silicon contaminants; however, most importantly, it confirms the incorporation of Cu into the adsorbent.

3.2. Batch Adsorption Studies

Effect of pH and Dosage

The pH where maximum adsorption occurs was investigated by varying the pH of the copper ion-containing solution from 1 to 8. The dose, initial concentration, and temperature were kept constant at 20 mg, 200 mg/L, and 25 °C, respectively. The pH studies determined that the optimum pH for maximum adsorption for BTHM was pH 6 at 25 °C, as shown in Figure 6a. The observed increase in adsorption in pH values above 6 is a result of precipitation that is known to occur to heavy metals in solutions such as Cu^{2+} ions at a higher pH where there is an increase in the concentrations of $Cu(OH)^{+}$ and $Cu(OH)_2$ as the pH of the solution increases. The amount of adsorbent for optimum adsorption of Cu^{2+} ions from solution was investigated. The ideal dosage of the adsorbent amount was between 10 and 100 mg, while the initial Cu^{2+} ions concentration was kept at 200 mg/L, at the solution pH of 4.3 at 25 °C. It was observed that the removal of the Cu^{2+} ions increases with an increase in adsorbent dosage for BTHM until 75 mg, and then the observed absorbance decreases. The optimum absorbance was determined to be 75 mg (Figure 6b). This can be explicated by the increase in active sites available at higher doses for the adsorbents to attach to.

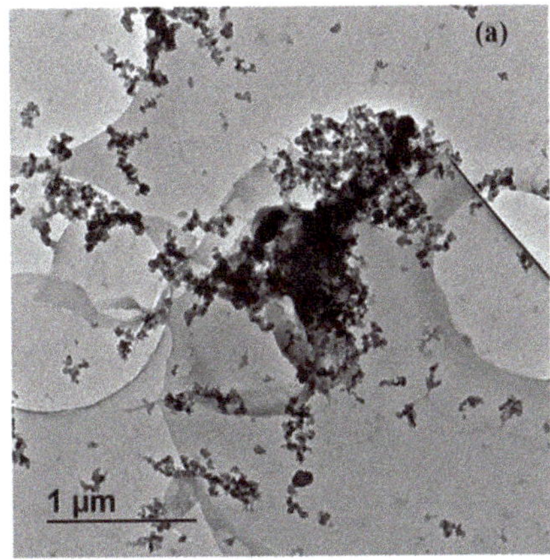

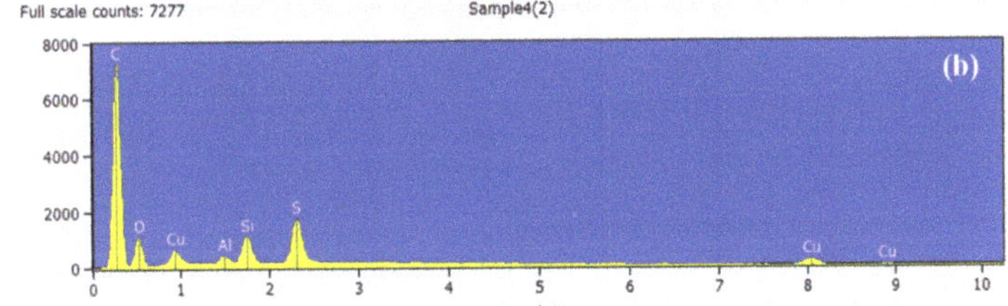

Figure 5. The images of (**a**) HR-TEM of Cu@BTHM and (**b**) SEM-EDS of Cu@BTHM.

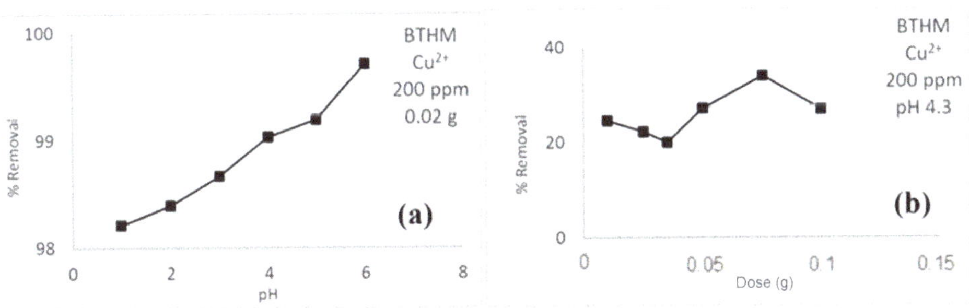

Figure 6. (**a**) pH and (**b**) dose studies for BTHM.

3.3. Adsorption Kinetics

The dynamic relationship between contact time and adsorbed Cu^{2+} ions at varying initial concentrations (50, 100, and 200 mg/L) at pH 6 and 25 °C was explored and is illustrated in Figure 7. The adsorbent amount was kept constant for BTHM as 1.5 g in 1 L solution. The adsorption kinetics were studied only at 25 °C. The pseudo-first-order

(PFO) model is based on the adsorption of Cu^{2+} ions onto BTHM following the first-order mechanism [37], the pseudo-second-order (PSO) model, which is based on the solid adsorbent capacity being proportional to adsorbate Cu^{2+} ions in the solution [38], and the two-phase pseudo-first-order model (2 phase PFO), which assumes the parallel rapid and slow adsorption processes taking place on the surface of the adsorbent [28]. Figure 7a–c show the fits of the kinetic data to the pseudo-first-order, pseudo-second-order, and two-phase pseudo-first-order kinetic models for Cu^{2+} removal by BTHM, respectively. In addition, Figure 7 shows the 95% prediction bands for the fitted models. The 95% prediction bands provide the area within which there is a 95% probability of future observations [39]; therefore, the greater the prediction band, the less accurate the model.

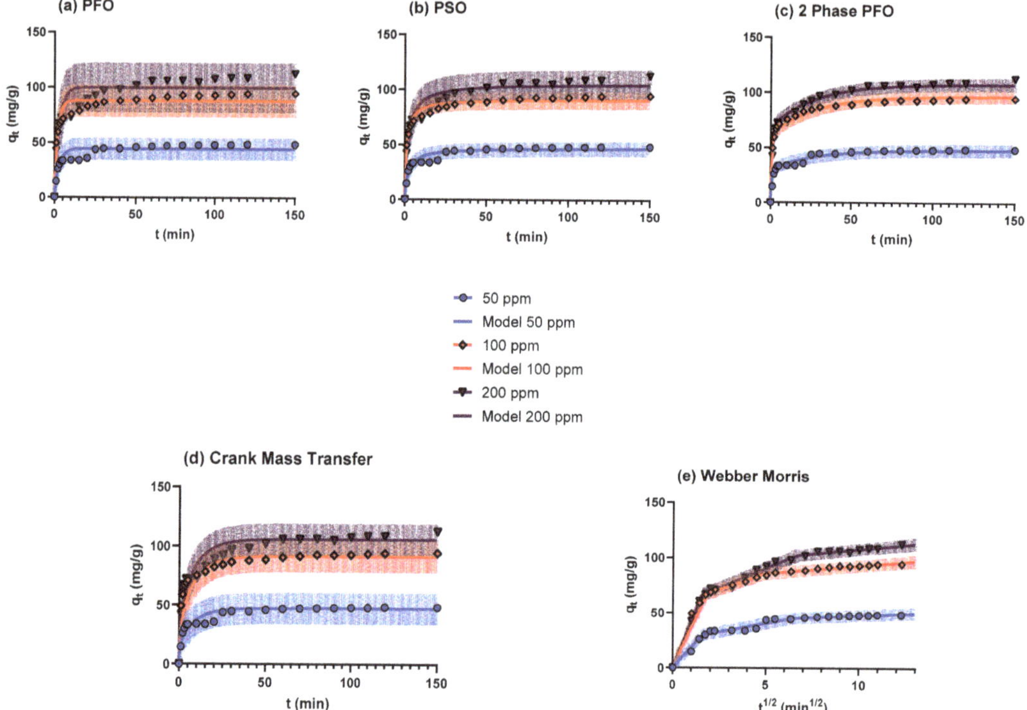

Figure 7. Graphical representation of the effect of contact time on Cu^{2+} adsorption by BTHM at 25 °C for different initial concentrations of Cu^{2+} ions and fits of kinetic data to the (**a**) pseudo-first-order (PFO), (**b**) pseudo-second-order (PSO), (**c**) 2 phase PFO, (**d**) Crank mass transfer and (**e**) Webber and Morris kinetic models.

The kinetic parameters obtained for the fittings are presented in Table 2. Table 2 shows that higher R^2 values and lower Sy.x and RMSE values were obtained for the 2 phase PFO model than either the PFO or PSO models, suggesting that the former model fits the data better; his observation is mirrored in the size of the 95% predictions bands shown in Figure 7. These results indicate that the adsorption shows a two-phase behaviour in which the adsorption takes place as two parallel adsorption phenomena, one fast and one slow, likely as a result of heterogenous surface interactions with the adsorbate. In addition, the 2 phase PFO model parameters are independent of the adsorbate concentration, with the qe predicted by equilibrium conditions and constant values for ϕ_{fast}, $k_{1,fast}$, and $k_{1,slow}$. Therefore, this model could provide the basis for the design and scaling of the system to continuous operation [40].

Table 2. Kinetics model constants and goodness of fit values for Cu^{2+} adsorption of BTHM.

Kinetic Model	Initial Cu Concentration		
	50 mg/L	100 mg/L	200 mg/L
Pseudo-first-order			
q_e	44.55	87.84	99.96
k_1	0.3418	0.5151	0.3477
R^2	0.8788	0.9105	0.8696
Absolute Sum of Squares	392.7	867.1	1999
Sy.x	4.546	6.755	10.26
RMSE	4.431	6.584	9.997
Pseudo-second-order			
q_e	47.40	91.42	105.9
k_2	0.009826	0.009487	0.004677
R^2	0.9455	0.9750	0.9504
Absolute Sum of Squares	176.6	242.4	760.3
Sy.x	3.049	3.572	6.326
RMSE	2.971	3.481	6.166
Two-phase pseudo-first-order			
q_e	47.89	96.74	107.7
ϕ_{fast}		0.6319	
$k_{1,fast}$		1.078	
$k_{1,slow}$		0.03732	
R^2		0.9922	
Absolute Sum of Squares		439.3	
Sy.x		2.776	
RMSE	2.662		
Crank mass transfer			
Q_{max}	47.40	91.42	105.9
D_e		3.7×10^{-10}	
R^2		0.9264	
Absolute Sum of Squares		1365	
Sy.x		6.531	
RMSE		6.531	
Webber and Morris			
D_{e1}		3.70×10^{-10}	
D_{e2}		6.15×10^{-11}	
D_{e3}		1.43×10^{-11}	
R^2		0.9934	
Absolute Sum of Squares		368.2	
Sy.x		2.587	
RMSE		2.437	

Units: q_e: mg/g, k_1, $k_{1,fast}$, $k_{1,slow}$: 1/min, k_2: g/mg.min, D_e, D_{e1}, D_{e2}, D_{e3}: m²/s.

Figure 7d and e show the resulting fits for the Crank and Webber and Morris models [27], respectively. The results from the fits are summarised in Table 2. From Figure 7d, it can be seen that a good fit can be obtained of the data exhibiting a relatively small 95% prediction interval—especially considering that a single parameter D_e was fitted for all three initial concentrations. Comparing the estimated D_e of 3.7×10^{-10} m²/s to the molecular diffusivity of Cu(II) ions in an aqueous solution at 25 °C–1.2×10^{-9} m²/s [41], it is clear that significant internal diffusional resistance is experienced by the Cu(II) ion as it diffuses to the adsorbent. This further results in the formation of thermodynamically stable structures due to the large time that Cu^{2+} must roam the surface of the adsorbents and attain minimum energy configuration before attaching to the growing island nuclei [42].

In Figure 7e, the first region is where the initial adsorption was observed to be fast and to interact with the boundary of the BTHM material. This was followed by a slower adsorption where the nanostructure pores took up more Cu^{2+}, and then, finally, the third region equilibrium was reached. It is interesting that the same D_e was estimated for the initial phase of adsorption by the Webber and Morris model as for the Crank model.

Using a plot of ln Kc vs. $1/T$, shown in Figure 8, the slope and intercept were $\Delta H°$ and $\Delta S°$, respectively. The thermodynamic parameters are presented in Table 3. The analysis of the plotted data shows that the adsorption process was endothermic and spontaneous, due to the positive $\Delta H°$ value and increasingly negative value of $\Delta G°$ with temperature, respectively [43]. The positive value of $\Delta S°$ shows that there was an increase in disorder at the adsorbent and Cu solution interface throughout the adsorption process [44].

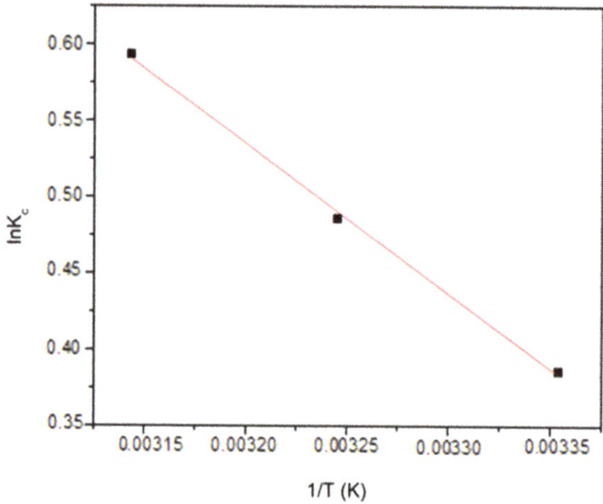

Figure 8. Thermodynamic plot of Cu^{2+} adsorption by BTHM.

Table 3. Thermodynamic parameter estimation for Cu^{2+} adsorption on BTHM.

Temperature (°C)	$\Delta G°$	$\Delta H°$	$\Delta S°$
25	−0.95175		
35	−1.25742	8.161679	0.030567
45	−1.56308		

Units: $\Delta G°$: kJ/mol, $\Delta H°$: kJ/mol, $\Delta S°$: kJ/mol.K.

3.4. Equilibrium Adsorption Isotherm

The effect of an initial concentration of Cu^{2+} ions on the adsorbent capacity was determined at fixed doses for BTHM, including 75 mg at pH 6. These adsorption isotherm studies were carried out at three different temperatures: 25 °C, 35 °C, and 45 °C. The initial concentration for Cu^{2+} varied from 20 to 200 mg/L. The fits of the equilibrium isotherm data for Cu^{2+} removed by BTHM to the Langmuir and Freundlich models are presented in Figure 9.

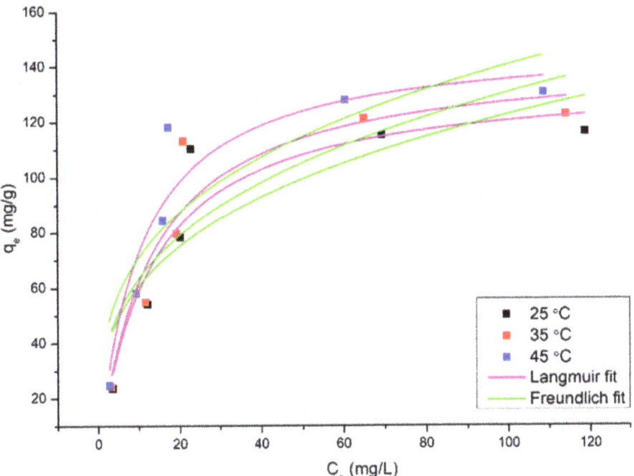

Figure 9. Adsorption equilibrium isotherms for Cu^{2+} adsorption by BTHM at different temperatures and fits of data to the Langmuir and Freundlich isotherm models.

The associated Langmuir and Freundlich isotherm parameters obtained for the fittings are presented in Table 4. The R^2 values are higher and they Sy.x and RMSE values are lower for the Langmuir model fitting, suggesting that this model fits the data better than the Freundlich model. The Langmuir maximum adsorption capacity of BTHM for Cu^{2+}, 136.1 mg/g (25 °C), is relatively high when compared to the other materials reported in the literature that were studied regarding Cu^{2+} adsorption, as shown in Table 5.

Table 4. Langmuir and Freundlich isotherm constants for Cu^{2+} removal by BTHM.

Isotherm Model	Temperature		
	25 °C	**35 °C**	**45 °C**
Langmuir			
Best-fit values			
q_m	136.1	143.9	149.5
b	0.07744	0.07693	0.09670
Std. Error			
q_m	15.98	17.39	15.86
b	0.03072	0.03068	0.03495
Goodness of Fit			
R^2	0.8923	0.8926	0.9067
Absolute Sum of Squares	783.8	880.6	856.3
Sy.x	14.00	14.84	14.63
Freundlich			
Best-fit values			
K_F	30.38	31.12	36.22
N	3.297	3.208	3.391
Std. Error			
K_F	12.08	12.18	12.89
N	1.088	1.023	1.061
R^2	0.7550	0.7678	0.7713
Absolute Sum of Squares	1784	1905	2098
Sy.x	21.12	21.82	22.90
R^2	0.7550	0.7678	0.7713

Units: q_m: mg/g, b: L/mg, K_F: $((mg \cdot g^{-1})(mg \cdot L^{-1})^{-1/n})$.

Table 5. Comparison of the qmax values for Cu removal by waste-derived activated carbon (AC) and modified activated carbon (MAC)-based adsorbents.

Adsorbents	q_{max} (mg/g)	Ref
Banana leave AC	66.2	[45]
Green vegetable AC	75.0	[46]
Date stone AC	31.3	[47]
Sawdust AC	2.23	[48]
Grape bagasse AC	43.5	[49]
Used tire AC	20.0	[50]
Dried tree fibre MAC	80.2	[51]
Waste wood MAC	84.5	[52]
Mill Waste MAC	4.4	[53]
Orange Peels, TiO_2 MAC	14.0	[54]
BTHM	136.1	Current study

3.5. Adsorption Mechanism

The before and after adsorption FTIR and XPS results were used to determine a credible adsorption mechanism for BTHM. As discussed, the BTHM material analysed using pH_{pzc} is negatively charged at isoelectric point of pH 4 and above; thus, the surface of the material at the selected optimum pH 6 is negative. The positively charged Cu^{2+} ions, therefore, have an opportunity for surface charge interaction. The high-resolution XPS spectra of various elements present in BTHM before adsorption (Figure 3) compared to the XPS spectra after adsorption (Figure 4) corroborate this proposed mechanism as the presence of a Cu energy peak in the survey scan of BTHM confirmed the successful adsorption of a Cu^{2+} ion onto the surface, along with the potential bonds indicated by the intensity and energy shift changes observed and discussed in the XPS spectra before and after adsorption.

3.6. Catalysis: Coupling Reactions

Finally, the spent adsorbent was then used as a catalyst in the coupling reaction to evaluate its ability to produce valuable coupled organic compounds. The catalyst denoted Cu@BTHM was air-dried in the fume hood overnight and stored in a glass container. The formation of the desired products was monitored using thin-layer chromatography (TLC) and the compounds formed were analysed using NMR. Coupling reactions are critical in organic synthesis of various pharmaceutical, cosmetic, and other important compounds. The synthesised compound was analysed using NMR and the spectra are illustrated in Figure 10. 1H NMR dissolved in deuterated DMSO, Bruker 400 MHz, had peaks appearing at 7.54 (CH3O-Ar-C-H), 7.50 (CH3O-Ar-C-H), 7.0 (C-H), 6.70 (CH3O-Ar-C-H), 3.85 (CH3), and 0.8–1.5 (O-Ar-C-H). These results confirm the successful coupling of iodobenzene with a phenol (as predicted in Scheme 1).